A CULTURAL HISTORY OF THE HUMAN BODY

VOLUME 1

A Cultural History of the Human Body
General Editors: Linda Kalof and William Bynum

Volume 1
A Cultural History of the Human Body in Antiquity
Edited by Daniel H. Garrison

Volume 2
A Cultural History of the Human Body in the Medieval Age
Edited by Linda Kalof

Volume 3
A Cultural History of the Human Body in the Renaissance
Edited by Linda Kalof and William Bynum

Volume 4
A Cultural History of the Human Body in the Age of Enlightenment
Edited by Carole Reeves

Volume 5
A Cultural History of the Human Body in the Age of Empire
Edited by Michael Sappol and Stephen P. Rice

Volume 6
A Cultural History of the Human Body in the Modern Age
Edited by Ivan Crozier

A CULTURAL HISTORY OF THE HUMAN BODY
IN ANTIQUITY

Edited by Daniel H. Garrison

Bloomsbury Academic
An imprint of Bloomsbury Publishing Plc

B L O O M S B U R Y
LONDON • OXFORD • NEW YORK • NEW DELHI • SYDNEY

Bloomsbury Academic
An imprint of Bloomsbury Publishing Plc

50 Bedford Square
London
WC1B 3DP
UK

1385 Broadway
New York
NY 10018
USA

www.bloomsbury.com

Hardback edition first published in 2010 by Berg Publishers, an imprint of Bloomsbury Academic
Paperback edition first published by Bloomsbury Academic 2014
Reprinted by Bloomsbury Academic 2015

© Daniel H. Garrison, 2010, 2014

Daniel H. Garrison has asserted his right under the Copyright, Designs and Patents Act, 1988, to be identified as Editor of this work.

All rights reserved. No part of this publication may be reproduced or transmitted in any form or by any means, electronic or mechanical, including photocopying, recording, or any information storage or retrieval system, without prior permission in writing from the publishers.

No responsibility for loss caused to any individual or organization acting on or refraining from action as a result of the material in this publication can be accepted by Bloomsbury or the author.

British Library Cataloguing-in-Publication Data
A catalogue record for this book is available from the British Library.

ISBN: HB: 978-1-84788-788-7
PB: 978-1-4725-5462-8
HB Set: 978-1-84520-495-2
PB Set: 978-1-4725-5468-0

Series: The Cultural Histories Series

Library of Congress Cataloging-in-Publication Data
A catalog record for this book is available from the Library of Congress.

Typeset by Apex CoVantage, LLC, Madison, WI, USA

CONTENTS

	Illustrations	vii
	Series Preface	ix
	Introduction *Daniel H. Garrison*	1
1	"The End is to the Beginning as the Beginning is to the End": Birth, Death, and the Classical Body *Valerie M. Hope*	25
2	Health and Disease *Patrick MacFarlane*	45
3	Sex *Marilyn B. Skinner*	67
4	Medical Knowledge and Technology *Brooke Holmes*	83
5	Popular Beliefs about the Human Body in Antiquity *Page duBois*	107
6	Reflections on Erotic Desire in Archaic and Classical Greece *Froma I. Zeitlin*	133
7	Marked Bodies: Gender, Race, Class, Age, Disability, and Disease *Brooke Holmes*	159
8	Marked Bodies: Divine, Human, and Bestial *Marguerite Johnson*	185

9 The Body of a Hero: Images of Herakles and Their
   Political Use in Antiquity   217
   Amalia Avramidou

10 The Self from Homer to Charlemagne   239
   Marc Mastrangelo

   NOTES   255
   BIBLIOGRAPHY   327
   CONTRIBUTORS   361
   INDEX   365

ILLUSTRATIONS

CHAPTER 1

Figure 1.1: A terracotta relief attached to the tomb façade (Tomb 100) of Scribonia Attice at the Isola Sacra Necropolis, Portus (near Ostia, Italy). Mid second century A.D.   29

Figure 1.2: Ancient Roman relief carving of a Roman midwife.   30

Figure 1.3: A Roman grave marker in the shape of an altar, Rome, late first or second century A.D.   33

Figure 1.4: Scene from a Greek vase (white ground lekythoi) depicting a funerary stele adorned with ribbons, with mourners bringing offerings to the grave.   34

CHAPTER 5

Figure 5.1: Greek open air shower baths for men.   110

Figure 5.2: Greek woman bathing.   111

Figure 5.3: Votive offering: pregnant woman wearing a binder, Greek.   116

Figure 5.4: A seated Greek woman on an obstetrical stool being held in position by her husband whilst giving birth aided by a midwife, another attendant dresses the first baby.   117

Figure 5.5: Woman with a tumor of the breast. Greek votive offering.   118

Figure 5.6: Asklepios/Aesculapius: The ancient Greek deity
of healing. 126

Figure 5.7: Engraving of a bust of the Greek physician Hippocrates. 128

Figure 5.8: Greek physician and patient, plaster cast in W.H.M.M. 129

Figure 5.9: Detail of the decoration of a red-figured Greek vessel
in the Louvre showing a satyr with a crippled leg. 130

Figure 5.10: Ancient Roman and Greek surgical instruments. 131

CHAPTER 8

Figure 8.1: Achilles dragging the body of Hector around Troy.
Mezzotint after G. Hamilton, 1794. 193

Figure 8.2: A performance of the play *Birds* by Aristophanes. 198

Figure 8.3: Ulysses [Odysseus] and the Sirens. Etching by P. Aquila
after G. Rossi after Annibale Carracci. 206

Figure 8.4: Three old hags surround a basket of newborn babies,
with bats in the distance. Etching by F. Goya, 1796/98. 213

CHAPTER 9

Figure 9.1: Herakles fights Gêras. Attic red-figure pelike,
ca. 480–470 B.C. 220

Figure 9.2: Marble relief showing Herakles mounted by
a young woman. 222

Figure 9.3: The struggle for the Delphic tripod. Pediment
of the Siphnian Treasury, ca. 525 B.C. 226

Figure 9.4: Herakles at the Garden of Hesperides. Attic red-figure
hydria by the Meidias Painter, ca. 420 B.C. 228

Figure 9.5: Apulian red-figure column-krater with artist painting
a marble statue of Herakles, ca. 350–320 B.C. 230

Figure 9.6: Marble statuette of the type of Herakles Mingens. 233

SERIES PREFACE

A Cultural History of the Human Body is a six-volume series reviewing the changing cultural construction of the human body throughout history. Each volume follows the same basic structure and begins with an outline account of the human body in the period under consideration. Next, specialists examine major aspects of the human body under seven key headings: birth/death, health/disease, sex, medical knowledge/technology, popular beliefs, beauty/concepts of the ideal, marked bodies of gender/race/class, marked bodies of the bestial/divine, cultural representations and self and society. Thus, readers can choose a synchronic or a diachronic approach to the material—a single volume can be read to obtain a thorough knowledge of the body in a given period, or one of the seven themes can be followed through time by reading the relevant chapters of all six volumes, thus providing a thematic understanding of changes and developments over the long term. The six volumes divide the history of the body as follows:

Volume 1: A Cultural History of the Human Body in Antiquity (750 B.C.E.–1000 C.E.)

Volume 2: A Cultural History of the Human Body in the Medieval Age (500–1500)

Volume 3: A Cultural History of the Human Body in the Renaissance (1400–1650)

Volume 4: A Cultural History of the Human Body in the Age of Enlightenment (1650–1800)

Volume 5: A Cultural History of the Human Body in the Age of Empire (1800–1920)

Volume 6: A Cultural History of the Human Body in the Modern Age (1920–21st Century)

General Editors, Linda Kalof and William Bynum

Introduction

DANIEL H. GARRISON

In the 130,000 to 200,000 years since homo sapiens evolved on this planet, the human body has had a more or less constant natural, anatomical construction. Its *cultural* construction is another matter entirely. The human body is the subject of its own master narrative in every culture and many subcultures. It is the task of this volume to reveal key features of that master narrative as it played out in the world of the Greeks and Romans and in Europe of the first millennium C.E.

Though to some extent grounded in the facts of human anatomy and physiology, the cultural body has a life and a construction of its own. In 1543 the anatomist Andreas Vesalius wrote,

> It is commonly believed that men lack a rib on one side, and that men have one rib fewer than women. This is plainly absurd, even if Moses did say in the second chapter of Genesis that Eve was created by God out of Adam's rib. Granted that perhaps Adam's bones, had someone articulated them into a skeleton, might have lacked a rib on one side, it does not necessarily follow on that account that all men are lacking a rib as well.

Well before the scientific era, it was increasingly evident that a gap was opening between traditional thinking and the biological facts of life. The modern period is in part defined by the need to distinguish between these two realities. A few other examples will suffice to illustrate features of the ancient body that (perhaps) differ from the body perceived in the modern world.

The belief that female nature is less stable than male was tied in Western antiquity—and long after—to the belief that the human womb was not anchored by tendons as it is in other animals but can wander about the body, attaching itself to the heart, the liver, the brain, and other unlikely places, causing a variety of disorders peculiar to women. It was thought that the womb was particularly susceptible to odors, and that it could be moved about by the application of pleasant or unpleasant odors to the vagina.

One of the most persistent constructs of the cultural body in the West, now embedded in our language, is the belief that character is determined by one of four fluids or humors produced in the body. Blood, bile or choler, phlegm, and black bile (a mythical product of the spleen invented to complete the tetrad) were not only natural fluids: a predominance of one over the other three produced what we still call sanguine, bilious or choleric, phlegmatic, and atrabilious or splenetic personalities. Originating in Hippocratic medicine, by the Middle Ages the humors had solidified into the four temperaments; by Shakespeare's time, they had become part of the language, beyond the reach of scientific refutation.

> Scarce can I speak, my choler is so great (2 *Henry VI* 5.1. 23)
> If she must teem,
> Create her child of spleen, that it may live
> And be a thwart disnatured torment to her. (*Lear* 1.4. 273–6)

Another cultural construct is the belief that certain parts of the body have a medical relationship with other parts to which they are not provably connected. The now largely discarded practice of cupping and drawing of blood through venesection (phlebotomy) developed in the West from the Greek belief that certain disorders were the result of an excess or plethora of blood. This ancient legacy became increasingly elaborate, not to say mystical: the Jewish Talmud ruled certain days favorable or unfavorable for bloodletting. Christians recommended certain saints' days. Astrological considerations also gained influence in its practice. Other lore emerging in the Middle Ages prescribed specific veins to be opened for certain conditions elsewhere in the body. The cephalic vein in the arm owes its name to the medieval belief that certain affects in the head could be treated by draining blood from that vessel, even though it does not go to the head. In the sixteenth century, pleurisy and related inflammations of tissues surrounding the lungs were treated by opening a tributary of the basilic ("kingly") vein of the arm, so called because it was also thought to communicate with the liver (right arm) or spleen (left arm). Therapy by acupuncture, still popular worldwide, is based upon similar doctrines of vital energy originating in Chinese traditional medicine. Such doctrines, East or West, are no less potent for being cultural constructs without a proven basis.

The examples we have given show what a tremendous range cultural constructs of the body have in history and why it is important to keep track of them if we can. We see what we have been told to see, and we believe what we have been told to believe about our bodies. Besides taking most of its terminology from ancient Greek and Latin, modern medicine is from time to time tripped up by some ancient conception that runs afoul of clinical reality. Our visual culture is no less conditioned by legacies from the ancient world.

PALEOLITHIC BODIES

There is not much reason to believe that our early ancestors took a great interest in the human body during the Upper Paleolithic when the first great art was produced in Europe. The most arresting masterpieces of cave art at Lascaux, Altamira, Trois Frères, Chauvet, and other sites in France and Spain depicted animals with great accuracy over a period of 25,000 years until about 10,000 years ago: bison, stags, aurochs, ibex, horses, and mammoths. Depictions of humans are relatively few and casual, especially when they represent the male. Three-dimensional representations of the steatopygous (fat-buttocked) female body are iconic in a different way from the animal art, displaying features of an idiom that lasted longer than any other in the history of art. Though they bear a slight resemblance to some women who can be seen in any American shopping mall, these figures are unlikely real-life specimens of hunter-gatherers who lived transient lives in the landscape of paleolithic Europe before the establishment of settled farming villages. Breasts, buttocks, hips, and thighs are prodigiously obese; feet and hands are typically missing, and there is usually no face. Shoulders are narrow, as of a person who does no work with the upper body, but the genital cleft is shown when not overlaid by fat. There is no sign of ornament or clothing. Though they come to us with no explanatory captions or primary informant to explain their meaning, it is easy to guess that they represent success in eating and reproduction, two privileges that did not come easily to Paleolithic man or woman.

These "Venus" figures, as they have been ironically (but perceptively) called, have been found all over Europe from France and Italy to Moravia and Russia, transcending the styles of any region or ethnic group. Because they are anatomical fantasies, they are the first appearance known to us of the human body as part of a symbolic and projective system, the first cultural construction of the human body. Variously carved in bas-relief on a limestone wall or molded in clay, the stylized, schematic "Venus" figure presents an instructive contrast with the sharp realism of animal figures belonging to the same period. It is the distinction between nature and culture.

NEOLITHIC IDIOMS

Naked female images, both of the "Venus" type and of less exaggerated features, are more numerous than male throughout the Upper Paleolithic. It would be a mistake to think that they represent a dead end in the history of art. Their lasting symbolic value is argued by a tendency to abstraction that survived the Paleolithic cultures and emerged as a theme in the Neolithic communities that developed in the upper Balkans, Asia Minor, Mesopotamia, Malta, and Greece. The repertory of forms grew into a visual idiom that survived into the Bronze Age and finds its most refined expression in the Cycladic figurines beginning in the Neolithic fifth millennium B.C.E. and reaching their peak in the third (Early Cycladic I–III), where at its most abstract the female figure is reduced to a "fiddle form" icon with a long narrow neck and no head, arms, or legs. Cycladic figurines, the great majority female, have been preserved in significant numbers because they were buried with the dead.

Much older than the Cycladic carvings are the Early Neolithic figurines of northern Greece that precede 6000 B.C.E. Here the idiom includes phallic heads and necks, hands upon or just below the breasts (a gesture that never died out), minimal facial features with slits for eyes, and heavy thighs. The Neolithic cultures of Sesklo (7000–3200 B.C.E.) and Nea Nikomedeia produced simplified female figurines with birdlike faces, others with pronounced buttocks, and the kourotrophos, a nursing mother with infant that was much later revived in Christian art. These types occur far beyond northern Greece: Neolithic Lerna in the northeastern Peloponnese, Hacılar and Çatalhöyük in southern Anatolia, Cyprus, and numerous other sites. Though they are part of the legacy of human representation in what is now Greece, they precede Greece and belong to a widely distributed koine that extended beyond the confines of the Aegean basin and included southeastern Europe, Malta, and some of the Adriatic basin. It is now thought that the early Greek Neolithic culture that employed these styles was influenced by a large-scale influx of pioneer communities from southwest Asia ca. 7500–6500 B.C.E., but it is unclear whether the idiom we have been describing was also indigenous to the entire Balkan peninsula and was familiar to the newcomers.

BRONZE AGE INTERNATIONALISM

Historical Greece, the homeland of classical civilization, was a distinct complex of traditions of which the neolithic substratum was only one element. The Aegean world of the third millennium benefitted from a new period of Early Bronze Age internationalism anticipated by Cycladic sea trading that linked the western Greek mainland with the Anatolian coast (3100–2700 B.C.E.). In the centuries that followed, Troy II, the distant ancestor of Homer's Troy at the south end of the Dardanelles, became prosperous; deep-hulled sailing ships displaced the Cycladic rowed canoes, and Minoan Crete began to assert its hegemony in the southern Aegean. The second millennium opened with a Cretan thalassocracy that extended its reach into southern Greece and was the first true civilization in what would become the Greek world. As suggested by the myth of Europa, the princess of Tyre carried by Zeus in the form of a bull to Crete to become the eponymous founder of Europe, Minoan Crete had strong ties with the Levantine east. The language of its still-undeciphered Linear A script is believed to be a Semitic dialect, and the script may be derived from an eastern Mediterranean cuneiform. Minoan trade relations extended to south Russia or Afghanistan (copper ingots, lapis lazuli), Attica (silver), and Egypt (fine stone and finished products such as stone vessels, scarabs, and statues).

A palace civilization with multiple centers in Crete and an important settlement at Akrotiri on the volcanic island of Thera about seventy miles north of Crete, the Minoans created art in several media including ivory, stone, faience, gold, clay, and wall painting. Human subjects included male figures such as athletes, harvesters, bull jumpers, and worshipers. Both male and female figures are shown clothed, but with greater attention to anatomical detail than was typical in earlier Neolithic or bronze age art of earlier eras. In what could be taken as an anticipation of the athletic ideal of Classical Greek art, Minoan male and female figures are notably narrow in the waist to the point of being wasp-waisted. Especially in contexts of worship, where they are often shown, women have well-developed, exposed breasts that are clearly distinct from the obese breasts of the Paleolithic "Venus" figures of Europe. The best-known type is the "snake goddess" shown with a snake in each hand and fully clothed in a long flounced dress that leaves only her breasts exposed. It is unclear whether this is a goddess in epiphany or a worshiper; multiple figures in cult scenes, similarly costumed, appear in gold rings and seal stones.

MYCENAEAN GREECE

Minoan civilization declined in the late fifteenth century B.C.E. and its palace sites were occupied by Greek Mycenaeans. It is unclear to what extent this decline was a result of the volcanic eruption of Thera not far to the north sometime between 1630 and 1550, as many Minoan remains are above the ash layer attributed to that eruption. It is clearer that the mainland Greek Mycenaean civilization, which had earlier been a subset of the Minoan, took on a life of its own in the last phase of the Greek Bronze Age. Its independence is signaled in myth by the story of the Athenian prince Theseus slaying the Minotaur and running off with Ariadne, the daughter of Minos, whom he later abandons on an Aegean island.

Mycenaean civilization was based on a decentralized set of sites in eastern and southwest Greece. Significantly, it is proven from their Linear B writing that the Mycenaeans were a Greek-speaking culture, dominated by a warrior elite probably descended from migrants who came from somewhere to the north or northeast. They took most of their language, some of their religion, and their distinctive character from the final ethnic component of the historical Greek people, the Indo-Europeans. Paradoxically, the arrival of this potent new group sometime near the beginning of the second millennium is marked in the archaeological record by a long period of relative stagnation (Middle Helladic—Emily Vermeule described its early stages as "spiritually very poor"). But contact with the Minoans early in the Late Helladic period (beginning ca. 1550 B.C.E.) stimulated a rapid development in Greece that lasted until the collapse of Mycenaean civilization about 1100 B.C.E.

Early in the Greek Bronze Age, designated in archaeology as the Helladic period, the stylized female idols that formed an artistic idiom in Neolithic and Cycladic art had begun to disappear, perhaps because of religious changes. By Mycenaean times, the bright colors and ornamental themes of figural art were wholly Minoan in derivation. The presentation of female figures conforms to Minoan style: short-sleeved jacket open at the breasts, tubular flaring skirts with rows of flounces, bare feet, and elaborately coiffed hair. Prominent breasts and erect posture remain the rule for women, and small, well-defined waists continue to represent the ideal for both sexes. A distinctly Mycenaean theme, appropriate to the people to be remembered for the sack of Troy, is the warrior, most famously represented in the "Warrior Vase" of the twelfth century showing a line of helmeted warriors, each with round shield, spear, and small sack of provisions, marching away from a grieving woman.

THE END OF THE BRONZE AGE

The Trojan War, traditionally dated at 1184 B.C.E., was the last great event of the Greek Bronze Age, which came to an end with a system collapse throughout the eastern Mediterranean between 1200 and 1100 B.C.E. The Greek warlords who were the nominal victors at Troy would soon have their own palaces reduced to a burnt layer. Athena, coming to Ithaca to rouse Telemachus to search for his father in Homer's *Odyssey*, disguises herself as a trader in iron—a mark in the later epic tradition of the beginning of what we now call the Iron Age. The archaeological record of the centuries that followed the legendary Trojan war suggests a "dark age" in which populations declined and many Greek emigrants settled east and south of the mainland, most importantly in Ionia on the southwest coast of Anatolia. In later centuries this outlying region would become the home of the first Greek philosophers and the home of Hippocrates of Cos, the father of Western medicine.

The Mycenaeans had maintained constant trade with the east, as the Mycenaean Greek community in Ugarit (Ras Shamra) in the northern Levant shows.

The system collapse of 1200 cut east-west trade routes and isolated Greece from western Asia for several generations. But this did not last forever. By about 900 B.C.E., contact with the Phoenecians revived to the extent that the Greeks adapted their Semitic alphabet, converting some of its characters to vowels. Thus began a pattern of adaptive reuse that would characterize the civilization that invented itself during the following centuries.

PANHELLENIC ATHLETICISM

Two events in the eighth century marked the emergence of a Panhellenic identity that accompanied the beginning of the Archaic period, the first era of historical Greece. The transformation of religious games in honor of Zeus at Olympia to Panhellenic games is nominally tied to 776 B.C.E. Very little is known of the early history of the games, as the earliest written record dates from the work of a local historian, Hippias of Elis, near the end of the fifth century B.C.E. The quadrennial Panhellenic Olympics were, however, important enough long before Hippias to become the basis for dating any event in Greek antiquity. For historians of the body, the important fact is that the main events were athletic, and that victories in the contests were supremely important for the cities and families of the victors. The revival of the Games in 1896 C.E., the colossal audiences they command today, and the prestige accorded to participating nations according to the number of victories achieved by their athletes demonstrate the durability of the underlying ideologies of the body in world culture.

Another significant feature of the ancient Olympics was the establishment of male nudity. The origin of this custom is unclear, but two ancient authors place it in the Games of the year 720 B.C.E. Vase paintings confirm the practice of athletic nudity. To some extent, this practice grew beyond the athletic arena to become a costume of sorts so that nakedness, a symbol of helplessness, was transformed to nudity, a symbol of power especially among the oligarchic classes who cultivated their athletic prowess in the *gymnasion*, literally a place of nakedness. Only the most powerful were secure enough to flaunt their unprotected condition. Thus begins the cult of the nude body as an icon of power and social rank. In athletic contests as in warfare, the body is implicitly not a cosmetic object as much as a means to *do* something. It was both an instrument of power as its abstract representation. It represented and made possible the supreme virtue of *aretê*, which is cognate with *aristos*, "best."

The second defining event of the eighth century was the composition of the Homeric epics, which re-created the Trojan War some 450 years after the actual event and became the defining literary projection of the Greek character. Probably set down in writing between 750 and 700 B.C.E., the *Iliad* and the *Odyssey* took full advantage of the distance in time and the absence of an exact written record to make the war and its aftermath an affair that bridges the gap between the Mycenaean and the Archaic eras, making the epics a kind of commentary upon both. Like the Olympic games, Homeric warfare in the *Iliad* was a competitive enterprise in which the Greek Achaeans are as busy competing among each other as they are occupied with the defeat of the Trojans. Students of the body and its representation have long ago noticed that Homer rarely describes the body as a whole. Bruno Snell once argued that he had no conception of it but could articulate only one part or another. Helen of Troy is famously left undescribed, as if Homer were allowing his readers to construct their own Helen from their personal imagination.

BODY AND SOUL IN THE EPIC WORLD

So far as the ideal Homeric body can be pieced together from his epics, it is not the short, dark-complexioned body of the Mediterranean genotype. Two of the greatest Achaean or Danaan heroes are fair-haired (Achilles, Menelaus), as is the goddess Demeter and the underworld judge Rhadamanthus. Odysseus is an untypical hero because he is short in the legs, and Agamemnon is lordly for his height. The fullest physical description in the *Iliad* goes to the deformed antihero Thersites, who rails against Agamemnon early in the poem:

> This was the ugliest man who came beneath Ilion. He was
> bandy-legged and went lame on one foot, with shoulders
> stooped and drawn together over his chest, and above this
> his skull went up to a point with the wool grown sparsely upon it.
> Beyond all others Achilleus hated him, and Odysseus.
>   *Iliad* 2.216–220, trans. Richmond Lattimore

Like slavery, deformity was hated by the Greeks who read Homer, but they had little compassion for its victims.

Without doubt the most important statement Homer makes about the body is dropped into the very beginning of the *Iliad* where the poet talks about the cost of Achilles's wrath, which is the theme of his story. This wrath "sent the strong souls of heroes to the house of Hades, but made *themselves* spoils for dogs and birds." As if inserting a polemic against any system of belief that overvalued the soul, the poet of the *Iliad* insists in this programmatic moment that the self is the body that is destroyed in death. Similarly in the *Odyssey* when Odysseus tries to embrace the spirit of his dead mother in the underworld but fails three times because she is only a wraith, she explains,

> [this] is only what happens, when they die, to all mortals.
> The sinews no longer hold the flesh and the bones together
> . . .

but the soul flitters out like a dream and flies away. Therefore
you must strive back toward the light again with all speed.
                    Odyssey 11.218–223, trans. Richmond Lattimore

It could scarcely be made clearer that anatomy is ontology in the Archaic mind. Your body is all you get. This helps to explain the high value the Greeks of later generations were to place upon physical conditioning, health maintenance, and medicine.

GENDERING NUDITY

A lesser writer of the same century as Homer was Hesiod, author of a narrative about divine origins, the *Theogony*, and a handbook of wisdom for farmers, the *Works and Days,* both composed in the same hexameter meter and epic dialect as the Homeric epics. Like the epics of Homer, Hesiod's poems were foundational in Greek thought, not least when they told of the creation of woman, a story recited in both Hesiodic poems. The centerpiece of both narratives about the creature sent by Zeus as the evil antidote to fire stolen by Prometheus and given to man, is the costuming of Pandora (the name is mentioned only in the *Works and Days* version) with a gown, a veil, a garland of flowers, and a gold crown. Thus at about the time (ca. 720 B.C.E.) male nudity was established as the uniform of physical excellence in the Olympic games, female nudity was being covered in Hesiod's account of the creation of woman. Like Homer's polemics about the centrality of the body to human identity and the unimportance of the soul, Hesiod's account turns out to be a kind of polemic against western Asian representations of Ishtar (Babylonian)/Astarte (Phoenician)/Inanna (Sumerian), the goddess of love and war who was sometimes represented in the nude. The literature and institutions of Greece were therefore in the first century of the cementing of Greek identity rejecting female nudity and elevating male nudity to heroic status where it constitutes what Larissa Bonfante has called "the chief distinction between Greek and barbarian." This distinction was to endure until the slow emergence of the female nude in the last century of the Classical period, which ended about 332 B.C.E.

Paradoxically, the human figure depicted in vase painting contemporary to Homer and Hesiod is not as nuanced or articulated as we should expect. The geometric style reduced humans to simple formulas: a triangular chest, simplified beaked head, and bulbous thighs tapering to spindly lower legs. There is no attempt at perspective or display of emotion except the formalized gesture of mourning with one or both hands placed on the head. It is not until the seventh century that Daedalic and orientalizing influences make their way into Greek art, introducing a naturalistic rendering of volume and an organic unity that anticipates the monumental kouroi of the late Archaic period: statues of powerfully erect young men that became the first great masterpieces of male nude art.

The opening of Greece to west Asian influences is an essential component of Greek thought and art from the beginning of its development in the eighth century. As we have seen, it was selective, rejecting such idioms as the female nude while embracing patterns of myth (Hesiod's *Theogony* is an adaptation of the Ancient Near Eastern succession myth) and entire narrative structures that are common to Homer and much older stories such as *Atrahasis* and the Gilgamesh Epic. This orientalizing revolution has so impressed itself on the current study of antiquity that the old conception of the "Greek Miracle" as a completely native, autonomous, and self-starting development has been abandoned outside nationalistic Greek circles.

A SCIENCE OF THE BODY

From that derivative beginning, Greek thought devoted itself to an ocularcentric regime peculiar to itself. Visual information was accorded such a privileged role that their verb *oida*, "I know," means literally "I have seen." This was tempered, to be sure, by a strong tendency to make the human body fit an intellectually constructed visuality, assimilating the body to abstract, archetypal forms. This tendency has been traced as far back as the Geometric Dipylon Amphora, where the schematic human figures are shown to conform to a kind of geometric progression in which the torso and arms form an isosceles triangle and the head, body, and legs fall into a 1:2:4 ratio. This habit would develop in the Classical period into the Canon devised by the fifth-century bronze caster Polyclitus (ca. 460–410 B.C.E.), in which the parts of the body were proportioned to each other and to the body as a whole according to mathematical ratios: "Perfection comes about little by little through many numbers."

Closely related to this belief in numerical proportion was the belief that the human body is a microcosm of the universe as a whole. It is not clear at what point in the ill-documented development of Greek philosophy this idea was first articulated, but it is deeply embedded in the cultural history of the human body. Partially expressed in Protagoras's saying "Man is the measure of all things," this seminal idea could as well be reversed to say, "All things are the measure of man," because it was never articulated whether man or the universe had priority. Beginning about 550 B.C.E. in Ionia, a series of philosophers sought to displace poetic *mythos* with reasoned *logos* about the universe. Moreover, the core object of this pre-Socratic inquiry was *physis*, nature, placing the body implicitly within the ambit of rational inquiry.

The basis of the new, rational study of the body, its functions, and its malfunctions was the belief that the body was subject to the same laws as the rest of the universe. When medicine began to emerge as a learned discipline around the middle of the fifth century B.C.E. it was a sister art to philosophy. The Latin saying *medicina soror philosophiae* preserves the memory of this ancient linkage, which came at the cost of a certain predisposition for a priori thinking that privileged general principles over specific observed facts about health and disease. While the link to abstract philosophy has weakened over time, the belief that medical knowledge is inconceivable without the elements of chemistry, biology, and physics is enshrined in every school's premedical curriculum.

The first philosophers of Miletus in southwest Anatolia transformed Greek thought by arguing that divine influence has no power over weather and the heavenly bodies. Less than forty-five miles to the south, a college of physicians under the leadership of Hippocrates of Cos made similar arguments about health and disease. The Hippocratic author of a monograph on epilepsy began *On the Sacred Disease* with a statement challenging the very name of his subject:

> The disease called "sacred" is not, in my opinion, any more divine or more sacred than other diseases, but has a natural cause, and its supposed divine origin is due to men's inexperience and to their wonder at its peculiar character.
>
> from the Loeb translation of W.H.S. Jones

In spite of all their weakness for postulates (such as the doctrine of humors previously mentioned), the Hippocratic authors had a largely pragmatic character that assumed the laws on nature work in a more or less regular way that becomes increasingly clear with experience and the study of case histories. It was a blend of skill (*technê*) and knowledge (*gnosis, logos*) whose proof could be seen not only in manual applications but also in teaching and discourse: diagnosis in the recognition and naming of a disease, etiology in explaining what caused the condition, and prognosis in predicting the course of a disease. Medical interventions, except for treatment of wounds, reduction of dislocations, and administration of home remedies, were rare, which led to reliance upon prevention by diet and exercise in health and regimen in acute diseases. Gynecology and obstetrics were important specialties: no fewer than ten of the sixty surviving Hippocratic works written between 420 and 350 B.C.E. are on these topics.

THE BLACK BOX

In the comparatively rational culture of learned medicine developed by the Hippocratic school of medical thought, one taboo about the body stands out in strong relief throughout most of antiquity until the late Middle Ages. Except for a short period in Hellenistic Alexandria, the human body was regarded as a "black box" containing no serviceable parts. This taboo continued after death, and it was an absolute wrong to mutilate a corpse, even if it were that of your worst enemy. The classic example of this rule comes at the opening moment of Greek literature, in the *Iliad* when Achilles drags the corpse of Hector behind his chariot and Apollo complains to the other gods about "this cursed Achilles within whose breast there are no feelings of justice . . . So Achilles has destroyed pity, and there is not in him any shame" (24.39–45, trans. Lattimore). This prohibition was powerful enough to resist the curiosity even of Aristotle (384–322 B.C.E.) and his contemporary Diocles of Carystus, who wrote an anatomy book based upon animals. To some extent, it made the inside of the body a mythical zone whose contents and processes were poorly understood and where fictions such as black bile and the wandering womb persisted long after they could have been put to rest.

Notwithstanding the provisions of the Hippocratic Oath, ancient medical practice observed few other prohibitions. "Do no harm" is not in the oath, contrary to popular belief: it comes from a collection of case histories involving epidemics and advice to the physician: "As to diseases, make a habit of two things—to help, or at least to do no harm" (*Epidemics* 1.11). This can be distinguished from the oath, in which the physician promises not to engage in assisted suicide, provide an abortive remedy, or practice surgery. All three of these were commonly practiced in ancient medicine, however, as Ludwig Edelstein long ago showed, arguing that the oath was a Pythagorean manifesto "uniformly conceived and thoroughly saturated with Pythagorean philosophy." This is not to say that the more representative Hippocratic works were unconcerned with ethics, decorum, and other aspects of medical deontology. The Hippocratic tradition did much to establish medicine as a profession free of religion and magic at a time when there were no schools of medicine and no certification or licensing to distinguish trained physicians from quacks, empirics, and faith healers.

THE SEXUAL BODY

Ancient medicine was disposed by its surrounding culture to take account of the sexual body. In this it was as conditioned by embedded tradition as by simple observation. From Homer onward, Greek writers took it for granted that women were more sexually driven and less capable of sexual self-control than men. The Hippocratics, less concerned with sexual deportment than with health, argued that intercourse along with menstruation and childbirth were as essential to the health of mature women as intercourse was to the health of men. The importance of sex had been so fundamental to the Greek view of the world that Hesiod accounted Eros one of the first gods in creation after Chaos, Gaia, and Tartaros. As in traditional Jewish culture, it was assumed that men would take sexual pleasure outside of marriage just as it was denied to married women. As on earth, so in heaven: in Greek myth from Homer onward, Zeus is sexually insatiable, and the wrath of Hera at his constant infidelities, sexual partners (such as Io), and love children (such as Herakles) is a recurrent theme of mythological narratives.

Greek vase painting glamorized the sexual orgies, real or imagined, that took place in the symposium or drinking party traditionally located in the *andrôn* or men's quarters of the upper-class private home. The women hired for these entertainments ranged from common *pornai*, the lowest class of sex workers, to elite *hetairai* or "companions" who were trained in the arts of exotic dance, musicianship, song, and conversation and who were free to dispense their sexual favors as they chose. Though public art avoided the female nude, vase painting was uninhibited, illustrating a variety of abusive and affectionate sexual acts and positions. Homosexual encounters were also a subject of vase paintings, though these followed a more limited protocol in which penetration of a male partner was avoided. The rules governing pederastic love are most famously explained in Plato's *Symposium* (ca. 385 B.C.E. with a dramatic date of 416), where the respectful sexual pursuit of a worthy youth becomes metaphoric for the pursuit of knowledge. The larger cultural history is related in K. J. Dover's *Greek Homosexuality* (1978, 1989), still the authoritative treatment of the subject. Pederasty, in which mature men were attracted to pubescent boys of the same social class, was an oligarchic tradition possibly originating in rites of passage. But homosexuality in general was not socially rejected or tabooed as it was in the Judaic world or as it came to be in monotheistic Europe.

Sexual scenes on black-figure and red-figure pottery are concentrated in the late Archaic period from the sixth to the early fifth century B.C.E., but the tradition remained active through the fourth century and moved into other media such as bronze mirror covers, suggesting that the market for erotic art may have included women. Representations in art and jokes in Aristophanes suggest that genital depilation was widely practiced by women and remained so into the Roman period.

The hetaera became a glamorous paradigm of the female body by the end of the Archaic period, with some well-known vases picturing them separately rather than in relation to their male clients: preparing for a symposium, washing their costume, or talking among themselves. Tall and svelte, the hetaera of vase paintings was a Mediterranean exotic, almost as much an anatomical fantasy as the grotesque "Venus" figures of Paleolithic art but of the opposite type, anticipating the build of the modern fashion model.

KOUROS AND KORE

Another side of the Archaic somatic ideal that made the Greek body famous was the *kouros*, the erect, youthful male nude whose early predecessors were the cast bronze nudes of the late eighth and early seventh centuries B.C.E., clad only in a belt with right arm raised to brandish a spear. Under the influence of Egyptian monumental statuary, the kouros has his hands lowered to his sides and the left leg slightly advanced, with weight evenly distributed. Unlike their Egyptian prototypes, Greek kouroi are almost always nude, set up as offerings to Apollo or as tomb markers and displaying the timeless athletic ideal represented by Apollo. Produced over a period lasting scarcely more than 150 years, the kouros became a sculptural ideal, marking an obsession with male youth and beauty that outlasted the form itself.

The female counterpart during the same period was the richly clad *korê*. Like the kouros, she often wore the enigmatic "Archaic smile" and was used as a sanctuary offering or grave marking displaying the aristocratic ideal. Produced between ca. 650 and 480 B.C.E., their body was almost entirely obscured by their elaborate costume, though in later years the powerful build of their buttocks and thighs was shown behind. A variant of the sturdily erect kore was the caryatid, a female sculpture employed as a column. These first appear about 650 B.C.E., but unlike the kore who was not produced after the Persian Wars (490–479 B.C.E.), caryatids became an established fixture of Greek and Roman architecture, symbolizing the stable, supportive virtue of the female ideal. They further suggest the relationship between architecture and the human body explained by the Roman architect Vitruvius and illustrated by the famous "Vitruvian man" drawn by Leonardo ca. 1487 C.E. This relationship survives in the names we have assigned to parts of the body. Anatomical dictionaries today name some thirty-six columns, twenty-six arches, seven labyrinths, six roofs, three vaults, and five floors in the human body.

POWER VERSUS GRACE

By the beginning of the Classical period (ca. 480 B.C.E.), the male nude was the established icon of power. This is nowhere better illustrated than in the Artemision Zeus (or Poseidon), a bronze statue found in three pieces recovered from the seabed off Cape Artemision at the northern end of Euboia. Slightly larger than life and with somewhat elongated arms, the god stands poised to hurl his thunderbolt or trident. Though based upon a sculptural type that goes back to the time of Homer, this statue demonstrates the meticulous observation of surface anatomy that became the celebrated legacy of Classical Greek art and sent numbers of art students to medical dissections in sixteenth-century Europe when that standard was being revived. Two more recently discovered specimens of this precisely detailed standard are the two oversize bronzes found underwater in 1972 near Riace Marina in southern Italy and now called the Riace bronzes. Their asymmetric posture anticipates the chiastic contrapposto position codified by Polycleitus (fl. ca. 450–415 B.C.E.) where the arm above the weight-bearing leg is relaxed and the one above the relaxed leg is tensed, with the head turned in the direction of the weight-bearing leg.

Though anatomically correct, the Polycleitan formula illustrated in his statue of a spear carrier, the *Doryphoros* (known only in Roman copies), and described in his manual, the *Canon* (also lost), is not entirely naturalistic. This was acknowledged even in antiquity: the Roman rhetorician Quintilian noted that the effect of the positioning was to reduce the weight of the figure and to give it "a grace beyond truth." Classical taste in representations of the body changed in the course of the fifth century from weight and power to lightness and grace. The evolution can be seen to begin as early as the Late Archaic Kritios Boy (ca. 480 B.C.E.), whose softer lines and reflective, inward appearance place it in a different world from the kouros and the early classical severe style. Later male nudes such as the Apollo Sauroktonos of Praxiteles (fl. ca. 375–340 B.C.E.) became more youthful and graceful, not to say effeminate. More important still, the female nude came out from behind wraps, first in the "wet drapery" style of the east frieze of the Parthenon (ca. 437–432 B.C.E.) showing Artemis adjusting the drapery of her chiton and in the relief sculpture of Nike adjusting her sandal on the Nike balustrade of the Acropolis (ca. 410–407 B.C.E.). By about 350 B.C.E., Praxiteles was able to carve the fully nude Aphrodite of Knidos, surely the most celebrated statue in antiquity, though the marble original is lost and is known only by way of later Roman copies.

Reputedly modeled after the sculptor's mistress the hetaera Phryne, the Knidian Aphrodite is anatomically correct with the sole exception of the genitals, which were routinely omitted from female nudes throughout antiquity. Most details of her appearance are impossible to know with certainty, as the original statue has disappeared and all that remain are Roman adaptations. She appears to have been of somewhat heavier build than the svelte hetaeras shown on earlier vases illustrating the world of the symposium or than the bronze caryatid mirror bases that have survived from the same period. All of those illustrate the archaic preference for a more boyish female body, with narrow hips and small breasts. The production of female nudes became a regular industry from the later fourth century until the end of antiquity. The male nude was never completely displaced, but the female body became the supreme paradigm of the ideal esthetic form.

OPENING THE BLACK BOX

The science of the human body (if such can be said to have existed in antiquity) was a Hellenistic development stimulated by the biological investigations of Aristotle (384–322 B.C.E.) and his contemporary Diocles of Carystus. The locus of the new knowledge shifted from Athens, Cos, and Cnidus to northern Egypt. Alexander the Great's general Ptolemy I Soter established himself as the ruler of Egypt in 323 B.C.E., and with his son Ptolemy II Philadelphus (308–246 B.C.E.) sought to make Alexandria an intellectual rival to Athens by founding a library and a museum, which was less a collection of exhibits than a state-sponsored research institution. Founded about 300 B.C.E., the library collected and standardized ancient texts such as the epics of Homer and the medical works attributed to Hippocrates, while the museum supported the research of geographers, critics, poets, and mathematicians. Among the medical researchers recruited to the museum were two anatomists, Herophilus of Chalcedon (ca. 330–260 B.C.E.) and Erasistratus of Ceos (ca. 330–255 B.C.E.), who were encouraged to forget the ancient prohibition against opening the human body and carry their investigations into the interior. The result of their work was epochal for knowledge of the body and its functions. Herophilus showed that the brain was the center of the nervous system (Aristotle had thought it was the heart) and distinguished motor from sensory nerves. He investigated the anatomy of reproduction, the importance of the pulse, the gross anatomy of the brain, and the mechanics of vision, among other things. Erasistratus, in some ways a more advanced thinker than Herophilus, was more controversial in his time and is less well known to us. He anticipated the views of seventeenth-century C.E. anatomists such as William Harvey by perceiving that the body is a machine explicable in terms of mechanics and physics and whose failures are sometimes the result of mechanical breakdown. He saw that the liver and kidneys function as filters, and that the heart is a pump for both venous and arterial blood. In another anticipation of modern science, he departed philosophically from Aristotle's teleological notion that purpose determines form. In this and other ways, Herophilus was a visionary many centuries ahead of his time and hence scarcely representative of antiquity. Had the fifty-year window into human anatomy not been closed by about 250 B.C.E., it is certain that knowledge of the body would have advanced much further in antiquity.

GALEN AND THE MEDICAL SECTS

As it turned out, little new understanding of the body developed before Galen of Pergamon (ca. 129 to 199 or 216 C.E.). The son of a successful architect in western Anatolia, Galen studied medicine in Athens and Alexandria and became imperial physician to the court of the Roman emperor Marcus Aurelius. The most prolific author of the ancient world in any language, his surviving works make up fully a tenth of everything that has survived in Greek from antiquity. Much of what he wrote was about medicine and anatomy, and he remained the best-recognized authority on the human body well into the seventeenth century C.E. even though he never dissected a human cadaver. An eclectic follower of the learned medical authors with a strong loyalty to the Hippocratics and Aristotle, Galen believed in teleologic design where "Nature does nothing in vain." His vague but determined belief in a supreme being whose intentions can be learned by study of the body's parts made him acceptable to Christianity and assured his authority throughout Christian Europe.

Roman medicine did not, however, adhere single-mindedly to the doctrines of the Hippocratics, Aristotle, Diocles, and Herophilus, which together came to be labeled the "Dogmatist" school because of their tendency to appeal to deductive first principles. Like medical practice in the United States prior to the 1910 Flexner Report's refusal to accredit schools teaching homeopathy, naturopathy, lay herbalism, nutritionism, reflexology, irridology, osteopathy, and vitalism, Roman medical practice was divided among sects or schools of medical thought that tried to reduce the complexity of medicine to a single approach. The most respected of these Hellenistic sects was the Empiricist, which rejected the study of causes and sought instead the single goal of effective treatment, learned in part by the study of case histories. Empiricists rejected the study of anatomy because it centered upon the dead rather than the living body. Not to be confused with mere empirics, self-taught healers who made up their practice as they went along, Empiricists combined firsthand observation with collections of data about symptoms, signs, and syndromes. They preferred to regard medicine as a craft rather than a science.

The most radical of the Roman departures from learned medicine was the Methodist sect, which became the most popular school of medical thought and practice in the period of the Roman Empire (after 32 B.C.E.). Methodism benefitted from a Roman instinct to distrust anything too intellectual, and in the second or early third century C.E. fell under the spell of Pyrrhonist radical skepticism recorded by Sextus Empiricus. But well before adopting those credentials, Methodist practice became an established system, probably by the time of Nero (ruled 54–68 C.E.), concentrating upon what the patient felt as a guide to treatment. It reversed the Hippocratic aphorism and declared that "Art is short, life is long," meaning that the skills needing to be applied to the healer's art are few. An earlier Roman practitioner, Asclepiades of Bithynia, provided Methodism with its most popular slogan, that the best doctor heals "safely, swiftly, and pleasantly": *tuto, celeriter, iucunde*.

Methodism was the most influential of the sects confronting the synthesis of learned medicine assembled in the extensive writings of Galen. It was not altogether the contemptible array of quackery that Galen made it out to be, and it claimed among its adherents the greatest writer on women's medicine before Galen, Soranus of Ephesus, whose *Gynecology* survives. Roman medicine was as pluralistic as Greek medicine had ever been, but it was specially shaped by a Roman distrust of intellectual complexity and by a more stubborn adherence to supernatural causes, making the boundary between magic and medicine fluid in a way it had not been since the Hippocratics. Though medicine was regarded by the Italians as an alien's profession and no fit occupation for a native, by 49 B.C.E. the demand for physicians was so great that Caesar conferred citizenship on any doctor practicing in Rome, and in 23 B.C.E. Augustus granted tax immunity to physicians.

ROMANITAS AND THE ROMAN BODY

Like its formal medicine, Roman art was so influenced by Greek models that it is extremely difficult to identify or authenticate peculiarly Roman or ancient Italian conceptions of the body. Roman identity, such as the *Romanitas* promulgated by Virgil and Augustan propaganda in the period of the Empire, had more to do with civic identity than with anything more internal or individual. Roman religion was also an official matter of symbolic and ritual duty to the state, and the early Christians were persecuted chiefly because they were unwilling to make the largely perfunctory gestures of obeisance that the state required. Its indifference to matters beyond state ritual was Roman religion's strength as well as its ultimate weakness, as this made room for every kind of exotic import. Such imports included the introduction of the cult of Aesculapius (Gk. Asclepios) in 291 B.C.E.; its long-term success is indicated by the growth of shrines to this healing deity as late as 100–150 C.E. Another example is the Roman cult of the Hellenistic Egyptian Serapis, which reached Italy in 105 B.C.E. and was officially recognized in 43 B.C.E. Both healing cults included the ritual of incubation, in which the god appeared to the patient in a dream and described a cure. Christianity would also be responsive to the need for a religion with healing powers. It has been pointed out that of the twenty miracles recorded in the Gospel of Luke, all but three are medical. The trend in later antiquity toward the desecularization of medicine was to have further consequences, as we shall presently see.

CARING FOR THE BODY POLITIC

An important development in Roman medicine during the period of the Empire was the appearance of the *valetudinarium*, an ancestor of the hospital dedicated to the care of slave workers on the large industrial farms or *latifundia* upon which Rome and other cities depended for their food supply. These care centers also became a fixture in the large military encampments on which the Empire increasingly depended for the stability of its frontiers.

Institutionalized care of the sick had been a part of Greek city life as early as 600 B.C.E., when some cities began to appoint a physician at least to remain in residence and supported him with tax money. With the growth of urbanism in the Hellenistic period this became a more common practice. The mandate is also heard in the Jewish Talmud: "It is forbidden to live in a city without a physician" (BT Sanhedrin 17b) The realization that health was a matter of public policy and not merely the private relation between doctor and patient was one of the most important legacies of the ancient world, and one that did not go into decline with the dissolution of the Roman Empire.

Though plague had been known as early as the beginning of Homer's *Iliad* (Apollo was the god of plague as well as healing), it became a growing threat in the congested cities of the Hellenistic and Roman world. There was no better conceptual framework than the general notion of pollution or miasma that lay on the frontier between religion and rational medicine. The Antonine smallpox plague of the mid-160s C.E., a "virgin soil" epidemic, came from Iraq via the Roman army. It found the Mediterranean populations as defenseless as the same smallpox would find the native population of North America. The pandemic found the great physician Galen so helpless against it that he had no choice but to leave Rome between 166 and 169. Another historic pandemic that ravaged the ancient world was the Justinianic Plague of 541–544, which introduced Europe to bubonic plague and recurred in 749 and in the pandemic of 1347–1349. Beginning in the eastern Mediterranean, it eventually spread throughout the Mediterranean and up into Europe as far as Great Britain. More than anything else in the ancient world, these plagues challenged traditional thinking about the body, disease, and care of the body politic as well as the physical body. The result was a revolution in the way medical care was distributed.

A consequence of epidemics and other conditions of urban and transient life was the development of hospitals (Lat. *hospitium*, "hospice"). A convergent product of the religious and social values of Judaism, Christianity, and later of Islam (first ca. 710 in Damascus, then ca. 805 in Baghdad) the first hospitals were not entirely medical establishments as much as refuges for the poor, the elderly, and others on the social edge. Where they were unable to cure disease, they served to contain it and bring a degree of comfort to its victims. As a consequence, by the late Middle Ages medicine had expanded from a profession to a civic institution, completing a development that had begun early in Greek antiquity.

Though Christianity had brought an increased reliance on thaumaturgy, the miracle healing featured in the Gospel of Luke, the newly dominant religion of Europe had no radical conflict with rational medicine. Galenism, a reliance upon the extensive medical writings of Galen, brought an end to the epistemic pluralism of Hellenistic and classical Roman medicine and the rival claims of the medical sects. From about 340 C.E., late Roman medicine abandoned its ancient taste for controversy and sought instead a rhetoric of certainty. The demand for consistency produced medical digests such as those written by Oribasius of Pergamon (ca. 325–400), Aëtius Amideus (fl. 530), and Paul of

Aegina (ca. 630). Alexandria, still a center of medical learning in 500, produced a Galenic canon of sixteen books (or 24 condensed into 16).

The end of the first millennium C.E. brought a kind of economic polarization to Europe and the Mideast. Christian Western Europe became more thinly populated, rural, and economically underdeveloped, while the Mideast became a place of flourishing cities, commercial economies, and lively intellectual life. One result was a decline in European learning and a growing demand for translations into Arabic and Syriac of the works of Hippocrates, Aristotle, Galen, and other Western authorities. The resultant translation movement that began in the early ninth century under Abassid patronage and lasted into the late twelfth century produced translations, digests, and commentaries that established Arabic-Islamic medicine and preserved some Greek medical writings that were lost in the original language. Much of European medicine of the later Medieval and Early Modern periods depended upon the translation into Latin of those Mideastern texts and digests.

DEVALUATION OF THE BODY

Attitudes toward the body changed in later antiquity under several converging influences. Orphic and Pythagorean ideas as early as the late Archaic period devalued the body, considering it a tomb that confined the soul. Though these were not organized religious movements, their influence took root in later thought. Platonic transcendentalism emphasized a distinction between the body and the soul that elevated the latter at the expense of the former. Plato's conception of the body's hierarchic division in the *Timaeus* placed the neck as an isthmus separating the immortal soul residing in the head from the mortal soul residing in the chest, "subject to terrible and irresistible affections" that included "pleasure, the greatest incitement to evil." The lowest nature is then situated below the midriff in the stomach, bowels, and liver. Plato's Socrates in the *Symposium* and other dialogues is a heroic figure whose supreme intellect makes it possible for him to disregard the demands of the body. Though Plato is not representative of mainstream culture of his own time, he was not the only person in the Classical period to subordinate the life of the body. The desire for purity had been manifested in Hippolytus's perverse rejection of sex in Euripides's *Hippolytus* (428 B.C.E.). The same instinct became the subject of one of Catullus's best poems before the middle of the Roman first century B.C.E., where Attis brashly castrates himself to serve the Asiatic goddess Cybele. This mania appears to have been the centerpiece of other Cybele poems in the same era. Though extreme, the theme marks a development that eventually subverted a culture that had been comfortable with its physicality and from Homer onward had made the body all but coextensive with identity.

The secular philosophies of the Hellenistic period that largely displaced traditional religion in the minds of educated people placed a premium on control of the body in caring for the self. Though the philosophy of Epicurus made pleasure the chief metric of the good, its prescriptions for the good life entailed a sharply curtailed indulgence in food, sex, and other bodily pleasures traditionally associated with happiness. With Epicureanism, pleasure migrated from a condition of the body to one of the mind. Stoicism, which became the standard moral compass of late antiquity and deeply influenced early Christianity, so emphasized indifference to pain that the word "stoic" remains identified in our vocabulary with that virtue. Roman stoicism was increasingly austere: sex should be only for procreation; diet should be vegetarian, preferably uncooked. Even the pleasure of bathing is questioned: "What do you see when you take a bath? Oil, sweat, greasy water, all nauseating. Every part of life is like that" (Marcus Aurelius *Meditations* 8.24).

When Christianity displaced Stoicism it was not so much by rejecting its precepts as by adding to them a happier message of ultimate salvation and everlasting life. The highest form of Christianity was the ascetic life, where Gk. *askesis* was literally preparation for the life after death. In a religion of fasting and prayer, mortification of the body took the place of care of the body.

THE BODY CLOAKED

The art that grew up with these statements about the body and its management was increasingly reluctant to idealize the body. This was partly because monotheist religion has a tradition of opposing any human representation. Judaism, Islam, and Byzantine Christianity all have traditions of iconoclasm that have led at one time or another to the destruction of religious imagery, and when Christianity overcame the official religion of the Roman Empire, images associated with classical paganism were sometimes destroyed. Christian bishops and authors opposed all representational art. Pre-Christian Romans had been avid connoisseurs of looted Greek art, murals, and mosaics illustrating themes in myth, and reproductions of Greek statuary. In the transition from late Roman to early Christian art there is no rupture more abrupt than the one illustrated by the schematized, symbolic style of the paintings in the Roman catacombs. But the ocularcentric tendencies of Western culture were too strong to be overwhelmed by a new religion that distrusted images. Moreover, the steep decline of literacy made it necessary to stimulate the unlettered imagination with pictures. In a letter written to the iconoclast Serenus of Marseilles in 600 C.E., Pope Gregory the Great justified the use of images: "What writing offers to those who read, a picture offers to those who look; in it they read who do not know letters." This justification is significant because it highlights an important feature of early Christian art: it served as the handmaid of religion and was closely allied to the written word. Pagan art in the West had existed largely by itself and for its own sake, and it suggested admiration rather than veneration. Classical art also represented its human characters as exemplars of the well-tuned body, whereas Christian art celebrated personages whose contempt for the body was central to their identity. Its human characters are clothed; only the crucified Christ is shown with nothing but a loincloth. Over time, reliance upon drapery that concealed the body led to an insouciance about anatomy in Medieval art that required correction in the early modern period when the disciplines of ancient art were reengaged.

The ancient West can be said to have a defining preoccupation with the human body for slightly more than a millennium, from the beginning of the Panhellenic games at Olympia (ca. 776 B.C.E.) to the edict of Theodosius I in 394 C.E. banning all such games as pagan rites. Connoisseurship of ancient Greek art, especially in the great cities, did not instantly disappear, but the values that had made it a living legacy gradually faded with the acceptance of Christianity.

The rise of monotheism had no such effect upon medicine. Though new knowledge and conceptual frameworks did not come to light in late antiquity, the popularity of religious pilgrimages combined with the religious acceptance of responsibility for compassionate care to make the hospital a fixture of urban life. This was especially true in pilgrim cities such as Rome, Antioch, Alexandria, and Jerusalem, where the capacity of a single hospital could reach as high as 500. Though medical care was at first tangential to other services, by the seventh century some hospitals had specialized wards for the treatment of surgical patients and patients suffering from ocular disease. Islamic hospitals first led the way in medical services. The medical writer Haly Abbas (Majusi, fl. tenth century C.E.) recommended that doctors go there for medical experience. But it was not until the high Middle Ages that the teaching hospital began to appear. The first in Christendom was the Pantokrator hospital in Constantinople (founded 1136), which anticipated teaching in its foundation charter. Thus the continuity that was lost in visual culture was maintained in medicine, thanks in large part to the Islamic world.

SUGGESTIONS FOR FURTHER READING

Boardman, John. 1993. *The Oxford History of Classical Art*. New York and Oxford: Oxford University Press.

Brown, Peter. 1988, 2008. *The Body and Society. Men, Women, and Sexual Renunciation in Early Christianity*. New York: Columbia University Press.

Campbell, Gordon. ed. 2007. *The Grove Encyclopedia of Classical Art and Architecture*. 2 vols. Oxford and New York: Oxford University Press.

Conrad, Lawrence I., Michael Neve, Vivian Nutton, Roy Porter, and Andrew Wear. 1995. *The Western Medical Tradition 800 B.C. to A.D. 1800*. Cambridge: Cambridge University Press.

Diebold, William J. 2000. *Word and Image. An Introduction to Early Medieval Art*. Boulder, Colorado: Westview Press.

Dover, K. J. 1978, 1989. *Greek Homosexuality*. Cambridge: Harvard University Press.

Fox, Robin Lane. 1987. *Pagans and Christians*. New York: Alfred A. Knopf.

Garrison, Daniel H. 2000. *Sexual Culture in Ancient Greece*. Norman: Oklahoma University Press.

Greek Ministry of Culture, Athens/National Gallery of Art, Washington. 1988. *The Human Figure in Early Greek Art*.

Guthrie, R. Dale. 2005. *The Nature of Paleolithic Art*. Chicago and London: University of Chicago Press.

Hampe, Roland, and Simon, Erika. 1981. *The Birth of Greek Art. From the Mycenaean to the Archaic Period*. New York: Oxford University Press.

Miller, Stephen G. 2004. *Ancient Greek Athletics*. New Haven and London: Yale University Press.

Nutton, Vivian. 2004. *Ancient Medicine*. London and New York: Routledge.

Stewart, Andrew. 1997. *Art, Desire, and the Body in Ancient Greece*. Cambridge: Cambridge University Press.

Vermeule, Emily. 1964. *Greece in the Bronze Age*. Chicago and London: University of Chicago Press.

CHAPTER ONE

"The End is to the Beginning as the Beginning is to the End"

Birth, Death, and the Classical Body

VALERIE M. HOPE

In the second century A.D. Artemidorus of Ephesus wrote a book titled *On the Interpretation of Dreams*. In this work Artemidorus collected stories about people's dreams, their predictive or prophetic elements and how dreams could be interpreted. The work is valuable not only as the sole surviving work on ancient dream analysis, but also for its social and religious insights. In one commentary on dreams about birth, Artemidorus noted that for a sick man such a dream foretold death, because, "The dead like babies are wrapped in cloth strips and are placed on the ground; and the end is to the beginning as the beginning is to the end."[1] This comment, little more than an aside, provides a central insight into both the rituals of birth and death in the ancient world and the symbolic manipulation of the ancient body—alive and dead. It suggests a parallel symbolism between entering and leaving this world, connections between the bodies of the newly born and the bodies of the dead, between human flesh and the earth and its fertility.

Artemidorus's brief comments on birth and death epitomize many of the issues and difficulties in reconstructing and interpreting ancient rituals, beliefs, and customs. Who was Artemidorus? Were his comments accurate? Did they apply across the ancient world or just in his locale and lifetime? Answering these questions about any ancient author (or source) is almost impossible. We face the challenge that what is loosely termed the ancient world or period stretched across the centuries and the miles. Thus we know little of Artemidorus's life beyond that he was from modern Turkey, spoke Greek, and lived in the second century A.D. under Roman rule. Whether his comments and insights were equally applicable to Archaic Greece, Classical Athens, Republican Rome, or every settlement in the Roman Empire is unclear. In understanding ancient rituals surrounding birth and death we are often forced to patch together snippets of information, comments, and insights that are disparate in terms of origin, period, and nature. This process can be far from satisfactory; we cannot assume that rituals, actions, and thought processes were static or unchanging across space and time.

The focus in this chapter largely falls on the city of Rome during the late Republic and the rule of the early emperors, that is, the first century B.C. and first two centuries A.D. However, the chapter will draw on the broader ancient context for birth and death rituals before taking a more detailed look at some of the embodied experiences that marked death and mourning in ancient Rome. The details of the expectations and experiences of the inhabitants of the ancient world may have varied, yet the themes that emerge here, centered on pollution, transition, transformation, inversion of norms, gendered experience, and performance, can be seen to characterize not just death in ancient Rome, but birth and death throughout the ancient world.

BIRTH

The arrival of a baby in the ancient world could be a joyful event, but it was also one marked by danger, anxiety, and a sense of pollution. Mortality rates for mothers and infants are difficult to estimate,[2] but the risks were real. Both Julius Caesar and Cicero lost their daughters, Julia and Tullia, respectively, following childbirth. Pliny the Younger, in one of his letters, tells how two sisters both died in childbirth.[3] Infant mortality may have been so high, at birth and in the months following, that it has been argued that emotional investment in babies was limited. Infants could be farmed out to other caregivers such as wet nurses and child minders, and infants were not afforded full burial rites if they happened to die.[4] The exposure of infants was also practiced, in both ancient Greece and Rome; a father could simply choose not to rear a baby and abandon it. How common such an action was and the circumstances that dictated it are much debated.[5] Adult interest in babies and children can be characterized as self-interested; the priority was for the continuity of the family name and the value of any offspring to provide for and support the parents. However, plenty of more emotive evidence from the ancient world suggests that children were not just viewed as a commodity, that the characteristics of children, including babies and toddlers, were cherished and that parents grieved deeply when their children died.[6]

Birth, midwifery, and early child care were primarily a female sphere, but a sphere within which men still expressed opinions and one they ultimately sought to monitor and control. The birthing mother would have been attended by women: family, neighbors, and wise women for the less well off, and experienced midwives for those who could afford them. Male doctors may have sometimes been present at the delivery, especially if there were complications, and may also have been involved at other stages of the pregnancy.[7] Texts written by men, and probably primarily read by men, record the practices and customs that surrounded childbirth, some of which were bound up with superstition and folklore and others with medical knowledge, however rudimentary.

Pliny the Elder, writing in the later first century A.D., described some of the practices used to hasten and ease labor and noted that boys were more easily delivered than girls.[8] Women could lessen labor pains by consuming various drinks: one made from powdered sow's dung; another from sow's milk, honey, and wine; and one concoction composed of goose semen mixed with water.[9] Pliny also mentioned the use of various herbs and amulets such as a snake's skins and a vulture's feather.[10] It is easy to be dismissive of such practices, some of which may have had a placebo effect, but childbirth with its inherent pain and risk to life was readily associated with superstition and religious belief. In the Roman world fifty-eight divinities were linked to childbirth,[11] covering various aspects of the pregnancy, labor, and delivery. Juno Lucina was the traditional main goddess of childbirth. The epithet Lucina may be derived from *lux* (light); birth brought the child from darkness to light and separated it from mother earth.[12] In ancient Greece women prayed to the goddess Artemis and god Asclepius (patron god of medicine), and also to specialized childbirth divinities such as Eileithyia or Lochia. Votive offerings might be made to these divinities following a successful labor.[13]

From a modern perspective it is tempting to give greater priority to medical writings on childbirth, but how widely these were read by those who actually delivered ancient babies, or whether they were written to reflect standard practices, is impossible to know. Particularly detailed is the work of Soranus, who wrote in the early second century A.D. and focused on midwives and procedures. Soranus argued that a midwife needed to be a competent woman, literate, respectable, of good memory, physically robust, with her wits about her; while the possession of long, slim fingers and short nails was also an advantage.[14] Soranus's list implies that some women may have fallen short of the ideal! The demand for literacy suggests that a good midwife should have read up on her subject—presumably including the works of Soranus—and not be duped by superstition. However, how educated midwives actually were is difficult to judge. In general there were few respectable professions for women to pursue in the ancient world. The evidence from Rome suggests that midwives were slaves and ex-slaves; in the eastern parts of the Roman Empire medicine may have been more widely respected, and some gynecological tracts were even written by women.[15] The reality may well have been, though, that educated midwives were in the minority and employed by only a wealthy few. The majority of midwives learned on the job; experience rather than formal training was what mattered most, experience that encompassed practical knowledge as well as folklore and superstitious beliefs.

Soranus also described procedures for the birth. He noted the equipment required and how the midwife should assist and support the woman.[16] Birth was expected to take place on a birthing stool with a crescent-shaped hole through which the baby was delivered and armrests for the woman to grasp. The stool needed a strong back, or alternatively a female assistant stood behind the stool to support the woman.[17] Reliefs depicting birth scenes provide some support for this (Figures 1.1 and 1.2). The production of such images showing childbirth is striking and can be seen as elevating the status of midwives and women, including those who died in childbirth.[18] Plutarch noted that the Spartans allowed the names of the deceased to be inscribed on tombstones only for men who had died in battle and women who had died in childbirth.[19] However, images of childbirth tend to suggest the passive, sometimes almost pitiable, role of women during the process, rather than elevate them to the equivalent status of warriors.[20]

After the birth the new mother and child could be regarded as vulnerable, and in some sense polluted. The exact cause and nature of the pollution is hard to pin down and encompassed spiritual and physical elements, folklore, and practicalities. The household was thought to need protecting from evil spirits; the mother and baby were at risk of infection and were also in a state of flux, due to movement from the prenatal world to this world for the child and to motherhood for the woman.[21] The sense of pollution marked an interim stage of transition before mother and child were integrated with the household. Indeed, the acceptance of the child was not automatic, and immediately after birth attention moved from the body of the mother to the body of the child. The baby was laid upon the ground for the midwife to assess its strength and viability.[22] The decision to raise the child ultimately rested with the father; the biological birth of the child needed to be followed by its social birth. It seems unlikely, however, that the father was the one to raise the newborn baby from the ground in a symbolic gesture of acceptance.[23] Instead, in both Classical Athens and Rome, ceremonies were held some days after the birth (infant mortality would have been particularly high in the early days) that gave an opportunity for the family, and to some extent the wider community, to see the baby and witness its acceptance into the household. In Athens, on the fifth or seventh day

Bas-relief antique en terre cuite, ornant le monument funéraire d'une sage-femme romaine, récemment découvert à Ostie. (Voir communication du professeur Capparoni, page 243). — Ce bas-relief représente une scène d'accouchement. La parturiente, nue, est assise sur le fauteuil obstétrical. Derrière elle, une assistante la maintient fermement en lui passant un bras sous l'aisselle. Devant, la sage-femme, vêtue d'une longue tunique, assise sur un petit tabouret, paraît s'opposer à la sortie trop brusque de la tête de l'enfant.

FIGURE 1.1: A terra-cotta relief attached to the tomb façade (Tomb 100) of Scribonia Attice at the Isola Sacra Necropolis, Portus (near Ostia, Italy). Mid second century A.D. A midwife delivers a baby with the laboring woman seated, and supported by an attendant. The relief suggests that Scribonia Attice was a midwife; a second relief from the tomb depicts a doctor and doctor's implements, suggesting that Scribonia Attice's husband was also in the medical profession. Wellcome Institute. M002646.

after birth a ritual called the *amphidromia* ("running around") was held; this involved animal sacrifice, with the father encircling the household hearth while carrying the child, and purification for the women, while friends and relatives sent gifts. The family might also decorate the doorway of the house with wool for a girl and an olive garland for a boy. For those who could afford a second ceremony, naming occurred on the tenth day after birth (*dekatē*), which could be a more festive occasion.[24] In Rome, the house doorway might be decorated with flowers, and the eighth day after birth for a girl and ninth after birth for a boy was called the *dies lustricus,* or purification day. The mother and child were ritually purified and the child was given a name.[25] These events completed the time of transition for the new mother, although the child was still perceived as vulnerable to physical and spiritual dangers and might be protected by amulets, such as the *bulla* worn around the neck of well-to-do male Roman citizens.

Childbirth in the ancient world focused attention on the bodies of both women and children. Both mother and child were vulnerable not just to death,

FIGURE 1.2: Ancient Roman relief carving of a Roman midwife. Wellcome Institute. M0003964.

but also to pollution and rejection. Birth transformed the woman into a mother and could thus elevate her status within the household as the provider of heirs, especially a son. But this transition to motherhood involved a sense of pollution, which could affect midwives and other women present at the birth. One of the reasons for midwives' low status, apart from accepting payment for work, may have been the polluting aspect of their role (compare with undertakers; see next section). Women played the central part in childbirth, yet men—husbands, fathers, and doctors—monitored the process. The male doctor through his written treatises assumed expertise over the female lore of childbirth, and in these writings, including the Hippocratic *Epidemics,* the birthing woman's role could be viewed as almost passive; it was the child who was active, struggling and forcing its way into the world.[26] The men of the household, while distanced from the biological and bodily aspects of childbirth, played a central role in the process of the acceptance of the birth and of the infant. The treatment of the baby's body—its physical handling and assessment—symbolized its place in the social hierarchy and gradual integration into the household. Women controlled the labor room and the biological aspects of birth, but it was men who controlled the social birth of the child.

DEATH

In the ancient world death, like birth, was ideally centered on house, home and family. A good death happened in one's own bed, with one's family and friends present, who would hear some final wise words and then perform the essential rites. The dying person would have time to put his affairs in order, check his will, entrust children to the safekeeping of others, maybe pray for an easy death and ready his appearance by washing.[27] Many people may not have achieved this ideal, dying away from home, or on the battlefield or unexpectedly without having time to prepare themselves. Pain and suffering must have also characterized many a deathbed, but idealized death scenes often promoted bravery in the face of death, serenity, and resignation; the bodily horrors of death were little acknowledged in the domestic ideal.[28]

In Greece at the moment of death the eyes and mouth of the dead person were closed. In Rome a relative would kiss the deceased to catch their last breath, close the eyes, and shout out the name of the dead person. In both cultures wailing and lamenting might follow, and the body may have been rested on the floor briefly to establish that death was certain. The body was washed by female relatives, then dressed and anointed. A coin could be placed in the mouth of the deceased to pay the underworld ferryman, and the jaws may have also been bound. The body was laid on a bed or couch (Greek *kline*, Latin *lectus funebris*), feet toward the door, for display and ready for visitors to pay their last respects. Hired mourners might sing and chant laments and dirges, and small gifts or offerings could be given to the dead. The body might be surrounded by torches and incense burners.[29]

Death brought a sense of pollution to the family and those who came into contact with the dead body. As with the pollution brought by birth, death pollution was a mixture of physical, spiritual, symbolic, and traditional factors. The corpse as decaying matter was viewed as unclean, but it was also betwixt and between this world and the next and vulnerable to supernatural powers.[30] In Rome the acknowledgment of a death meant that the family became a *familia funesta*, obliged to undertake the funeral and also prohibited from its usual activities, that is, symbolically marked out by death.[31] However, in both Greece and Rome the direct handling of the corpse primarily by women meant that the pollution may have affected the men of the house least of all. In Athens the preparation and touching of the body may have been limited to women closely related to the dead, especially the elderly.[32] A death in the family may have been particularly problematic for men with a public religious role. In Greece those who had recent contact with the dead were banned from temple precincts.[33] In 82 B.C. when the Roman dictator Sulla was presiding at the festival of Hercules, he was unable to visit his dying wife and was forced to divorce her and have her removed from his house. Sulla was honoring religious scruples and apparently subsequently grieved for his wife deeply, honoring her with a lavish funeral.[34] The emperor Tiberius, a stickler for traditional behavior, also took his religious role seriously and delivered the eulogy for his son Drusus in A.D. 23, with a veil placed between him and the body, "so that the eyes of a high priest might not look upon a corpse."[35] The use of paid undertakers, especially in Rome, to perform the menial tasks associated with washing, preparing, and dressing the body may have also created a sense of remove between the well-to-do elements of society, both male and female, and dead bodies. Undertakers, generally slaves, were maligned and stigmatized for their close contact with the dead and also the fact that they received payment.[36]

On the day of the funeral, probably within three days of the death (perhaps longer in Rome), the body was carried out for burial. The procession (Greek, *ekphora*; Latin, *pompa*) could be an important aspect of the funeral ritual, with musicians, hired mourners, family, and friends creating a noisy and eye-catching pageant. There is some evidence for the use of carts to transport the body in Archaic Greece,[37] but in both Greece and Rome a bier carried by up to eight pallbearers was probably the norm. In Rome during the late Republic and early Empire the procession for members of the aristocratic elite could involve actors wearing masks (*imagines*) of other deceased relatives, creating a scene that united present, past, and future.[38] A funeral speech could also be delivered in the Roman Forum, which praised the deceased and thereby elevated the status of surviving family members.[39]

Once the body arrived at the cemetery or tomb it could be either inhumed or cremated. The rites performed at this stage are unclear but probably involved animal sacrifice, the use of formulaic expressions, and the calling out of the name of the deceased.[40] It is also unclear why there were shifts in use between cremation and inhumation. In Greece inhumation and cremation were practiced concurrently from the eighth to the fourth century B.C. In the Hellenistic period inhumation became more popular. In Rome inhumation was often regarded as the oldest rite, although cremation was most popular from the late Republic through the first century A.D., with inhumation gaining ground in the second century A.D. and then becoming almost universally employed across the Roman Empire. These shifts in burial rite are perhaps best related to changes in fashion rather than alterations in deep-seated religious belief.[41] It is worth noting that other rites such as exposure and mummification were known of and practiced in some parts of the Roman Empire but were not widely exported. Ancient authors could view such rites as a cultural indicator.[42] How dead bodies were treated was part of cultural identity, what differentiated the Greeks and Romans from others.

On returning from the funeral, the mourners were purified (by fire and water in Rome) and may have shared a funerary meal. Nine days after the death and funeral in Rome, a feast at the grave and further purification rituals were performed.[43] The exact timing of the equivalent rituals in ancient Greece is unclear, but these may have involved the symbolic sweeping out of the house and feasting.[44] Graves whether for inhumed or cremated remains were generally marked. Grave indicators could be simple—for example, terra-cotta pots, upright stone slabs, cinerary urns, and funerary altars (Figure 1.3)—or highly elaborate architectural flights of fancy.[45] The grave or tomb was the focus for regular visits.

In Greece the tombstone (stele) could be decorated, wrapped with colorful ribbons, and adorned with flowers (Figure 1.4). Flowers—especially violets and roses—were brought to Roman tombs, and offerings of food and wine were common in both cultures. Annual festivals—the *Genesia* in Greece and the *Parentalia* in Rome—were set dates for visiting and honoring graves, but other festivals, anniversaries, and birthdays also provided occasions for tomb cult.[46] The physical bodies of the dead might be gone, but they could still be regarded as requiring sustenance (if more symbolic than actual) and remembrance.

Any account, whether of birth or death rituals, that covers such a large time span and several cultures is bound to be a composite one. There were

FIGURE 1.3: A Roman grave marker in the shape of an altar, Rome, late first century A.D. or second century A.D. The epitaph was set up to Cornelia by her husband and another man and depicts the dead woman in a portrait bust (*Corpus Inscriptionum Latinarum* VI, 16362). Wellcome Institute. L0013594.

FIGURE 1.4: Scene from a Greek vase (white ground lekythoi) depicting a funerary stele adorned with ribbons, with mourners bringing offerings to the grave. Wellcome Institute. M00518.

differences and variations in the details of the general picture painted here of Greco-Roman funerary rites. Common themes did unite the death rituals of Greece and Rome—pollution; respect for the dead; tradition; custom; superstition; and the articulation of social differences according to wealth, gender, and status. But it remains difficult to reconstruct precise details and to understand some of the religious and spiritual aspects to the customs. Did people make offerings at graves because they really believed that the dead needed sustenance in the afterlife, or did they do these things because it was accepted tradition or a way of honoring and remembering the dead? Were those who came into contact with the dead actually thought to be impure or polluted, or was this a matter of convention, little questioned or understood? What we can say is that the dead body was the focus of a series of accepted rites that transferred it from home to cemetery, from the family to the kingdom of the dead, that removed the inanimate and potentially threatening corpse from this world to the next. The body was at the center of these rituals: on the one hand an inanimate and passive symbol, but on the other hand driving the actions of the survivors.

ROMAN CORPSES

Funeral rituals center on the body, on the corpse. The rituals transfer the corpse to its new abode, removing it from house and home, or at least the realm of the living, to the cemetery. The corpse is thus transformed, both symbolically and physically, from its former living, breathing status to that of an inanimate and distant ancestor. The human corpse, during and after these rituals, is a source of anxiety and ambivalent emotions. It is all that is left of the person who has been loved and lost, but it is also associated with decay, disease, and pollution as well as symbolizes the mortality and vulnerability of the living survivors.[47] The corpse is inanimate but retains a certain power or hold over the living.

In the Roman world bodies defined essential identities based on gender and age, but bodies could be further codified by dress, adornment, gesture, brands, and tattoos. Such distinctions did not cease with death. How dead bodies were treated often mirrored and reinforced social distinctions that had pervaded life. Dead bodies were separated from the living by a series of rituals that fulfilled emotional, spiritual, and practical considerations (see preceding section). However, the proper disposal of the dead also sprung from a sense of shared compassion and humanity; it was decent and civilized to dispose of the dead appropriately.[48] For a body to lie unburied had implications for the living and the dead. Not only was this a way to dishonor the memory of the dead (see below), but it could also reflect badly on the status of the living survivors, suggesting failure in duty or family responsibility; there were also religious and superstitious implications. Those inadequately buried were believed to be unable to rest in peace. Ghost stories often centered on those who haunted the living until they were given decent and proper burial.[49] The ghost of the assassinated emperor Caligula (Gaius) was said to haunt the place where he had been rapidly cremated and then buried, until his body was properly laid to rest.[50] The extent to which people actively believed in ghosts, spirits, and the afterlife is difficult to judge; elite authors could be dismissive of common beliefs, but many others may have kept an open mind about what happened after death,[51] and at the very least traditions and tales about the discontented dead enforced the idea that the dead should be buried properly and treated with respect.

For the great and the good, the treatment that a dead body received was expected to reflect life. For the wealthy elite, especially men, the ideal was for the corpse to be displayed in its finery and then to be carried on an elaborately decorated bier, accompanied by a lengthy procession, which would include a display of the family masks (*imagines*). A flattering eulogy would be delivered, before the bier was placed on an impressive pyre that might be painted and decorated. The collected ashes (or inhumed remains) would be marked by a suitable memorial tomb that could, through words and images, summarize the notable life of the deceased, acting in some respects almost as a substitute for the living body. During the funeral rituals the corpse took center stage and was in some respects objectified: decorated and perfumed, a silent and immobile witness to the unfolding events. Yet in other respects these grand funerals blurred the distinctions between the inanimate corpse and living bodies. The *imagines,* including one of the dead man, might be worn by actors, bringing the dead back to life and intermingling the generations, the long dead and the more recently dead, of the family. The mime employed to impersonate the emperor Vespasian at his funeral was particularly convincing, toying with the emperor's reputation for stinginess.[52] Effigies of the dead were also used at some funerals, especially if the body had already been cremated elsewhere. This suggests the symbolic importance of the body and the role of funeral rituals in the politics of power. In A.D. 193 the funeral of the emperor Pertinax was organized by the emperor Septimius Severus. Pertinax had died some months earlier, and his successor had been overthrown by Septimius Severus, who now needed to legitimize his rule as the rightful heir to Pertinax. With no body available, a richly dressed effigy of Pertinax was made and displayed in the Forum; it was even fanned by a slave as if to keep away the flies. People processed by the effigy, which was surrounded by images of distinguished Romans, as if paying their last respects. Septimius Severus read a eulogy, and then with the crowd wailing and lamenting the effigy was borne to a pyre, which was three stories high and adorned with ivory and gold. The consuls set fire to the pyre and an eagle flew out and away, symbolizing that Pertinax's soul was to be placed among that of the gods. Thus Pertinax was made immortal, and Septimius Severus emphasized that the seat of power was rightfully his.[53] This was not so much a funeral as a ceremonial institution that was integral to the safe transfer of power.[54]

The converse of this pomp and ceremony, and the elevated status associated with some bodies, was how the bodies of the poor and destitute could be treated. Most people hoped for and probably achieved a decent burial, with the family doing the best that they could, or the individual took out an insurance policy, joining a burial club (*collegium*) that, for a monthly subscription, would pay out a lump sum on the demise of a member.[55] Even slaves, who in theory had no income or property, could join burial clubs or at least expect their remains to be interred in the tomb of their owner. The funerals of such people may have been relatively simple but performed the basic function of ensuring decent disposal and some form of commemoration. There were those, however, who must have had no one to provide for or care for their bodies. John Bodel has estimated that for Rome as much as 5 percent of the population would have fallen into this category and that as many as 1,500 corpses annually had to be disposed of at public expense.[56] These bodies may have been dumped in mass communal burial pits. The author Varro refers to *puticuli* (pits) into which the bodies of the poor were thrown,[57] and during the late Republic such paupers' graves may have been located in the Esquiline area of Rome.[58] Whether these pits continued under the early emperors, when cremation was the normative burial rite, we simply do not know,[59] but we can say that the disposal of the extreme poor, the treatment received by their bodies, provided a striking contrast to that received by the wealthy and honored members of society.

Disposing of the dead might be regarded as an essential human act, but elaboration or simplification of the rite reflected social status. Equally, transgression of certain social boundaries and conventions could also affect how dead bodies were treated. In Rome certain groups were thought to degrade, dirty, and demean their living bodies through their profession; these could include actors, prostitutes, gladiators, and undertakers. Recruited from people of low status, including slaves, these people sold, displayed, or soiled their bodies for the entertainment or use of others.[60] There is some evidence that the stigmas of life followed such groups to the grave; they may not have been denied burial but could perhaps be buried separately, placed at the edge of the cemetery, just as they had lived at the edge of society.[61] The means of death could also affect access to funerary rites and burial areas. Some suicides, especially those who hanged themselves or were guilty of crimes, and thus made a dishonorable exit from life, may have been denied full burial rites.[62]

The ultimate indignity was corpse abuse. Some bodies were singled out for attack; so great were the transgressions of the living body that the dying and the dead body did not deserve to receive the expected spiritual, moral, and practical norms. Common criminals met horrific ends: crucifixion, burning alive, or being thrown to the beasts. There were no humanitarian concerns about speedy and discrete executions; death was a public spectacle.[63] The bodies of criminals became objects of display. Following the slave revolt of Spartacus (71 B.C.), 6,000 captives were crucified along the via Appia from Capua to Rome and their bodies left to rot as a grisly reminder of the price for defying Rome.[64] The author Petronius described how in Ephesus the crucified bodies of thieves were guarded by soldiers to prevent their removal.[65] These bodies were crucified at the edge of the settlement and thus near a cemetery. The contrast between the graves of the respectable on the one hand and the exposed rotting corpses of the condemned on the other must have been a striking one.[66] The living bodies of criminals were humiliated and tortured; then their dead bodies were mutilated, abused, and displayed; and finally they were denied a funeral and burial. In Rome what was left of the body was most probably dumped in the river Tiber, which washed away the waste and purified the city.[67]

Criminal or treacherous members of the elite might be allowed a more dignified exit, being given the opportunity to commit suicide or a relatively swift dispatch by means of the sword. In dangerous times, such as under the reign of terror of the emperor Tiberius, people implicated in the numerous treason trials often chose suicide to ensure that their body received decent, if often rapid, disposal.[68] However, for those captured as traitors and public enemies, their fate was even worse than that of common criminals. The emperor Vitellius was tortured, executed, and then his body was dragged through the streets to the river Tiber.[69] Sejanus, the captain of the Praetorian Guard and onetime right-hand man to the emperor Tiberius, was hauled to prison, tortured, and executed, and his corpse was then abused by the crowd for three days before it was thrown into the river.[70] The statues of Sejanus were also toppled and destroyed, paralleling the demise of his actual body. Statues and portraits of the condemned were often removed, defaced, and abused, underlining once more the power of images and effigies as stand-ins for real bodies.[71] In honored funerals the *imagines* brought the dead back to life; by contrast the parade of abused corpses and statues underlined the destruction of identity and reputation. Indeed, the fate of such men reversed all the normal elite expectations for honored treatment of the corpse, a grand funeral, praising eulogy, and respectful treatment of lasting reminders of their body and reputation such as statues, inscriptions, monuments, and tombs. To see these things denied and destroyed, or the prospect of seeing this, was horrific. In *damnatio memoriae*, a series of penalties formulated against public enemies and traitors, the name of the damned could be erased from inscriptions and their portrait images, including the *imagines*, destroyed.[72] Instead of their portrait image being placed among the ancestors, the contorted features of these victims' heads might be displayed and abused. Decapitation and head-hunting could be seen as barbaric, a practice of the Celts rather than the Romans,[73] but with its easily recognizable features the head could become a trophy.[74] The great orator and politician Cicero was beheaded and his head and hand then displayed in the Roman Forum for a long time.[75] The body of the emperor Galba was decapitated by a soldier, paraded on a stick, then sold and thrown to the ground.[76]

The Roman body could be punished after death for the crimes of life, or the body could be elevated and praised for the successes of the living person, or demeaned and stigmatized as the living body had been. In many ways the corpse, and also by association portraits, images, and statues, was open to manipulation. The dead body was inanimate and powerless, and in the hands of the living, corpses could be powerful symbols to be honored and respected, or mutilated and destroyed. Yet corpses were not completely passive; the corpse still retained the essential features and some aspects of the identity of the living person. Corpses could act on the senses, emotions, and fears of the living. The sight, smell, and touch of the corpse affected the living and framed their lasting memories of the dead person, the funeral, and how the dead were finally disposed of.[77] In some respects the corpse was the primary participant in the funeral performance; it was adorned, processed, and displayed; it mingled with other dead ancestors, temporarily brought back to life; it sat as if listening to the eulogy and laments; and the latter in some ways may have given the dead a voice. The corpse was an actor, or at least the central prop extraordinaire in funeral theatre, and the parody of this was corpse abuse and denial of burial.

ROMAN MOURNERS

Death did not just affect the bodies of the deceased, it also marked and altered the bodies of the bereaved. Mourning was a physical as well as an emotional journey; it entailed the bodily expression of internal feelings of anger, loss, and grief and thus marked out the bereaved from other members of society. In Rome the house of the bereaved family was signaled by a cypress branch, and the mourners were in some senses isolated from the rest of society, identified by their demeanor, gestures, and clothing.[78] The *familia funesta* was polluted by its close contact with the dead, and the associated funerary rituals were as much about reintegrating these survivors as disposing of the corpse.

In the initial stages of grief mourners might use their bodies dramatically to display their loss and suffering. Grief was noisy; the human voice gave expression to loss through wailing, shouting, and more organized laments. It was traditional to call aloud the name of the dead person and articulate what the dead had been deprived of. If the body was displayed by the well-off, this lamenting may have been undertaken by hired mourners and dirge singers. We do not know the precise content of what was sung, shouted, and cried out, but there may have been a standard repertoire of poems and songs. Musicians, such as flute players and trumpeters, may have also been present. All this could be accompanied by gestures—raised arms, the beating of breasts, the scratching of cheeks, the pulling (perhaps even cutting) of hair, and the use of ashes to dirty hair and faces. The behavior of the bereaved inverted norms for the body, such as dirt as opposed to cleanliness, dark rather than light clothing, untidiness as opposed to an ordered appearance.[79] The ancient sources refer to bodies altered, almost temporarily scared through grief, made dirty, bloodied, and unkempt to give physical expression to loss.[80]

Mourners also wore distinctive clothing, dark or black as opposed to white. How long this clothing was worn for is uncertain. Paulus, a writer on legal matters during the later second and early third centuries A.D., noted that those in mourning were to dress plainly, with no purple or white clothes or jewelry, and they were to avoid dinner parties. Paulus also indicated how long mourning was supposed to last: parents and children aged over six could be mourned for a year, children under six for only one month, a husband for ten months, and close blood relatives for eight months.[81] However, such stipulations seem to have been aimed at women more than men. In general men were admired for not expressing their grief in public, for putting public duty before personal misfortune, and for getting back to work as quickly as possible. Men played a significant role in public funerary ritual, acting as pallbearers and making laudatory speeches, but these were controlled and measured ways of expressing the loss to the wider community. Dramatic gestures of grief were not for men, who were expected to show their loss with decorum and then, in public at least, move on. In contrast the female role in funerary rituals was more dramatic, sensory, and emotional. Open mourning could be described as "womanly"; it was behavior that particularly characterized a bereft mother.[82] The subtext was that a woman's dramatic reaction to a loss was understandable and to some degree acceptable, and thus in contrast a similar reaction was not acceptable for a man.

Tradition dictated that women play a prominent role in public mourning and funeral ritual. Mourning was one of the rare occasions during which women had an accepted place and role in the public gaze. Women's actions and appearance gave expression to the family's and the community's grief. It was women who wailed, cried, and sang laments; women who beat their breasts, scratched their cheeks, and pulled at and dirtied their hair; and it was women who would usually give the last kiss to the dying, wash the dead body, and finally receive the ashes.[83] These were accepted and important roles; however, they could place women in an ambivalent position. On the one hand women could be seen as taking on the dirty and polluted tasks associated with death; on the other hand women were playing a vital role in the important transition from life to death that was essential for the well-ordered running of society.[84] It could be argued that death was both degrading and empowering for women, and as a result men sought to monitor and ultimately to restrain the female role in public mourning.[85] Mourning, when done on a grand or dramatic scale, had the potential to become political and competitive; the behavior of women reflected on their fathers, husbands, and sons and thus ultimately on the state. The legal restrictions on mourning periods, as noted previously, were aimed particularly at women. The early laws of the Twelve Tables (c. 450 B.C.) also sought to stop excessive displays and prohibited women from tearing at their cheeks and wailing too loudly.[86] We can also note that after Rome's overwhelming defeat by Hannibal at the Battle of Cannae (216 B.C.) mourning was restricted since there was a real fear that the grief of women, symbolizing the suffering felt throughout the city, was becoming self-destructive and crippling.[87] Women also could be prevented from mourning those who were condemned or dammed; a decree issued to condemn Cnaeus Calpurnius Piso following his suicide in A.D. 20 stated that he was not to be mourned by the women of his house.[88] Mourning was one of the duties of a woman, but the nature and extent of that mourning could be controlled by men.

However, to polarize male and female mourning roles may be to accept ancient stereotypes too readily. Expressing grief may have been regarded as "womanly," but some men did struggle with their losses, while women, especially elite women were expected to conform to male standards, controlling their emotion and retaining dignity. Gender was not an absolute given; in their mourning men could be womanly, women manly. The dominant association of women with death and mourning appears to have been eroded during the late Republic and early Imperial period. The increasing use of undertakers and hired professional mourners reduced the need for the female relatives of the deceased to make dramatic displays of grief themselves or to be so intimately involved in the preparation and tending of the corpse before disposal. Extreme female reaction to loss could now be seen as demeaning, as something that characterized the lower orders and those paid to do such jobs. Elite women could find themselves in an uncomfortable and often contradictory position; as mothers, wives, and daughters they were expected to lead the mourning, to express their loss, but simultaneously grief was now more to be suppressed than flaunted.[89]

In the eyes of many male elite authors, guided by philosophy, mourning (as opposed to grief) was a performance. Mourning was acted out at funerals by people hired for the job, and thus for men, and women of a certain standing, to act so and to give physical and audible expression to their loss in public was false and at best melodramatic. By contrast, some poets did take a softer line, acknowledging that controlling grief in public was not always so easy, and that setting limits to mourning according to gender and social class was unfeeling.[90] However, it remains clear that real pain was perceived to be the preserve of the bereaved and could be contrasted with that of hired mourners who enacted grief that they were not suffering.

The conflicts and debates in Roman society about how one should mourn, differences between male and female roles, between the elite and nonelite, between the bereaved and hired mourners are suggestive of the centrality of the essential rites, and their emotional and physical manifestation. Even with changing perspectives, however, the place of women in the rituals, whether family or hired help, remained important. It was women who guided the dead away from the living, just as it was women who brought new life into the world. In many respects Roman funerary rituals both mirrored and inverted the rites associated with birth. The dead were placed on the earth, like the baby; the dead were washed and clothed, as was the baby; the eyes of the dead were closed, those of the baby opened; at death and birth women might expose their breasts, and milk was one of the traditional offerings made to the dead, just as it was the food of life for the newborn.[91] At the grave or pyre women, hired professional singers, are thought to have sung a traditional lament, the *nenia,* to encourage the dead to leave this world. The precise content of these laments is largely unknown, but the word *nenia* is also associated with children's rhymes and some magical incantations, suggesting that the different types of *nenia* may have shared repetitive and soothing aspects.[92] Hired mourners were like midwives, selling their skills, abilities that guided the living into this world and guided the dead out of it. Both were marred by their physical intimacy with human biological processes, by their closeness to bodies that were in transition. Female mourners may have acted "as macabre wet nurses of sorts, guiding their charges in the direction of the tomb."[93]

CONCLUSION

Birth and death represent the ultimate transitions in human life. People seek to control, monitor, and ritualize the biological processes that allow bodies to enter and leave this world. Reconstructing the details of ancient birth and death rituals can be a complex process, one involving assembling pieces of evidence to create a composite picture. It remains difficult to know how widespread certain practices were and how people regarded these practices. Were certain rituals dismissed as little more than superstition or regarded as essential and central; were they seen as anachronistic procedures that were little respected or vital traditions to be honored? What does emerge in studying birth and death in the ancient world are common strands and themes centered on pollution, performance, and transformation. The established rituals guided and structured behavior; their performance allowed the changing status, the separation and incorporation, of the newly born, the dead, and the bereaved.

The body defines the individual and identity. Bodies can be objectified, treated as symbols to be manipulated, and this may be especially the case with bodies in transition—for the dead and the newly born. But people also live through their bodies, and those closest to the dead and the newly born—mothers, midwives, undertakers, hired mourners, and the bereaved—had their bodies transformed sensually, emotionally, and physically by the bodies of others. Bodies can become an interface or site of intersection for the biological and the social, for the collective and the individual, for the emotional and the physical.[94] Studying ancient birth and death highlights the role of the inanimate body in dialogues of power and acceptance (and the converse), but also the sensory aspects of the embodied experience. Focusing on the treatment of the dead in ancient Rome suggests how the corpse could become a powerful symbol in the hands of the living, to be honored or abused, but also that the corpse had a value only because it continued to represent the identity of the individual; it was in some respects the central performer in the rituals. Roman mourning was also centered on performance and for some was an extremely physical and emotional bodily experience, yet ancient discussions about gendered roles in mourning are suggestive of underlying discourses of power that sought to control and codify (not always with success) bodily display and excess.

In the ancient world the bodies of babies and the bodies of the dead could be treated in a similar fashion; both could be objectified, seen as items that were to be controlled by others, elevated, abused, or abandoned according to the needs of the living. But these bodies also acted upon the bodies of others, changing them physically or emotionally or sensually, providing them with points of reference, comparison, and memory. Women in particular were affected by birth and death, and their experience or knowledge in these areas could enhance their status[95] and contribute to the definition and negotiation of gendered roles. Female bodies became the medium by which ancient bodies in general entered and left this world.

CHAPTER TWO

Health and Disease

PATRICK MACFARLANE

Offering a survey of the concepts of health and disease in classical antiquity is a somewhat forbidding enterprise.[1] We have countless reflections on these notions from many sources, including medical writers (such as the writers of the treatises that comprise the Hippocratic Corpus,[2] non-Hippocratic medical writers such as Diocles of Carystus and Praxagoras of Cos and, later on, Galen), philosophers (the pre-Socratics, Plato, Aristotle, and Cicero), historians (Thucydides and Herodotus), and poets (Homer, Hesiod, and Lucretius).[3] Considering these eclectic sources also reveals that the ancients thought of health and disease in a number of different contexts beyond their typical place in relation to particular human bodies. To give just a simple example, Plato, in his famous dialogue *Republic,* has Socrates refer to his first sketch of a just city as the healthy (*hygiês*) city, while the next city that Socrates and his interlocuters construct in speech is referred to as the feverish city (*phlegmainousan polin*).[4] Thus, the concepts of health and disease become metaphorical in Socrates's evaluation of *political* arrangements.[5] This metaphorical usage continues up through the classical era and into the first centuries of Christianity, when the bodily notions of health and disease are applied to the health and salvation of the soul and the concomitant disease of sin. Roman Catholics still respond today, after the body and blood of Christ are offered on the altar at Mass during the liturgy of the Eucharist, "Lord, I am not worthy to receive you, but only say the word (*verbo*) and my soul (*anima*) will be healed (*sanabitur*)."[6]

Despite the rich metaphorical usage that the concepts of health and disease have in ancient literature, their primary sense applies to the good or bad condition of the human body. But even this rather common and obvious application of health and disease is problematic. Health appears rather simple: it is the good condition and proper functioning of the human body. Disease, on the other hand, is complex and multifarious; its forms are legion.[7] Thus the picture that emerges from studying the concepts of health and disease in the literature of classical antiquity is decidedly lopsided: health appears as a state or condition of the body to be achieved through careful management of regimen (including proper diet and exercise), while disease is a constant danger that threatens the health of the human being from all quarters—from within and without the body. This may explain why in the Hippocratic Corpus so many passages (even entire treatises) are devoted to disease, while only a few are given to health.[8] We are able to conclude from this disparity that disease was, paradoxically, the natural condition of many human beings in antiquity, while health was fleeting, and therefore prized for its rarity.

In what follows, then, I explore health and disease as they are discussed primarily in the medical and philosophical literature of antiquity, while occasionally making reference to important poetic, historical, and theological reflections. The dialogue between medical and philosophical writers in antiquity is quite fruitful since medicine, even in its earliest days, emerged as a counterpart to philosophical reflection.[9] This is because ancient medicine shared much of its subject matter, the body and its parts (organs and tissues), including its elements (earth, air, fire, and water), qualities or powers (hot, cold, wet, and dry), and fluids (blood, humors, urine, etc.), with the philosophy of nature (*phusikê*). Thus Tertullian, the early second century C.E. Christian apologist and polemicist, refers to medicine as the "sister of philosophy" (*sororem philosophiae*).[10]

THE PRE-SOCRATICS AND HIPPOCRATICS

Many of the pre-Socratic philosophers show a remarkable curiosity in questions of health and disease as these manifest themselves in the relationship of human beings to nature (*phusis*).[11] In accord with their rational inquiry into nature, the pre-Socratic philosophers generally deemphasized the relationship of the gods to disease, a relationship prevalent in the poetry of Homer and Hesiod.[12] This mode of explanation characterizes the work of the Hippocratic writers as well.[13] If human beings are part of nature, then they and their bodies must be composed of natural elements, whether these elements are the classic four (earth, air, fire, and water), or some mixture based on these four. Empedocles of Acragas (flourished ca. 460 B.C.E.), who declared himself to be a healer of some sort, thought that the main tissues of the human body (bone, sinew, flesh, and blood) were all certain ratio-mixtures of the four basic "roots" of all things, earth, air, fire, and water.[14] One of the earliest discussions of health and disease comes from the pre-Socratic philosopher Alcmaeon of Croton (flourished ca. 500 B.C.E.), possibly connected to the Pythagoreans, who is notable for making the health of the body analogous to the "health" of the political community:

> Alcmaeon holds that what preserves health (*hygieias*) is the equality (*isonomian*) of the powers—moist and dry, cold and hot, bitter and sweet and the rest—and the supremacy (*monarchian*) of any one of them causes disease (*nosou*); for the supremacy of either is destructive. The cause of disease (*noson*) is an excess (*hyperbolêi*) of heat or cold; the occasion of it surfeit or deficiency of nourishment; the location of it blood, marrow or the brain. Disease may come about from external causes (*tôn exôthen aitiôn*), from the quality of water (*hydatôn poiôn*), local environment (*chôras*) or fatigue (*kopôn*) or violence (*anagkês*) or similar causes. Health (*hygieian*) on the other hand is the harmonious (*symmetron*) blending (*krasin*) of the qualities (*tôn poiôn*).[15]

What emerges from this fragment from Alcmaeon, and the notion that will go on to predominate in ancient Greek notions of health and disease, is the idea that health is a "harmonious blending" (or we could say "proportionate mixture") of everything found in the body—but not only a balance within the body, a balance that the body has with its environment as well.[16] Disease is the opposite of health, a disproportionate separation of these constituents of the body from one another. Later on, Aristotle returns to this important notion of proportion, particularly the proportion between hot and cold, as largely determinative of health and disease, much like Alcmaeon.[17]

Alcmaeon's association of health with a proportionate mixture of bodily qualities or powers (moist and dry, hot and cold, bitter and sweet) is similar to the well-known Hippocratic conception of health as constituted by a thorough mixture of the four humors (*chymoi*) or bodily fluids: blood (*haima*), yellow bile (*cholê xanthê*), black bile (*cholê melaina*), and phlegm (*phlegma*).[18] We find this notion in the Hippocratic treatise *On the Nature of Man*, a treatise that, thanks to the influence of Galen, became the *locus classicus* of the Hippocratic teaching on the principal causes of health and disease.[19] The author of *On the Nature of Man* writes that human beings "enjoy the most perfect health (*hygiainei malista*) when these elements are duly proportioned (*metriôs*) to one another in respect of compounding, power and bulk, and when they are perfectly mingled (*malista memigmena*)."[20] Disease, on the other hand, occurs when "one of these elements is in defect (*elasson*) or excess (*pleon*), or is isolated (*chôristhêi*) in the body without being mixed (*kekrêmenon*) with all the others. For when an element is isolated and stands by itself, not only must the place which it left become diseased (*epinoson*), but the place where it stands in a flood must, because of the excess, cause pain (*odunên*) and distress (*ponon*)."[21] Again, similarly to Alcmaeon, we find the Hippocratic writer speaking of health as a due proportion or measure, this time, of fluids in the body. Disease occurs from the separation and assertion of one humor at the expense of others, upsetting this mixture.[22]

It is not difficult to imagine where the doctrine of the four humors came from. Doctors and medical writers must have seen that certain fluids were vital for the normal operation of the body: blood that seeped from wounds or from menstruation, phlegm coughed up or sneezed out, and yellow bile that was vomited as a result of more serious illnesses.[23] Black bile was perhaps more difficult to observe, though its presence might have been inferred from seeing decomposed blood, which takes on a darker hue, in urine and stools.[24] Although the four humors are given a central role in *On the Nature of Man*, they are by no means used to explain health and disease in all the treatises that comprise the Hippocratic Corpus. Indeed, some treatises explain both health and disease primarily as the result of other factors while speaking of the humors as "secondary and subordinate causes" (*sunaitia kai metaitia*).[25] For example, in the Hippocratic treatise *Airs, Waters, Places*, the author argues that health and disease are largely the result of geography, including the predominating winds, waters, and soil conditions of the places that patients inhabit.[26]

In another treatise of the Hippocratic Corpus, *On Breaths*, the author writes:

> Now bodies, of men and of animals generally, are nourished by three kinds of nourishment (*tropheôn*), and the names thereof are solid food (*sitia*), drink (*pota*), and wind (*pneuma*). Wind (*pneuma*) in bodies is called breath (*phusa*), outside bodies it is called air (*aêr*). It is the most powerful of all and in all, and it is worth while examining its power (*dunastês*).[27]

The author then gives an elaborate encomium to the power of air—how it is able to tear trees from the ground, create large waves that toss ships about on the sea, and even cause the seasons to change.[28] From this he concludes that air is the most vital element for human life because humans can live for a time without food and drink, but hardly at all without air.[29] And yet, this author argues that the air we breathe can also give rise to fearsome diseases like fever,[30] while the air that is generated within us due to a poor diet is the cause of all other pathological conditions, minor (such as headache) and major (such as dropsy, apoplexy, and even epilepsy).[31]

The author of *On Breaths* attaches significance to the air we breathe for our overall health, but also to food and drink. He highlights environmental factors, such as air quality, and other external factors that become internal, that is, the food and drink that we consume. Other Hippocratic writers speak of the importance of exercise too, as necessary for maintaining health and preventing disease.³² Together, diet and exercise fell under the category of regimen, the proper management of which the ancients thought crucial for health and well-being.³³

Given ancient emphasis on appropriate regimen, it is unsurprising that the author of the Hippocratic treatise *On Ancient Medicine* claims that it was through the development of the art of cooking that the art of medicine was discovered. Offering something like a conjectural history, he argues that initially human beings suffered much "from strong and brutish living (*hypo ischurês te kai thêriôdeos diaitês*) when they partook of crude foods, uncompounded and possessing great powers—the same in fact as men would suffer at the present day, falling into violent pains and diseases (*ponoisi te ischuroisi kai nousois*) quickly followed by death."³⁴ In those dark times, reminiscent of the Hobbesian state of nature, where life was "nasty, brutish, and short," only those humans with strong constitutions survived while the weak perished.³⁵ This grim state of affairs provoked humans to search for "nourishment that harmonized with their constitution" (*trophên harmozousan têi phusei*).³⁶ Little by little through experimentation, they were eventually able to subdue the powers of foods so that instead of harming them and causing diseases, it would be properly nutritious:

> Experimenting with food they boiled or baked, after mixing (*emixan*), many other things, combining (*ekerasan*) the strong and the uncompounded with the weaker components so as to adapt all to the constitution (*phusin*) and power (*dunamin*) of man (*anthrôpou*), thinking that from foods which, being too strong, the human constitution cannot assimilate when eaten, will come pain (*ponous*), disease (*nousous*), and death (*thanatous*), while from such as can be assimilated will come nourishment (*trophên*), growth (*auxêsin*), and health (*hygieiên*). To this discovery and research what juster or more appropriate name could be given than medicine (*iêtrikên*), seeing that it has been discovered with a view to the health (*hygieiêi*), saving (*sôtêriêi*) and nourishment of man (*trophêi tou anthrôpou*), in the place of that mode of living (*ekeinês tês diaitês*) from which came the pain, disease and death?³⁷

In other treatises of the Hippocratic Corpus we find various prescriptions not only for preparing foods,³⁸ but also for what types of foods should be eaten in the different seasons of the year. For example, the author of *Regimen in Health*, addressing specifically laypeople (*tous idiôtas*) as opposed to other medical experts, suggests that the proper diet should be contrary to the seasons of the year. In winter, a wet and cold season, one ought to drink as little as possible and eat breads and roasted meats so that "the body (*to soma*) will be most dry (*xêron*) and hot (*thermon*)."[39] In spring the author recommends taking more drink, substituting barley cake for bread, decreasing the amount of meat consumed (boiling it instead of roasting it), and increasing the amount of vegetables in the diet, making the body "cool and soft" (*psychron kai malakon*) in order to prepare it for summer, since this season is hot and dry. The cycle continues as summer gives way to fall and winter: "By opposing opposites (*ta enantia poieonta*) prepare for the change from summer to winter."[40] By changing the diet gradually, the bodily constitution balances itself with and counteracts the change in climate.

Parallel to the attention paid to cooking foods so that they were suitable for digestion was the internal process of digestion (sometimes referred to as "concoction") itself. Digestion (*pepsis*) was important because it was largely through concocting food that the body was nourished, important fluids or humors (e.g., blood) created, and waste materials excreted (feces and urine). That the body is busy cooking its food under the agency of its innate heat[41] in order to provide a proper mixture of its contents and to nourish itself is an idea found in a number of Hippocratic treatises.[42] But eating also brought with it the possibility of indigestion (*dyspepsia* or *apepsia*), where the innate heat fails to master and assimilate the food consumed. Indigestion might yield harmful residues such as gases or fumes (causing flatulence or sour eructations),[43] or, if there was too much waste material produced (feces), intestinal blockages could occur, leaving the waste to putrefy in the body, resulting in a host of illnesses.

Disease and illness ravaged the ancient world, and ancient doctors did their best to combat it.[44] In G.E.R. Lloyd's lapidary formulation, "The ancient Greeks were plagued by plagues real and imaginary."[45] As we have seen from our brief look at some of the formulations of health in the Hippocratic Corpus, Hippocratic writers had a variety of approaches toward disease as well. Disease could be the result of poor diet (or eating too much or too little), improper exercise, imbalance of the body's constituents, and unsuitable environmental conditions. Diseases might also be caused by the very air we breathe[46] or even digestive gases that enter the body, getting trapped in the tissues, clogging vital conduits (vessels, *phlebes*) between organs, or settling in and occupying places where they did not naturally belong.[47] The author of the Hippocratic treatise *Diseases* 1 says:

> Now all our diseases (*nousoi*) arise either from things inside the body, bile and phlegm, or from things outside it: from exertions (*ponôn*) and wounds (*trômatôn*),[48] and from heat that makes it too hot, and cold that makes it too cold.[49]

Outside of this rather general explanation of the causes of disease, certain Hippocratic writers were prepared to list a number of additional factors. For example, in *Nutriment*, a tract composed of aphorisms, the author writes:

> Differences of diseases depend on nutriment, on breath, on heat, on blood, on phlegm, on bile, on humours, on flesh, on fat, on vein, on artery, on sinew, muscle, membrane, bone, brain, spinal marrow, mouth, tongue, oesophagus, stomach, bowels, midriff, peritoneum, liver, spleen, kidneys, bladder, womb, skin. All these things both as a whole and severally. Their greatness great and not great.[50]

Thus the ancient physician was faced with a staggering myriad of possibilities to take into consideration for the cause of disease.[51] What this passage suggests is that disease was not only limited to some sort of humoral imbalance (as the writer of *On the Nature of Man* might have it), but could also be located in the malfunction of or injury to specific parts of the body. To this already burgeoning list of potential causes for illness and disease, we could also add other factors, such as "the custom (*etheos*), mode of life (*diaitês*), practices (*epitêdeumatôn*) and ages (*hêlikiês*) of each patient."[52] And let us not forget to add the *sex* of each patient (since there was great emphasis on treating diseases peculiar to women in the Hippocratic Corpus),[53] along with attention to the environmental conditions of the patient including the winds, waters, seasons, and air quality.[54] Unlike health, disease admitted of no easy account or classification.[55] What is important to understand, generally speaking, is that the Hippocratics saw disease and its potential causes as ubiquitous, and its ubiquity called out for philosophical consideration.[56] It is to that consideration that we now turn.

PLATO AND ARISTOTLE

We find a number of reflections on health and disease in the philosophical dialogues of Plato (427–347 B.C.E.). In the *Republic*, Socrates, Glaucon, Adeimantus, and others seek to understand the nature of justice. To discover the nature of justice in the soul, Socrates first proposes that they find it in a city, so they build an imaginary "city in speech." The first political arrangement they create Socrates refers to as the healthy (*hygiês*) city:

> They'll produce bread, wine, clothes, and shoes, won't they? They'll build houses, work naked and barefoot in the summer, and wear adequate clothing and shoes in the winter. For food, they'll knead and cook the flour and meal they've made from wheat and barley. They'll put their

honest cakes and loaves on reeds or clean leaves, and, reclining on beds strewn with yew and myrtle, they'll feast with their children, drink their wine, and, crowned with wreaths, hymn the gods. They'll enjoy sex with one another but bear no more children than their resources allow, lest they fall into either poverty or war.

It seems that you make your people feast without any delicacies, Glaucon interrupted.

True enough, I said, I was forgetting that they'll obviously need salt, olives, cheese, boiled roots, and vegetables of the sort they cook in the country (*en agrois*). We'll give them desserts, too, of course, consisting of figs, chickpeas, and beans, and they'll roast myrtle and acorns before the fire, drinking moderately (*metriôs*). And so they'll live in peace (*en eirênêi*) and good health (*meta hygieias*), and when they die at a ripe old age, they'll bequeath a similar life to their children.

If you were founding a city for pigs, Socrates, he replied, wouldn't you fatten *them* on the same diet?[57]

Glaucon famously complains that this city is hardly appropriate for human beings, the sort of cultivated Athenian gentleman that he himself represents. So, the healthy city is allowed to transform by opening up people's desires for luxuries, until we have the second city, which Socrates refers to as the feverish city (*phlegmainousan polin*). The feverish city is complete with every luxury a civilized and cosmopolitan Athenian could wish for, including perfumes, prostitutes, and pastries. Importantly, in the feverish city, people will eat meat. Notice that Socrates links the health of the first city with their diet: it is simple and vegetarian, the fare of country folk.[58] The fact that people will be carnivores in the feverish city requires that more territory be annexed for raising animals. Socrates argues that this will eventually lead to war and conflict with the city's neighbors. In other words, the unhealthy lifestyle brought on by desire for luxuries leads to social and political illness.[59]

Plato uses the themes of health and disease in a number of his other dialogues as well, such as the *Gorgias*, *Charmides*, *Symposium*, and *Phaedrus*.[60] We find the fullest discussion of them, however, in his magisterial dialogue on natural philosophy, the *Timaeus*. Near the end of that dialogue, the character Timaeus explains the many ways in which disease can destroy the human body, signaling not only the prevalence of disease among human beings, but also the precariousness of health.[61] Disease, it seems, is closer to the normal way of things for human beings. Just as in the *Republic*, where in the allegory of the cave Socrates explains that most humans live in a psychic condition of ignorance and falsehood, so in the *Timaeus* the greater part of human bodily experience is one of ill health.[62] In fact, the first word Timaeus utters in the dialogue is "illness" (*astheneia*).[63]

Plato describes three sorts of disease in the *Timaeus*: first, diseases that result from an imbalance of the body's primary elements (earth, air, fire, and water); second, diseases that attack the body's tissues (blood, flesh, sinew, marrow, and bone) that the primary elements compose; and third, the "class of diseases" (*nosêmatôn eidos*) that result from air (*pneumatos*), phlegm (*phlegmatos*), and bile (*cholês*).[64] Timaeus begins his account of disease commenting on the first sort:

> How diseases originate (*to hothen tôn nosôn*) is, I take it, obvious to all. Given that there are four kinds of stuff out of which the body has been constructed—earth, fire, water and air—it may happen that some of these unnaturally (*para phusin*) increase (*pleonexia*) themselves at the expense of others. Or they may switch regions, each leaving its own and moving into another's region. Or again, since there is in fact more than one variety of fire and the other stuffs, it may happen that a given bodily part accommodates a particular variety that is not appropriate for it. When these things happen, they bring on conflicts (*staseis*) and diseases (*nosous*). For when any of these unnatural (*para phusin*) occurrences and changes take place, bodily parts that used to be cold become hot, or those that are dry go on to become moist, and so with light and heavy, too. They undergo all sorts of changes in all sorts of ways.[65]

One can easily discover in Timaeus's formulation the influence of both Empedocles and Alcamaeon.[66] While Timaeus is unspecific concerning which diseases arise from an imbalance of the four basic elements, he holds that diseases attacking the tissues constituted from them (blood, flesh, sinew, marrow, and bone) are caused when the normal formation (*genesis*) of these tissues is reversed (*anapalin*), leading to their degeneration (*tauta diaphtheiretai*).[67] The older part of the flesh dissolves, and the waste material produced as a result floods back into the blood vessels, filling them with "acidic and salty qualities" (*oxeiais kai halmurais dunamesi*) and "bile (*cholas*) and serum (*ichôras*) and phlegm (*phlegmata*) of every sort. These are all back-products (*palinaireta*) and agents of destruction (*diephtharmena*)."[68] What seems to be different here from some Hippocratic accounts is that bile and phlegm, instead of being natural constituents of the body, are described by Timaeus as unnatural (*para phusin*) products generated from the disintegration of diseased tissue. Once in the blood vessels, these dangerous materials travel throughout the body, waging war (*polemia*) and destroying the parts of the body "that have stayed intact and kept to their posts."[69] Timaeus lists a number of differently colored biles (black, green, and yellowish orange) that are forged in the fires of inflamed flesh. Rotting flesh mixes with air (*aeros*)[70] and produces bubbly, foamy white phlegm (*leukon phlegma*).[71] Timaeus seems to attribute the manufacture of these noxious agents to an improper concoction (*dyspepton*) of the body's tissues.[72] This might occur because the body is heated too much, perhaps by fever, or because the flesh of unhealthy people is too dense to allow proper ventilation, in turn setting the body's heat out of balance.[73] Somehow or other the body begins a course of consuming itself instead of its food and drink.[74] While bile and phlegm figure prominently in Timaeus's nosology, air (*pneuma*) is also named as the cause of serious diseases such as tetanus.[75]

Compared with disease, health gets relatively short shrift in the *Timaeus*. This, however, is to be expected, given Timaeus's theme that health is simple while disease is complex, a theme that is also evident in the fragment of Alcmaeon.[76] In a sense, the entire *Timaeus* is one sustained reflection on the tendency of the natural world to change, corrupt, and perish. Human beings, because they are composed of the same basic material that constitutes the visible cosmos, must surrender to this tendency as well. This somber realization might be a Platonic echo of the famous fragment of Heraclitus: "Nature loves to hide" (*phusis kruptesthai philei*).[77] Pierre Hadot observes that the Greek also makes sense if rendered into English this way: "What is born tends to disappear."[78]

But there is hope that our bodily suffering will be redeemed, according to Plato, because the soul is immortal. Socrates and his friends discuss this on Socrates's last day, portrayed in the *Phaedo*, a dialogue rife with medical imagery.[79] Many times throughout the dialogue, Socrates speaks of the trouble that comes from the soul being "infected" by the nature of the body.[80] All the many desires that come from our embodiment make concentrated philosophical inquiry, and therefore happiness, nearly impossible.[81] After presenting numerous arguments for the immortality of the soul, in an effort to console his friends that he will be all right even though his body will perish, Socrates bravely drinks the poison (*to pharmakon*) prepared for him by the jailer.[82] As he feels the poison enter his belly, Socrates utters his famous last words:

> "Crito, we owe a rooster to Asclepius; make this offering to him and do not forget."—"It shall be done," said Crito, "tell us if there is anything else." But there was no answer. Shortly afterwards Socrates made a movement; the man uncovered him and his eyes were fixed. Seeing this Crito closed his mouth and his eyes.[83]

Many have wondered why Socrates asks that Crito make a sacrifice to Asclepius, the Greek god of health and healing, from whom Hippocrates himself claimed descent.[84] What had Socrates been healed from? Nietzsche supposed that this was the ultimate sign of Socrates's weakness and decadence:

> Concerning life, the wisest men of all ages have judged alike: *it is no good.* Always and everywhere one has heard the same sound from their

> mouths—a sound full of doubt, full of melancholy, full of weariness of life, full of resistance to life. Even Socrates said as he died: "To live—that means to be sick a long time: I owe Asclepius the Savior a rooster." Even Socrates was tired of it. What does that evidence? What does it evince? Formerly one would have said (—oh, it has been said, and loud enough, and especially by our pessimists): "At least something of all this must be true! The consensus of the sages evidences the truth." Shall we still talk like that today? *May* we? "At least something must be *sick* here," *we* retort. These wisest men of all ages—they should first be scrutinized closely. Were they all perhaps shaky on their legs? Late? Tottery? Decadents? Could it be that wisdom appears on earth as a raven, inspired by a little whiff of carrion?[85]

Nietzsche sees in Socrates's final words a plea for release—the weakling's way out of the fullness of life on earth. Like a raven circling over a bloated corpse, philosophy is a scavenger that lives on the putrid scraps of a civilization unhealthy and in decline. Philosophy, in Nietzsche's estimation, is the manifestation of decadence. For Plato, *medicine* is the mark of decadence.[86]

But did Socrates and Plato really resent human life and the body, as Nietzsche supposes? By taking account of the many admonishments for care of the body along with care of the soul that we find scattered throughout the dialogues of Plato, he interestingly is not the extreme hater of the body that he often appears to be. We have already seen Socrates's emphasis on a simple vegetarian diet for human beings in the *Republic*. Consider also the proper regimen of diet and exercise he prescribes for his guardians later in the *Republic*, and seemingly for all human beings in the *Timaeus*.[87] Another Socratic writer, Xenophon, tells us that Socrates

> strongly urged his companions to take care (*epimeleisthai*) of their health (*hygieias*). "You should find out all you can," he said, "from those who know. Everyone should pay attention (*prosechonta*) to himself throughout his life; what sort of food (*brôma*) and drink (*pôma*) and what sort of exercise (*ponos*) are beneficial (*sumpheroi*) to him, and how he should use (*chrômenos*) them in order to enjoy excellent health (*hygieinotat*). For it is difficult," he said, "to find a physician (*iatron*) who knows those things conducive (*sumpheronta*) to one's own health (*hygieian*) better than the one who pays attention (*prosechontos*) to himself."[88]

The Socratic injunction to know yourself here includes knowledge of your own bodily nature, in addition to his usual emphasis on self (or soul) knowledge.[89] In this vein, we might think about the one member of Socrates's circle of intimates whose absence is conspicuous in the *Phaedo*: Plato himself. Plato has

Phaedo say toward the beginning of the dialogue that Plato was not there to share in Socrates's sorrow on his last day. Why? He was ill (*êsthenei*).⁹⁰ Plato, because of his illness, misses the final lesson of Socrates. This suggests that without proper attention to our bodily condition, our soul's well being could also be jeopardized.

Aristotle (384–322 B.C.E.), the most prominent of Plato's students, was the son of a physician.⁹¹ As I alluded to earlier, Aristotle provides a statement of what constitutes health in the *Physics*:

> Further, we say that all excellences (*aretas*) depend upon particular relations (*pros ti*). Thus bodily excellences such as health (*hygieian*) and fitness (*euexian*) we regard as consisting in a blending (*krasei*) of hot and cold elements in due proportion (*summetriai*), in relation either to one another within (*tôn entos*) the body or to the surrounding (*to periechon*); and in like manner we regard beauty, strength, and all the other excellences and defects.⁹²

This passage firmly establishes Aristotle within the Greek medical tradition of viewing health as a balanced mixture, this time of hot and cold, either in the body or between the body and the environment. Within the body, Aristotle explains that the organs of the heart and brain govern the relation of hot and cold, respectively. The heart houses the vital heat of the animal, while the brain, since it is cold, moderates this heat.

> The brain (*encephalos*), then, makes temperate (*eukraton*) the heat and seething of the heart. In order, however, that it may itself have a moderate (*metrias*) amount of heat, branches run from both blood vessels, that is to say from the great vessel (*megalês phlebos*) and from what is called the aorta (*aortês*), and end in the membrane (*mênigga*) which surrounds the brain; while at the same time, in order to prevent any injury (*blaptein*) from the heat, these encompassing vessels, instead of being few and large, are numerous and small, and their blood scanty and clear, instead of being turbid and thick. We can now understand why fluxes (*ta rheumata*) have their origin in the head, and occur whenever the parts about the brain have more than a measured mixture (*symmetrou kraseôs*) of coldness. For when the nutriment (*trophês*) evaporates (*anathumiômenês*) upwards through the blood-vessels (*tôn phlebôn*), its residue (*perittôma*) is chilled by the influence of this region, and forms fluxes (*rheumata*) of phlegm (*phlegmatos*) and serum (*ichôros*). We must suppose, to compare small things with great, that the like happens here as occurs in the production of showers (*huetôn*). For when vapor (*atmidos*) steams up (*anathumiômenês*) from the earth and is carried by the heat

into the upper regions, so soon as it reaches the cold air that is above the earth, it condenses again into water owing to the refrigeration, and falls back to the earth as rain. These however, are matters which may be suitably considered in the *Principles of Diseases* (*en tais tôn nosôn archais*), so far as natural philosophy (*phusikês philosophias*) has anything to say about them.[93]

In Aristotle's explanation of digestion or concoction (*pepsis*), vapors stream up into the head from the stomach, and when they are cooled in the region of the brain, they fall back through the body, chilling the vital heat in the heart, knocking out its power for a time.[94] This is his explanation of sleep.[95] We wake up when digestion is finished (*pephthêi*).[96] If the brain fails to perform its cooling function, and the heat of the heart increases disproportionately, illness could result.[97] In this passage we can see Aristotle assimilating some of the most prominent medical notions of his predecessors: the importance of balance (in this case between heart and brain), the effects of digestion,[98] meteorological explanations, and the close connection between natural philosophy and medicine.

Aristotle speaks of other pathological conditions throughout his own work. Some that figure prominently are melancholy and fever,[99] epilepsy,[100] tetanus,[101] leprosy,[102] abscesses,[103] gangrene,[104] gynecological issues,[105] disabilities,[106] and consideration of the causes of failure to generate or reproduce, mostly in human beings.[107] Interestingly, Aristotle even speaks of old age (*gêras*) as a "natural disease" (*noson phusikên*).[108] Of course, Aristotle's remarks about these various ailments are not isolated from the rest of his natural philosophy. Most of these discussions are found scattered throughout his natural works, particularly in his writings on animals.[109] His discussions of illness and disease are integrated into his natural philosophy. Indeed, Aristotle often explains normal physiological processes by referring to a pathological condition. For example, he explains the beating of the heart by likening it to the throbbing of an abscess.[110] Similarly, Aristotle compares the condition of sleep to epilepsy, noting that sleep actually *is*, "in a way" (*tropon tina*), epilepsy.[111] It remains unclear to what extent Aristotle employs humoral explanations in his work. When he offers lists of the body's fluid contents, he sometimes mentions humors, but he mainly speaks of "residues" (*perittomata*) that result from digestion.[112] One important residue that figures in many of Aristotle's nosological explanations is *pneuma*.[113] On various occasions, however, he does speak of certain kinds of people using the humoral language of constitutional types.[114]

Aristotle also engages in anatomical description and physiological explanation. For example, in the *History of Animals*, that compendium of relentless observation of animals and their parts, ways of life, activities, and characters, Aristotle offers first an overview of the vascular anatomy of Syennesis, the physician of Cyprus,[115] Diogenes of Apollonia,[116] and Polybus.[117] After citing the views of these predecessors on vascular anatomy, Aristotle then gives his own scheme, which stretches over some four and a half pages.[118]

The more important question to consider is why Aristotle believes this foray into the vascular anatomy of animals is worth the careful attention he pays it. The arrangement of the vessels in the animal body, as well as their source or origin in the heart, are actually crucial elements for understanding the comprehensive scope of the medical themes in Aristotle's philosophical biology.[119] They underpin the physiological aspect of Aristotle's psychology.[120] This is because Aristotle locates both the nutritive soul and the sensitive of soul of animals in the heart.[121] The heart houses the vital heat of the animal that is responsible for transforming half-digested food into blood, which, as the final stage of nutriment, truly feeds or nourishes the rest of the animal body, thereby allowing the animal to "maintain (*sôzei*) its substantial being (*tên ousian*)."[122] Aristotle speaks of the heart as the "supreme organ (*kyrion*) of the sense faculties (*tôn aisthêseôn*)," arguing that the heart is the location of the "common sensorium (*koinon aisthêtêrion*) belonging to all the sense organs (*pantôn tôn aisthêtêriôn*)."[123] The medium of both taste and touch is the flesh, which is just a conglomeration of finer and finer ramifications of the blood vessels that branch out from the heart. This makes the heart itself the actual organ of taste and touch.[124] Thus the blood vessels of the body transmit sensory information to the central sense organ located in the heart. Additionally, Aristotle argues in *Motion of Animals* that the heart initiates motion by using hot and cold to heat or chill the blood and *pneuma* that are present in the vessels and thus communicate with the body.[125]

The heart and the blood vessels are therefore vital in Aristotle's account of animal life, so vital that at one point in the *Parts of Animals* he refers to the heart as "the acropolis of the body."[126] Since the body is so dependent on the heart for its health and vitality, any pathological agent that threatens it threatens the life of the animal.[127] It is clear then that close study of Aristotle's anatomy and physiology gives us a sense of the importance of medical themes in his work, along with an idea of how robust his hylomorphism is in terms of his philosophical psychology.[128] Having outlined the main contours of notions of health and disease among the medical writers and philosophers of classical Greek antiquity, we turn now to a consideration of these themes after Aristotle and into later antiquity.

AFTER ARISTOTLE AND BEYOND

Lloyd raises the interesting question of whether or not much changed after the researches of the pre-Socratics, Hippocratics, Plato, and Aristotle. His answer is instructive: "In certain respects or in some sense the answer can be a straightforward yes. But those respects turn out to be peripheral and unimportant compared with those in which the similarities between earlier and later periods are more striking than the contrasts."[129] In Aristotle's own generation, and in the generation immediately after his death, there were a number of philosophers and medical writers who wrote on health and disease: Mnesthius and Dieuches (both Athenians), Diocles of Carystus, and Praxagoras of Cos.[130] Diocles of Carystus (on the Euboean peninsula, close to Athens) was regarded by some as a follower (*sectator*) of Hippocrates, and he was called by the Athenians a "younger Hippocrates" (*iuniorem Hippocratem*).[131] Apparently he based his explanations of the causes of disease on an imbalance of the elements in the body, but he was more specific about certain conditions.[132] For example, both he and Praxagoras held that the cause of paralysis was due to blockages in the vessels that branch off of the heart:

> But Praxagoras and Diocles say that it occurs because of thick and cold phlegm (*phlegmatos*) gathering around the offshoots (*apophuseis*) growing out from the heart (*kardias*) and the thick artery (*pacheias artêrias*), [offshoots] through which voluntary movement (*hê kata prohairesin kinêsis*) is distributed over the body.[133]

Praxagoras and Diocles both seem to agree (similarly to Aristotle) that the heart housed the soul's perceptive and motive powers.[134] Diocles is perhaps most notable for his emphasis on proper regimen, which he particularizes for each time of day: after getting up from bed in the morning, during the time prior to lunch, after lunch, at dinner, after dinner, and before bed at night. He also specifies the appropriate regimen (much like the Hippocratic writer of *Regimen in Health*) for each time of year.[135] In this context Diocles even speaks of proper oral hygiene: in the morning after rising, one should

> brush the gums (*oula*) towards the teeth (*odontas*) and the teeth [themselves] either simply with the fingers themselves or with rubbed pennyroyal (*glêchônos*) of even smoothness, both on the inside and on the outside, and wipe off the remains of food (*sitiôn*) that are attached to them.[136]

Praxagoras, a celebrated physician in his day, based his nosology on humors, identifying eleven separate humors instead of the canonical four.[137] According to Galen, Praxagoras (along with Hippocrates, Aristotle, Phylotimus, "and many others") argued that "when the nutriment becomes altered in the veins by the innate heat (*emphutou thermasias*), blood (*haima*) is produced when it (the heat) is in moderation (*summetrias*), and the other humors (*chymoi*) when it is not in proper proportion (*ametrias*)."[138] Thus, a disproportionate amount of innate heat gives rise to humors that can cause illnesses such as melancholy and fevers.[139]

The two greatest medical theorists of the Hellenistic period, roughly 323 B.C.E. (the death of Alexander the Great) to 30 B.C.E. (the Roman takeover of Alexandria), were Herophilus of Chalcedon and Erasistratus of Ceos, and both were possibly connected to the Lyceum, the Peripatetic school established by Aristotle.[140] While Herophilus seems to have based his nosology on humors,[141] Erasistratus favored mechanical explanations for bodily processes, and although pseudo-Galen claims that he explained all diseases resulted from blood entering the arteries (*paremptôsis*), which Erasistratus believed contained only *pneuma*,[142] there is evidence to suggest that his nosology was more sophisticated.[143] And yet, like Mnistheus, Dieuches, Diocles, and Praxagoras, the evidence upon which our knowledge of the doctrines of Herophilus and Erasistratus depends is fragmentary.[144] This fragmentary knowledge mirrors the political fracturing of the world in the wake of Alexander's imperialism, and the subsequent intellectual fracturing of the medical and philosophical schools of the Hellenistic period into sects. For philosophy, these sects were the Epicureans, Stoics, and Skeptics;[145] for medicine, the Rationalists[146] (sometimes referred to as Dogmatists or Logical physicians), Empiricists, and Methodists.[147] As one can see just by glancing at the names of these sects (both medical and philosophical), they divided primarily on account of their epistemological differences, though their physics played an important role as well.[148] It wasn't until the world was made whole again under the dominion of the Roman Empire that medicine and philosophy were reunited in the all-encompassing medical philosophy of Galen of Pergamum (129–200/216 C.E.).[149]

Galen saw himself as continuing the medical tradition inaugurated by Hippocrates and continued by philosophers such as Plato and Aristotle; he even wrote a treatise titled *The Best Doctor Is Also a Philosopher*.[150] Like his many predecessors, he explains health and disease primarily as a result of different sorts of mixtures found within the body:

> It would thus seem plausible that the best constitution (*aristê kataskeuê*) of the body is that in which all the homeomerous parts (*ta homoiomerê*) (this, of course, is the name given to those which appear single in nature to the senses) retain their proper mixture (*oikeian krasin*). The composition of the organic parts (*tôn organikôn*) from these homeomerous ones (*ek toutôn*) is then a matter of the best-balanced constitution (*emmetrotata kateskeuastai*) of them with regard to size, amount, construction, and relationship between each other. That body in which all the functions (*energeiais*) are working their best can easily be observed also to be the body least subject to illness (*dyspathestaton*). For the part which functions best is a product (*ekgonon*) of both good mixture (*eukrasias*)

of the homeomerous parts (*tôn homoiomerôn*) and well-proportioned constitution (*symmetrou kataskeuês*) of the organs (*tôn organôn*); but this is the nature of the body described above; therefore its activities will be the best of all.[151]

What Galen is referring to in this passage is really a conglomeration of Hippocratic and Aristotelian notions that combines emphasis on proportion and mixture with the material constituents of the tissues and organs. For Aristotle, the elements formed the fundamental level of all material reality (a mixture of earth, air, fire, and water), which, along with their qualities or powers (hot, cold, wet, and dry) combined to form the homeomerous parts of the body (blood, flesh, bone, vessel, and sinew; i.e., the tissues generally). Out of these homeomerous parts were formed the body's organs (heart, liver, eyes, brain, arms, legs, bladder, stomach, etc.), which Aristotle called the anhomeomerous parts.[152] Ultimately it is the soul that creates, orders, maintains, and exercises the functions of the anhomeomerous parts (the organs themselves), along with the homeomerous parts, and the elements and their powers, which compose the organs.[153] For Galen, then, health is equivalent to the natural functioning of the body and its parts, while disease is something that disrupts the body's natural functions by upsetting the fundamental elemental balance that underlies the body and its parts.[154] This is why Galen is so hostile to medical writers, such as Erasistratus and Asclepiades of Bythinia, who seem not to take account of the soul and the teleological arrangement of the body and its organs but instead propound mechanistic nosologies inspired by ancient atomism.[155]

Because Galen, similarly to Plato and Aristotle, discerned such a close connection between the soul and the body, he argued that the doctor was in the best position to treat the soul by treating the body; for him, psychic health was an extension of somatic health:

> The whole of Hippocrates' discussion of waters and of the mixture of the seasons is designed to show that the faculties (*dunameis*) of the soul (*psychês*) depend (*hepesthai*) on the mixtures of the body (*sômatos krasei*)—and not just the faculties within the spirited and desiderative parts of the soul, but those in the rational part too. . . . Anyone can see that the body and soul of people in northerly regions are quite opposite from those near the tropics. And anyone can see, too, that those who live in between the two, in a well-balanced (*eukraton*) region, have better bodies, characters, intelligence, and wisdom than those men. . . .
>
> We, at any rate, know that everything eaten is first of all "drunk down" into the stomach, where it undergoes a preliminary process of transformation, then received by the veins which lead from the liver to the stomach, and that it then produces the bodily humors, by which all

other parts, including the brain, heart, and liver, are nourished. But in the process of nutrition these parts become hotter than normal, or colder, or wetter, in accordance with the nature of the humors which predominate.

So, it would be wise of my opponents—those men who are unhappy at the idea that nourishment (*trophê*) has this power (*dunatai*) to make men more or less temperate, more or less continent, brave or cowardly, soft and gentle or violent and quarrelsome—to come to me even now and receive instruction (*mathêsomenoi*) on their diet. They would derive enormous benefit from this in their command of ethics; and the improvement in their intellectual faculties (*logistikou dunameis*), too, would have an effect on their virtue, as they acquired greater powers of understanding (*sunetôteroi*) and memory (*mnêmonikôteroi*). Apart from food and drink (*trophais kai pomasi*), I would teach them about winds (*anemous*) and mixtures in the environment (*periechontos kraseis*), and places (*chôras*), instructing them which to select and which to avoid.[156]

To support his own view that the condition of the soul is determined by the condition of the body, Galen cites passages from the Hippocratic treatise *Airs, Waters, Places*, Plato's *Timaeus* and *Laws*, and Aristotle's *Parts of Animals* and *History of Animals*. But what is truly remarkable about this passage, as Lloyd observes, is that here we see Galen, primarily a physician, "infiltrating the domain of philosophy and taking over the traditional role of the philosopher as the source of moral advice and guidance."[157] This is the fulfillment of the conception of the role of the physician announced by the author of the Hippocratic treatise *On Ancient Medicine*:

> I also hold that clear knowledge (*gnônai*) about natural science (*phusios*) can be acquired from medicine (*iêtrikês*) and from no other source, and that one can attain this knowledge (*katamathein*) when medicine itself has been properly comprehended (*perilabêi*), but till then it is quite impossible—I mean to possess this information (*historiên*), what the human being is (*anthrôpos ti estin*), by what causes (*aitias*) he comes to be, and similar points accurately (*akribeôs*).[158]

The Hippocratic writer of *On Ancient Medicine* argues that it is medicine alone that will not only set the research agenda for natural philosophy, but will also provide the best insight into the nature of the human being, an area of inquiry traditionally supposed to be the most important province of the philosopher.

This notion of Galen's, that the best physician is also a philosopher, nicely links him up to the tendency of philosophers, especially Hellenistic philosophers, to cast philosophy as a guide to the best sort of life—a "philosophy for life."[159] Because ancient philosophers such as Plato and Aristotle sought comprehensive understanding as their goal, they felt themselves able to speak wisely on all aspects of human life, including on matters of health and disease, since these matters comprised an integral part of leading the happy life. This tendency already had ancient precedents. The pre-Socratic philosopher Democritus of Abdera drew an analogy between philosophy and medicine, saying that "medicine cures (*akéetai*) the body of diseases, and wisdom clears (*aphaireitai*) passions (*pathôn*) from the soul."[160] The author of the Hippocratic treatise *Decorum* tightens this analogy, exhorting his listeners to transplant (*metagein*) wisdom into medicine (*sophiên es tên iêtrikên*) and medicine into wisdom (*iêtrikên es tên sophiên*), "For a physician (*iêtros*) who is a lover of wisdom (*philosophos*) is the equal of a god (*isotheos*)."[161] By the time we get to Socrates and Plato, we witness that the philosopher's search for wisdom leads him to understand that the health of the body is largely dependent on the health, that is, the good order, of the soul.[162] This notion is perhaps most forcefully expressed by Socrates in the *Crito*, where Socrates's lifelong friend, Crito, comes to his prison cell to help him escape the night before his execution. Socrates must face the decision of whether to stay or to flee, to die or to live. This is an artful Platonic application of the medical notion of *krisis* (the moment in an illness that determines whether the patient lives or dies), to a moral (psychological) choice that remains in the control of the person instead of in the hands of the disease.[163] Ironically, by choosing to remain in prison, preserving his psychic (moral) health, Socrates dies. For Socrates, one must live well (*eu zên*) and not just live (*zên*), and sometimes living well might be, paradoxically, equivalent to dying.[164] Some four hundred years after the death of Socrates, Cicero echoes these sentiments when he writes,

> Where men are carried away by desire of gain (*pecuniae cupiditate*), lust of pleasure (*voluptatum libidine*), and where men's souls (*animi*) are so disordered (*ita perturbantur*) that they are not far off unsoundness of mind (*insania*) (the natural consequence for all who are without wisdom [*insipientibus*]), is there no treatment (*curatio*) which should be applied to them? Is it that the ailments of the soul (*animi aegrotationes*) are less injurious than physical (*corporis*) ailments, or is it that physical ailments admit of treatment (*curari*) while there is no means of curing souls (*animorum medicina*)? . . . Assuredly there is an art of healing the soul (*animi medicina*)—I mean philosophy (*philosophia*), whose aid must be sought not, as in bodily diseases (*in corporis morbis*), outside ourselves, and we must use our utmost endeavor, with all our resources (*opibus*) and strength (*viribus*), to have the power to be ourselves our own physicians (*ut nosmet ipsi nobis mederi possimus*).[165]

Here Cicero alludes to the Stoic attitude that conceived the passions or emotions (*pathê*) of the soul as more threatening to the health of the individual than any bodily sickness.[166] However, where we usually have to seek the help of a physician to heal our bodily illnesses, by studying philosophy we are able to become our own physicians to that most important part of ourselves: our soul.[167] The sentiment we hear in Cicero's words will later be referred to by Jesus Christ as a proverbial expression (*parabolên*): "Physician, heal thyself (*Iatre, therapeuson seauton*)."[168] Indeed, inspired by the many healings of Christ throughout the Gospels, Christian writers raise the themes of health and disease to the level of eschatology. I conclude this chapter with some brief observations on the notions of health and disease found in some Christian writers.

Tertullian memorably asked,

> What is Athens to Jerusalem? The Academy to the Church? The heretic to the Christian? Our institution is the porch of Solomon, who himself taught that "the Lord should be sought with simplicity of heart" (Wisdom 1:1). Let them seek it, those who prefer a Stoic, Platonic, or dialectical Christianity. We have no need for curiosity after Christ Jesus; no need for research after the Gospel.[169]

For Tertullian, the revelations of Christianity had brought something entirely new onto the scene, marking an irreconcilable rupture from the learning of pagan antiquity.[170] Nevertheless, the Gospels offer many instances of health and disease that echo pagan Greek medical writers. This seems especially evident after a careful study of some of the images of disease that are presented by the Gospel writer Luke, whom Paul calls "the beloved physician" (*iatros agapêtos*).[171]

The Gospel of Luke contains a number of scenes where Jesus cures various illnesses. He heals lepers,[172] a paralytic,[173] a woman afflicted with hemorrhages,[174] a crippled woman,[175] and a man with dropsy,[176] to cite instances of conditions that would have been commonly seen by ancient physicians. In one passage, Jesus heals the fever of Peter's mother-in-law.[177] This case is notable because in parallel passages from the other Synoptic writers (Matthew and Mark), the fever is simply named; however, Luke gives a somewhat more descriptive account of the fever itself.[178] Luke's passion narrative also differs slightly from the other Synoptic writers. Luke includes an interesting medical detail from Christ's agony in the garden, during the night before he is handed over to the authorities. Luke writes: "He was in such agony (*en agôniai ektenesteron*) and he prayed so fervently that his sweat (*hidrôs*) became like drops of blood (*thromboi haimatos*) falling on the ground."[179] That Christ sweats blood as a result of his agony is left out of Matthew and Mark, but the phenomenon of sweating blood had been noticed by philosophers.[180] And yet, while there may be some Greek philosophical or medical antecedents to descriptions of illness, sickness, and disease in the Gospels, what must constantly be kept in mind is that "disease is viewed primarily in the light of its purpose and meaning, and not generally with regard to its etiology or causality" by the Gospel writers.[181] Christ heals, when he heals, so that God and his power may be glorified.[182]

Christ's healing of blind, mute, lame, sick, mutilated, and impaired bodies is also a sign, harbinger, or promise of his ultimate power to heal the human soul from the curses of sin and death and to restore humanity to eternal life, the life that was lost after the Fall. It is within this context that Matthew interprets Christ's healing work: "When it was evening, they brought him many who were possessed by demons, and he drove out the spirits by a word and cured all the sick, to fulfill what had been said by Isaiah the prophet: 'He took away our infirmities (*astheneias*) and bore our diseases (*nosous*).'"[183] But Christ's ultimate healing work will be accomplished only once he has taken on the pain, suffering, and despair of human death on the cross. It is this intense bodily and spiritual suffering that Christ's defeat of death ultimately cures.[184]

If there is one truth about the human condition that emerges from the New Testament, it is that human beings suffer and die. One gets the same sense from reading the texts of the Hippocratic *Epidemics*.[185] It is almost as if we are there at the beside with the ancient physicians as they carefully observe their patients, marking their symptoms, seeing if they rally from their illness or fade into death. One is often dumbstruck when considering the awesome amount of human suffering that these physicians must have seen, and in many cases were powerless to help. This is very much Plato's view in the *Timaeus*—the whole cosmos is like a hospital ward, and we are the patients. Several centuries later, St. Augustine offers a similar image: "All humankind ails, not from bodily sickness, but from sin. A diseased giant, it languishes over the surface of the entire world, from east to west."[186] To what end all this suffering? It was Christianity perhaps that first introduced the notion that suffering had some positive aspect, a salvific value, that would be redeemed by God.[187] Christians suffering from sickness and disease could take comfort in joining their suffering to Christ's own, in the hope for eternal life.

Finally, despite Christ's contrarian views of ancient Jewish dietary prescriptions, he insists, particularly in the Gospel of John, on the salutary effects that will come from consuming his own body and blood, a teaching that became the ritual of the Christian Eucharist.[188]

> "I am the bread of life. . . . Whoever eats my flesh and drinks my blood has eternal life, and I will raise him on the last day. For my flesh is true food, and my blood is true drink. Whoever eats my flesh and drinks my blood remains in me and I in him. Just as the living Father sent me and I have life because of the Father, so also the one who feeds on me will

have life because of me. This is the bread that came down from heaven. Unlike your ancestors who ate and still died, whoever eats this bread will live forever." These things he said while teaching in the synagogue in Capernaum. Then many of his disciples who were listening said, "This saying is hard; who can accept it?" Since Jesus knew that his disciples were murmuring about this, he said to them, "Does this shock you?"[189]

It is shocking indeed, and perhaps we may sympathize with the disciples who thought that Jesus' words were difficult to accept.[190] Mysteriously, the body and blood of Christ provides Christians with true food and true drink for the ultimate health of their souls, for eternal life. Thus, some of the early Christian fathers refer to the Eucharist as "the medicine of immortality and the antidote against dying that offers life for all in Jesus Christ."[191] In effect, we have come full circle. Just as the ancient medical writers and philosophers emphasized the salutary benefits of proper regimen, so too does Christ emphasize the saving power that comes from the consumption of his own flesh and blood. The Eucharist nourishes and gives strength to the entire body of Christ, the church, with Christ as its head, as they make their way through their earthly pilgrimage to eternal life.[192] In regimen, as in so many other things, Christianity truly made all things new.[193]

It goes without saying that I have not given a comprehensive account of ancient notions of health and disease in this chapter.[194] Nevertheless, if I have given the reader some idea of the reach of these concepts in antiquity, and if I have indicated sources for further study, then I will be satisfied.

CHAPTER THREE

Sex

MARILYN B. SKINNER

Because the ancient world had notions about human bodies, male and female, that do not resemble the ways in which we conceive of our bodies today, its perceptions of gender and its codes of sexual behavior will also seem foreign to us. This chapter takes body history as its point of departure for examining selected aspects of ancient gender and sexuality.[1] One fundamental premise underlay the ancient sex/gender system: the importance of bodily integrity. Adult male citizens might penetrate another sexually but their own bodies were in theory impenetrable, while the bodies of sexual objects, being open to physical incursion, were de facto stigmatized as somehow lesser. Viewed from this perspective, the sexual protocols of antiquity may be hard to reconcile with our own.[2]

MEN, WOMEN, AND REPRODUCTION

The agricultural world of prehistoric Greece assimilated human reproduction to the production of crops: in an economy of scarcity each was necessary for survival.[3] Since farming is labor-intensive, children were a vital resource in a small, self-sufficient rural household. The existence of an heir also assured parents of care in old age and the orderly transmission of property. Given a high infant mortality rate, procreation was a major concern, and discourses on gender focused upon the part each sex played in the reproductive act. Women's bodies were seen as analogous to the fecund, receptive earth itself. Accordingly, the Greek language developed a complex set of metaphors drawing on that analogy: woman is field to be plowed, furrow to be sown with seed, and storage jar made of clay.⁴ While this paradigm ensured wives and mothers of citizen males a central function in religion as intercessors with powerful fertility goddesses, it also diminished their supposed contribution to the child's heredity. In abstract terms, man was regarded as the agent of reproduction, the sower of seed, while woman was only the passive matrix in which it developed and grew.⁵

Concentration on male generativity, to the exclusion of the female, is epitomized in a famous passage of Aeschylus's *Eumenides* (658–61) in which the god Apollo attempts to mitigate the heinousness of Orestes's matricide by denying mothers any genetic share in parenthood:

> The so-called mother of the child is not its parent,
> but rather the nurse of the newly-sown embryo.
> The male who mounts begets. She, like a stranger
> for a stranger, preserves the shoot, should the god not harm it.

Furthermore, the definition of marriage as a contractual union between father and son-in-law, not man and wife,⁶ and such cultural practices as identification solely by patronymic and public registration of infant boys, but not girls, indicate that a "fantasy of descent from male to male" prevailed in the popular mind.⁷

Athenian law may reflect a different understanding of heredity. Cornelius Nepos explicitly states that it permitted the union of half-siblings born to the same father (*namque Atheniensibus licet eodem patre natas uxores ducere*, *Cim.* 1.2). Unions between full siblings, however, were regarded as incestuous (Pl. *Leg.* 838b–c). Two passages attesting half-sibling marriage emphasize that the partners had different mothers.⁸ It is a reasonable inference that marriage with a matrilateral half-sister was prohibited,⁹ which could imply recognition of the mother's genetic contribution to her offspring. Yet close-kin marriage in Athens (of matrilineal and patrilineal first cousins, and even of uncle and niece) was extremely common, owing to kinship loyalty and the desire to consolidate property within the extended family. Nevertheless, squabbles over inheritance were frequent. It is possible, then, that marriage between matrilateral half-siblings was discouraged not for genetic reasons but to prevent inheritance disputes between the fathers of the married couple, an issue that would not arise in the case of patrilateral half-siblings.

The philosopher Aristotle subsequently grounded his explanation of biological difference on the theory that the father is the sole genetic parent. The essential biological purpose of the male is to generate in another, while that of the female is to receive the generated being, in the form of semen, and clothe it in matter. In each sex, special organs exist to serve those purposes—for males, the penis and testes, for females, the uterus.¹⁰ Semen, Aristotle later postulates, is a residue of the blood produced by nourishment (726b.5–15), which has been condensed by the heat of the body to such a degree that it is able to transmit the principle of movement, or *psychê*. Only the hotter male body has the capacity to concoct blood to the point where it is capable of activating matter; as its counterpart, the colder female body merely produces menstrual fluid. This is a less thickened form of blood emitted as a monthly discharge, which in a pregnant woman is absorbed by the fetus and so contributes substance to its growth (729a.25–33). Thus woman's nurturing function in the reproductive process, though necessary to the child's survival, is ancillary to the prime act of generation performed by the man. Because of her lesser body heat, she herself can be regarded as "a defective male, as it were" (737a.28).

Under the Roman empire, Aristotle's line of thinking was pursued to its logical end, the "one-sex" model of human physiology. Male and female reproductive organs were seen as essentially the same, differing only in their internal or external positioning. Galen, the great second century C.E. physician, asks his readers to imagine the male parts turned inside out and placed within the body and the female parts, also inverted, located outside it. The scrotum, he claims, would take the place of the uterus, the penis the vagina; meanwhile, the uterus, facing outward, would contain the ovaries in the manner of a scrotum, and the cervix would become the male member. Like the eyes of a mole, which exist in the animal but cannot open, the female organs, already formed within the fetus, are not able to emerge and project outward due to lack of vital heat (*UP* 14.6). While this is an imperfection in the woman, he concludes, it is perfect for human reproduction, since it provides a safe place for gestation. Preserved by Byzantine and Arab physicians, Galen's treatises were handed down to students of medicine in Renaissance and early modern Europe. Until well into the eighteenth century, that anatomical model remained the standard one for Western medicine, not only consulted for diagnosis and treatment but also controlling cultural assumptions about gender.[11]

BODIES AND SEX IN ANCIENT GREECE

In antiquity the premise that men were the warmer sex was originally contested; for example, the Hippocratic treatise *Diseases of Women* asserts that women have a higher body temperature because they retain more blood (1.1.40–41).[12] However, the authority of Aristotle and Galen finally prevailed, and the notion that men were warmer became standard medical doctrine. Popular wisdom also regarded men as drier, in contrast to the damp earthiness of women. Health was thought to depend upon a balance of bodily heat and cold, dryness and moisture, proportionate to the natural constitution of the individual and the surrounding climate. Although dryness was the ideal male physiological condition, in excess it withered the frame, inducing weakness and even bringing on old age prematurely. Hence sex could have bad physical consequences: expulsion of sperm involved expenditure of a man's liquid strength, or *menos*, and postcoital lassitude was a sign that the body was parched. In the *Works and Days*, an epic compendium of conventional lore, the archaic poet Hesiod remarks that "women are horniest, but men most debilitated" during the dog days of summer, when heat draws moisture from the flesh (586–88).[13] Men, accordingly, were urged to adopt a regimen of moderation in respect to sex: although the act was natural and necessary for the continuation of the species, the intense pleasure associated with it posed the risk of addiction, the impulse of ejaculation gave a jolt to the system, and generation itself involved transmitting a portion of one's own life force to another.[14] Sex, for a man, was but one step removed from death.[15]

Women, on the other hand, required sexual activity in order to maintain health. The surplus moisture in their bodies, stored as blood, had to be discharged regularly, either through menstruation or, better, by nurturing a fetus during pregnancy. After childbirth, too, blood was converted into milk and expelled through lactation. Thus pregnancy and nursing kept the female body in balance and was prescribed by Hippocratic medicine as the cure for many gynecological ills. Moreover, a woman's womb, if empty, might become excessively dry and contract. Since it was believed to be a separate organ, unconnected to the rest of the body, it would then rise upward toward the moister organs, causing hysterical suffocation. Intercourse dampened the womb, preventing contraction and displacement, and it also heated the blood, allowing it easier passage during the menses.[16] These physiological demands provoked a corresponding ravenous sexual appetite: an experienced woman supposedly had a boundless capacity for sex, coupled with a lack of rational control over her desires. With her inexhaustible procreative energy, she needed to be kept in check, for otherwise she posed a concrete threat to masculinity.[17]

In traditional Greek thought, then, female was to male not merely as matter was to life and spirit but also as raw nature was to culture. Women's untamed reproductive vitality had to be channeled into meeting the civilized needs of the household economy—production and rearing of children, food preparation, manufacture of clothing and other domestic goods, conserving property, caring for the sick and performing the last services for the dead. All those functions were assigned to the domestic sphere, and actions having a liminal quality, such as the ones connected with birth and death, were in addition considered inherently polluting. Corollary ideas about women's temperament reinforced those basic gender assumptions: women were "naturally" inclined to love children and make sacrifices on their behalf;[18] they were thought to be innately fearful, which better suited them to life indoors;[19] and, far from resenting their seclusion, they were believed to prefer it.[20] Such instinctual predispositions aligned them with the animal, rather than the human, world. Consequently

Aristotle theorizes that women, while possessing a rational faculty, are by nature less capable of using it to govern themselves[21] and on that basis justifies their lifelong guardianship by males. In cases of adultery, moreover, Athenian law treated the man as the more responsible party, on the grounds that he had the greater capacity for self-restraint. Such notions of women's inferiority in mind and temper carried over into the construction of the "feminine" role in the act of sex.

ACTIVITY AND PASSIVITY

In the contemporary Western world, sexuality is perceived in terms of object preference. The distinction between those who, by natural inclination, are drawn to persons of the same sex and those attracted by persons of the opposite sex is fundamental to our thinking about sexuality: it defines personal identity on a very basic level; establishes separate categories of erotic behavior; and, in the eyes of many people, fixes the morality of a given act. While the tendency to fancy one sex over the other certainly existed in the ancient world and was acknowledged as such, in the majority of individuals erotic attraction was probably more diffuse, whereas Western culture, with its either/or mentality, has traditionally exercised a stronger narrowing effect upon innate propensity.[22] Thus every head of household was expected to marry and have regular intercourse with his wife in order to beget heirs. Private preference could be indulged only in the domain of extramarital sex.

"Normal" sexuality was based upon the notional genders of active and passive, respectively associated with the "masculine" and "feminine" roles in the sex act. Yet "active" and "passive" were not the same as our biologically based genders of "male" and "female," for the class of passive partner might include persons of either sex—girls and women, boys, and even adult men, though the latter were despised for it. Women were passive by nature, but male individuals who fell into that category, owing to age, status, or moral character, were regarded as "feminized." Criteria of beauty for youths and girls were more or less uniform: in early Attic vase painting, the attractive *hetaira* or courtesan is slim and oddly boyish, while in later art the beautiful boy is girlish.[23] The antithesis of passivity, the active sexual role, was on the other hand the prerogative of just one group, adult citizen males.

The ideal in ancient sexual relations was not, as a rule, reciprocity.[24] Instead, sex was hierarchically structured and conceptually integrated with power. Adult males of high status were seen as having the power to protect their bodies from invasive assault. Both in Athens under the radical democracy and in Rome the body of the citizen male was legally inviolable; even the Roman soldier, subject to beating by his military superiors, might still kill an officer attempting to violate him sexually.[25] All sexual encounters in which at least one of the participants was an adult male of citizen status were therefore expected to conform to a dominance-submission pattern: the man played the active part by penetrating the mouth, vagina, or anus of the other person, who in submitting to penetration revealed his or her lack of corporeal autonomy.[26] Status was thus reaffirmed under the most intimate circumstances, since a virile male demonstrated his intrinsic advantage over members of other groups by making use of their bodies for his pleasure.[27] Although the broad imposition on antiquity of this "winner-take-all" model of ancient sexuality, commonly referred to as the "penetration model," has been challenged as too simplistic,[28] much current scholarship has accepted it, since it does underscore fundamental differences between the ancient and the modern worlds—though possibly at the cost of minimizing resemblances.

PEDERASTY

In certain parts of archaic and classical Greece, and notably in sixth- to fourth-century B.C.E. Athens, for which we have the most detailed evidence,[29] the authoritarian paradigm of sexual dominance and submission was tempered by the social institution of pederasty. Pederasty involved the formal courtship of an elite citizen youth, the *erômenos* or "beloved," by an older male lover, or *erastês*. While sexual gratification was the lover's recognized aim, apologists for the custom surrounded it with an elaborate rhetoric of training in virtue, involving the testing of both participants.[30] The adolescent was supposed to withhold his favors until the potential lover had proved himself worthy as a mentor; he then obliged the *erastês* out of gratitude for attentions rendered and the hope of receiving further instruction in manly skills. Courtship rituals required the hopeful suitor not merely to give gifts and provide services, but virtually to abject himself before the beloved, who was therefore portrayed as the controlling figure in the relationship.[31] This simulated inversion of power roles encouraged the boy to develop self-possession, maturity, and good judgment and meanwhile checked hubristic tendencies on the part of his lover. Fostering wisdom, courage, and temperance through intense social bonding, pederasty could thus be defended as a source of benefits for the entire Athenian democracy. Evidence in the form of texts and vase paintings shows that this ideology permeated aristocratic discourse and was more or less accepted by ordinary citizens, although class tensions induced some suspicion of it.[32] The fact that Socrates, famously abstentious in his personal life, could ironically mine pederastic rhetoric for metaphors encapsulating his quest of the Good indicates that the conventions of this upper-class pursuit were part of the operative knowledge of the whole culture.

Classical Greek society was nevertheless prudish about the facts of pederastic coition. The verb most commonly applied to the youth's role in the act was *kharizesthai*, a euphemism literally meaning "to do [one's lover] a favor"; in other words, he was expected to derive no actual enjoyment from serving as passive partner. Xenophon preserves a fragment of Socrates's banquet conversation voicing what was apparently a common assumption: "A youth does not share in the pleasures of sex with the man, as a woman does, but soberly looks upon the other drunk with passion."[33] On a small number of Athenian vase paintings, intercourse with a boy is depicted, but the position assumed is normally intercrural and, with rare exceptions, penetration itself does not occur. Outside of Old Comedy, where obscenity was sanctioned by the Dionysiac festival, mention of anal sex was taboo. Anxiety over the adolescent's function as sex object took the form of concerns about his capacity to make the "ephebic transition"—that is, to shift from the passive to the active role, maturing into an adult male capable of achieving pleasure through erection and orgasm while still exercising due control over his passions.

FAILED MASCULINITY

"Masculinity in the ancient world was an achieved state, radically underdetermined by anatomical sex,"[34] and hence not spontaneously attained with the passage of years. Frequently in Attic tragedy the protagonists are youths whose lives are cut short through their own immature decisions (Euripides's Hippolytus and Pentheus, for example). Worse, from a classical Greek perspective, was the case of the grown man, the effeminate *katapugôn* or *kinaidos*, who had formed a craving for anal sex and thus abandoned himself to a dissolute life of submission to other men. *Problemata*, a treatise falsely attributed to the philosopher Aristotle, furnishes a medical explanation for the *kinaidos*'s unfortunate condition: because the vessels that in normal men conduct semen from the upper body to the penis and testicles are blocked, the secretion flows into the anus and collects there, triggering insatiable sexual excitement (4.26.29–30). Yet Greek science further recognized the role of early habituation in creating a lifelong bent toward sexual passivity. The same chapter of *Problemata* also suggests that repeated penetration at the time of adolescent "ripeness" (*hêbê*) could produce a quasi-natural desire for this experience through the recollection of pleasure.[35] In the *Nichomachean Ethics* Aristotle himself mentions passivity (along with other morbid propensities such as pulling out one's hair, biting one's nails, and eating dirt) as a practice that might result from nature or also habit, "as in the case of those violated since boyhood."[36]

Surprisingly, the philosopher exempts those *kinaidoi* whose condition is owed to nature from the charge of *akrateia* or lack of restraint, which, he says, is no more appropriately applied to them than it is to women.[37] Sexual difference is consequently the model to which all other forms of social inferiority—whether determined by status, class, ethnicity, or perceived failings of character—can be assimilated. Whoever falls short of the ideal, embodied in the person of the self-restrained free male citizen in the prime of life, is automatically feminized.

THE RISE OF A HETEROSEXUAL ETHOS

Athenian pederasty was an aristocratic institution that indirectly served class interests: ostensibly it benefited the community by training youths to be exemplary citizen-soldiers, but it also forged cross-generational bonds between elites to help insure that the nobility would retain its privileges, especially in the face of challenges posed by the radical democracy. With some exceptions, chiefly the comedies of Aristophanes and forensic speeches directed to juries, our literary sources for sexual attitudes in classical Athens are biased toward an upper-class audience and predictably focus upon homoerotic relations. More sensitive, however, to the tastes of the larger buying public, vase paintings indicate that a no less romantic and idealized model of heterosexual *erôs* coexisted alongside the pederastic model of sexuality and over the course of 150 years steadily gained ground. Courtship scenes involving youths, which reached the height of their popularity during the last half of the sixth century under the oligarchic regime of the Pisistratids, virtually disappear after 510 B.C.E., a phenomenon that has been attributed to the expanding power of the democracy.[38] Even during their heyday, if the surviving corpus is at all representative, such homoerotic vases were vastly outnumbered by those featuring images of heterosexual love. During the next decades we observe a growing interest in respectable women as erotic objects, culminating, after the second half of the fifth century, in depiction of the bride as an icon of romantic desire. In such wedding images "sexuality is shown in a polite but unmistakable manner as the bond that ties together the basic unit of the *polis*."[39] By the late fifth century, there was also a widespread vogue for interior scenes of domestic life featuring the tranquil occupations of women, which possibly served as vehicles of imaginative escape from the uncertainties caused by the long-drawn-out Peloponnesian War.

During the Hellenistic period (323 B.C.E.–30 C.E.), when mainland Greece passed under the rule of Macedonian monarchs and its city-states lost their independence, institutionalized pederasty ceased to have any real political function. Decreased opportunity to participate directly in civic government reduced its importance as a vehicle of socialization and cross-generational bonding. Consequently the boy's conduct was no longer implicated in a web of communal concerns about his present self-restraint and future moral development. With those larger ethical components removed, the pederastic love affair had no ramifications for any facet of society beyond the lives of the two principals. Literature of this period consequently takes a trivializing or sentimental stance toward pederasty: it empathizes with the misery of the *erastês*, but no longer raises questions about the honor or merit of either partner. In Hellenistic epigram, the beloved youth is stereotyped as fair but likewise fickle and greedy. Being the unattainable object of desire, he is chiefly a peg on which to hang a conceit.[40]

Attentiveness to female beauty and feminine psychology became dominant artistic preoccupations during this era. During the late fifth century, though, Athenian playwrights had already responded to the emerging interest in women and the domestic sphere with their memorable tragic heroines and fantasies of women exercising power. Contemporary sculptors were meanwhile experimenting with the partially clothed female body by depicting victims of violence with one breast partially revealed[41] and employing diaphanous drapery to give an illusion of nudity, as on the young Nike who adjusts her sandal from the parapet of the temple of Athena Nike on the Acropolis (411–407 B.C.E.).[42] Developments in the literature and monumental art of the high Hellenistic period may be therefore regarded as the culmination of long-standing trends.

Praxitiles's Aphrodite of Cnidus (ca. 350 B.C.E.) broke new ground by representing the goddess of love fully unclothed: putting the viewer unapologetically in the position of voyeur, it exploited the sensuous appeal of the nude female figure. Poetry simultaneously developed a special fascination with the emotional awakening of the inexperienced girl—Theocritus's Simaetha (*Id.* 2) or Medea in Book 3 of Apollonius of Rhodes's *Argonautica*—and the predicaments she would face. Widespread social changes taking place in the fourth and third centuries B.C.E. gave impetus to these artistic movements. The early Hellenistic age was a time of great cultural disruption. Political unrest, wars between city-states, Macedon's conquest of mainland Greece, and the subsequent drain on manpower and resources occasioned by Alexander the Great's invasion of the Persian empire combined to cause extensive rural poverty and a weakening of family ties. Contemporary accounts speak of considerable numbers of exiled and displaced persons.[43] When men left their ruined farms to seek a living as mercenaries, many never returned; tombs and ancestor cult were abandoned.[44] Isolated individuals, who could no longer find personal satisfaction through kin networks and civic institutions, invested emotionally in private relationships. Unlike boys, women could be valued as responsive erotic partners and equal (or almost equal) companions. Marriage consequently began to be regarded as a source of fulfillment for husband and wife as well as a strategy for family continuity. At Athens, the sentimental New Comedy of Menander and his contemporaries responded to this change in attitudes with dramatic plots that defended romantic love as a sound basis for choosing a mate. Rape of a citizen girl, with a resulting pregnancy, is sometimes a plot device, but in that case the sexual assault is treated as a peccadillo on the part of the ardent young man, and the play ends happily when he marries his former victim.[45]

Emphasis on private subjectivity coincided with an improvement in women's social and economic status. Even affluent parents were raising fewer children. Yet daughters were more cherished and indulged than they had been before, with many upper-class girls receiving the kind of liberal education formerly reserved for boys. Not surprisingly, we find evidence of females moving into the arts and professions, gaining recognition as physicians, painters, and writers.[46] Women who had inherited considerable wealth increasingly used it to promote their own standing within the community by serving as public priestesses, making costly religious dedications, and financing public works.[47] Neo-Pythagoreanism, a popular intellectual movement, promoted a single standard of marital chastity for spouses and celebrated the educated woman's capacity to direct the moral formation of her children. Stoicism and Epicureanism, the two major Hellenistic philosophical schools, likewise recognized women's equal potential for achieving wisdom and virtue. Thus a number of separate factors, all reinforcing each other, together contributed to the materialization of a more pronounced heterosexual ethos.

Romantic love between men and women was endorsed and sanctioned at the highest levels of Hellenistic society. The defining literary works of the period were written in Egypt at the court of the Ptolemies, the Macedonian ruling dynasty, by scholar-poets attached to the museum and library at Alexandria founded by Ptolemy I Soter. These poets eulogized the religious and cultural benefactions of the powerful queens Arsinoë II, full sister as well as wife to Ptolemy II Philadelphus, and Berenice II of Cyrene, married to her cousin, Philadelphus's son Ptolemy III Euergetes. Both royal unions were portrayed by the court poets as passionate love-matches. As successors to the Pharaohs, Arsinoë and Philadelphus had adopted the practice of sibling marriage, a sacred tradition of Egyptian kingship imitating the relations of the paradigmatic divine couple Isis and Osiris. Since Isis's devotion to her brother-spouse was a central theme of native Egyptian myth and ritual, Ptolemaic ruler ideology emphasized the harmonious love between king and queen as well.[48] Arsinoë also furthered worship of Aphrodite as patroness of the mutual desire of husband and wife: Theocritus's *Idyll* 15 commemorates an elaborate ritual tableau she staged honoring the goddess's consort Adonis, which featured an effigy of Aphrodite embracing her mortal bridegroom (*gambros*, 129). Callimachus paid tribute to Berenice's wifely affection in his *Lock of Berenice*, which tells how the queen's tress, offered to Aphrodite in thanksgiving for Euergetes's safe return from battle, was transformed by the gods into a constellation as an enduring memorial to her vow. This elegy was later translated into Latin by the Roman poet Catullus (c. 66). These eulogistic compositions profoundly influenced the subsequent literary tradition, but the cult of romantic heterosexual love and marriage was not so much a fad invented by court poets of the Ptolemies as it was a deep-seated response to the temper of the times.

DEVIANT FEMALES

Alongside promotion of the wife as emotional partner, we find a mounting hostility toward female homoeroticism. This phenomenon is not unexpected. In societies like that of archaic and classical Athens, where male and female spheres of activity were sharply divided and bonding was chiefly homosocial, what women might get up to among themselves would cause little anxiety for men. On archaic Lesbos, women's desire for other women had been given canonical expression in Sappho's songs, which were reperformed by men at Athenian drinking parties and highly admired.[49] Plato's Aristophanes matter-of-factly mentions female "companions" (*hetairistriai*) attracted to other women but does not elaborate.[50] There are a few uncertain allusions to sex between nonrespectable women on Attic vase paintings. Otherwise the issue is disregarded. Now we observe opposition to love between women articulated on several fronts. Fourth-century Attic comedy "heterosexualizes" Sappho, portraying her as the object of pursuit by male poets and the rejected lover, in turn, of the beautiful ferryman Phaon, for whom she commits suicide.[51] Biographical remarks contained in a papyrus from Imperial Roman times[52] may allude to scholarly writings from the third and second centuries B.C.E.: Sappho, it is said there, has been accused by some of being *ataktos* ("undisciplined") and a *gynaikerastria* or "lover of women." Together with efforts to normalize Sappho's erotic life or condemn it as licentious, we find female same-sex bonding branded as unnatural not only by Plato[53] but also by the normally tolerant epigrammatist Asclepiades. In *Anth. Pal.* 5.207 he denounces two courtesans who have developed an erotic attachment to each other:

> The Samian women Bitto and Nannion are unwilling to frequent
> the school of Aphrodite according to her rules,
> and desert to other things not good. Mistress Cypris, abhor
> the fugitives from that intercourse you preside over.

The activities of Bitto and Nannion do not fall under Aphrodite's sphere of influence; by engaging in them, the women turn outlaw and merit her hatred. This judgment illustrates the ancient tendency to regard *ta aphrodisia* as synonymous with asymmetrical power relations and consequent penetration.[54] What falls outside that framework is no longer ignored but has instead become a problem for males. There is a nervous curiosity about what pairs of women actually do; Asclepiades's vague expression *ha mê kala* ("things that might not be nice") invites prurient speculation. Metaphors of desertion, carrying a heavy charge in a culture that identifies masculinity with virtuous deeds on the battlefield, drive home the point that Bitto and Nannion have forsaken their obligations as courtesans. Unless we are missing some humor here, a considerable degree of apprehension over female sexual autonomy is reflected in this quatrain. Far worse than betraying the speaker with another man is betraying him with each other.

SEX AS METAPHOR

In the literature of classical Rome, largely produced during the last two centuries of the Republic and the first two centuries of the Empire (200 B.C.E.–200 C.E.), the conceptual pattern of sexual relations seems much the same as the one prevailing in Greek art and literature from the archaic through the Hellenistic periods: sex is, in essence, an asymmetrical relationship between the insertive and the receptive partner. Overall, there are many more references to deviations from the norm, but, even so, the hierarchy of male dominance and female submission is assumed as a firm principle and alternative arrangements are viewed as perverse. When the poet Catullus attacks Julius Caesar and his associate Mamurra as *cinaedi* ("gender deviates") who take turns servicing each other (57.1–2), when the emperor Caligula is described by his biographer as interchanging active and passive roles,[55] or when the philosopher Seneca expresses horror at monstrous manly women who forget their sex so far as to penetrate men,[56] such practices are represented as bizarre in the extreme.

Frequent mention of abnormal behavior, however, by no means indicates that flesh-and-blood Romans were regularly violating the rules of sex and gender. In fact, knowledge about the actual sexual lives of individuals, whether emperors or slaves, cannot easily be recovered from the historical or archaeological record—in the case of slaves, because even material remains such as tombstones mostly provide only scanty, formulaic information, and, in the case of certain emperors, because the admittedly copious reports are top-heavy with conventional slanders. In Roman culture, furthermore, allegations of sexual misbehavior—especially in oratory and other forms of political discourse, or in satire and invective—are largely metaphoric. That is, the hierarchical power dynamics of "normal" sex are symbolically employed to indicate the way things should be in a properly organized and functioning society, while accounts of irregular sex encode perceptions of alarming transformations in the social milieu caused by such impersonal forces as fluctuations in the economy, changes in the structure of government, demographic trends, and the threatening prospect of lower-class upward mobility.

Why did Roman culture draw upon sex as a metaphorical vehicle for discussing perceived societal irregularities? In Rome, as in Greece, claim to the active or insertive position was understood to be synonymous with established social supremacy. However, Roman social stratification was exceptionally complex and hard to fathom, even by insiders, because lines of demarcation were fuzzy. The democratic Athenian *polis* maintained an ideological fiction of political equality among heads of household, despite actual differences in birth or wealth. Though Roman ideology looked back to an imaginary golden age of rugged yeomanry, when riches and luxury had not corrupted the population, it did not proclaim the virtues of egalitarianism. Class disparity was in fact normalized through a patronage system in which goods and services were exchanged between those of higher and lower standing. In calibrating social position, birth, rank, and respectability all played a part, but, according to the satirists,[57] money might confer economic power that trumped all other factors, even when its possessor was a former slave. Particularly under the Empire, freedmen found ready opportunities for financial advancement, often by doing work that elites despised. Meanwhile, citizen males of limited resources, dependent for support upon the caprices of a patron, were vulnerable to psychological humiliation if not physical insult.[58] Literary texts consequently display an obsession with the conflict between inherited privilege and the clout exercised through wealth. Civil wars and, later, the erosion of upper-class political power and authority likewise produced status anxiety. Because the dominance-submission grid of ancient sexuality, with its clear-cut distinction of masculine and feminine roles, was thought to be ordained by nature, mapping sexual misconduct upon status incongruities implied that any departure from the traditional strict nexus of birth, wealth, and consequent privilege was a crime against the natural order. Manliness was solely an elite prerogative, while the marginal but upsetting outsider, such as the former slave of Oriental extraction, was rendered effeminate.

ROMAN WOMEN

In Rome women did not occupy the conceptual space assigned to them in the Greek symbolic universe. While the proper Greek wife performed her obligations chiefly in the domestic as opposed to the public sphere, the well-born Roman woman had communal, financial, and familial responsibilities that took her outside the home, sanctioned her public visibility, and conferred male traits: she might, for example, manage property and exercise patronage in the same way a man would. The highest-ranking Roman priestesses, the six Vestal Virgins, were endowed with many of the prerogatives of magistrates. Working-class women, slave or free, labored alongside men, performing, as artisans or shopkeepers, identical tasks. Consequently, the Roman male conceptualization of the female sex was bipartite: Roman women were gendered as both "Same" and "Other," a mode of thinking that permitted them to display praiseworthy male attributes without overstepping boundaries and allowed for vacillation between masculinity and femininity in their impact upon observers.[59] Gender boundaries were actually more permeable for Romans than for Greeks, more prone to destabilization. This enclave of latent ambiguity within the sex/gender system gave poets and artists added room to represent power relations obliquely.

Women's freedom of movement, however, aroused fears about the chastity of the *matrona* (the married woman of respectable status) within the community as well as among her kinfolk. The prosperity and power of the state depended upon the continued goodwill of the gods, who were profoundly offended, it was thought, by the wickedness of the ruling classes. Principal blame for adultery was placed not on the lover but on the irresponsible, pleasure-seeking woman,[60] who, pursuing gratification, abandons her moral duty to raise god-fearing and obedient sons.[61] Perhaps unease over potential adultery was one way of displacing vague misgivings about wives' independent conduct of their business and social affairs onto a tangible area of concern. Loss of confidence in the patronage system and the weakening of class divisions appear to find concrete expression in tales of well-born adulteresses attracted to social inferiors.[62] Whatever the reality behind these worries, they were serious enough to persuade Augustus, the first Roman emperor, to put forward a law making the act of consorting with a married woman a public offense tried before a standing criminal court, with severe penalties for both the convicted adulterer and the adulteress. Passage of this legislation, the *lex Iulia de adulteriis coercendis*, in 18 B.C.E. was followed by a series of notorious trials recorded in the pages of Roman historians, many with political ramifications. Problems arising from the vagueness of the statute generated an enormous body of judicial literature on technical issues, but it nevertheless remained in force until very late antiquity.[63]

ROMAN HOMOEROTICISM

Jurists distinguished adultery proper, defined as intercourse with a married woman, from the crime of *stuprum*, or illegitimate sexual congress with an unmarried girl or respectable widow or divorcee.[64] Nevertheless, both kinds of offense were punishable under the *lex Iulia*.[65] However, the concept of *stuprum*, which, properly speaking, consisted in "the violation of the sexual integrity of freeborn Romans of either sex,"[66] also applied to relations, consensual or forced, with citizen boys. Sexual attraction to youths was considered normal, just as it was in Greece, but the chastity of freeborn boys was guarded by law and custom. Below the age of puberty, such boys wore special clothing, a purple-bordered toga and a golden amulet (*bulla*) to mark their status. Laws protected them from unwanted attention; following them on the street, suborning their attendants, or pestering them was legally actionable.[67] By the third century C.E., capital punishment was prescribed for the accomplished seduction of a citizen boy.[68] Even after a youth of sixteen or seventeen had received the *toga virilis* ("man's toga") that marked his adulthood, his morals were an object of concern; defending his client M. Caelius from charges of adolescent unchastity, Cicero explains to a jury that the young man had been placed by his father for oratorical training under Cicero's own supervision and assures his listeners that during that precarious time of life Caelius had been as rigorously sheltered as in his paternal home.[69] Slaves and male prostitutes were therefore the only legitimate objects of masculine homoerotic desire. Anecdotal evidence indicates that seductions of boys occurred, but concerns about adultery with matrons seem far more prevalent in our sources.[70]

In the Roman scheme of sexual morality, then, there was no provision for romantic attachments to free citizen youths. Pederasty, defined in Greek terms as the institutionalized cross-generational relationship of two male citizens, was unthinkable, and the word, applied to Roman practices of boy-love, is misleading.[71] There is plenty of evidence, artistic, literary, and even inscriptional, for liaisons with slaves and social inferiors, but, because the power dynamics were so distinct from those courtships figured on Greek vases and imagined in the dialogues of Plato, it is hard to relate one kind of homoerotic relationship to the other. Slave boys who were special favorites were known as *delicati*, "pets"—yet, no matter how indulged, they were at the mercy of their owner. As for the prostitute, his client bought what services he wanted. Some pantomime actors were much in demand, like Bathyllus, reputed to be the favorite of Augustus's friend Maecenas,[72] but even so were stigmatized and suffered legal disabilities because of their profession.[73] In such relations there was no room for the flexibility and negotiation theoretically possible in the encounter of the Greek *erastês* and *erômenos*, with its ethical demands on both citizen participants. Roman poets can therefore adopt Hellenizing conventions in their love lyrics, giving the beloved a Greek name to signal that they are writing in the sympotic tradition, praising his beauty and lamenting his coyness, avarice, and infidelity. Yet these motifs seem artificial. Perhaps one attraction of Roman pederastic poetry was its evocation of an imaginary setting that glamorized the sordid mechanics of actual sexual transactions with slaves and hired partners.

CONCLUSION

In this chapter we have surveyed the corollary schemes of sex and gender arising from assumptions about male and female physiology that originally emerged as folk beliefs in pre-classical Greece. Men, because of their greater body heat, were believed to be the sole agents of generation; the female was considered merely the passive recipient of seed. Gender stereotypes that associated masculinity with dominance and femininity with submission consequently informed the conceptual structuring of the sex act. Under the "penetration model" the female role did not necessarily correspond to anatomical sex. Yet in classical Athens (and no doubt elsewhere) the etiquette of homoerotic boy-love permitted a temporary and artificial reversal of positions. Conventionally, the boy occupied the "feminized" role as the passive partner, but as the prize of courtship he was allowed to set the rules of engagement.

Ancient sex/gender patterns are not entirely uniform, though: they admit of variations across time and space and display new features in the civilizations of Hellenistic Greece and Rome. During the Hellenistic period romantic heterosexual love came to dominate art and literature. Even female desire outside of marriage might be depicted in a positive light. Still, the androcentric orientation of sex remained unchanged: women's erotic impulses were regarded as good only when directed toward men, while desire for other women was pilloried. In Rome, social hierarchies were complicated by patronage networks that paradoxically blurred differences of rank, class, and even gender. Sexual discourses, in which an unequivocally asymmetrical framework of power relations could still be taken for granted, became a means of expressing social anxieties metaphorically. At the same time, legal provisions criminalized two forms of actual sexual offense: adultery, which intruded on the marital rights of the husband, and injury to the chastity of citizens, whether female or male. All such protocols trace their origin back to one deep-seated assumption: sex comprises the union of two bodies, one that preserves its integrity during the encounter and one that yields to the other. To the ancient world, our modern Western ideal of sex as a free, mutual, and *desirable* surrender of self would be incomprehensible.

CHAPTER FOUR

Medical Knowledge and Technology

BROOKE HOLMES

The question of whether a physician knows what he is doing when he acts on the body is not a straightforward one. In asking it, are we trying to discover if the physician is familiar with, and has the skill to implement, the practices prescribed by the relevant medical tradition in a given clinical situation? Or are we interested in determining whether he has a grasp of the causes of the patient's illness and the reasons his therapy will bring about the desired outcome? What if the physician is working in a culture whose ideas about disease and therapeutic practices our own medical authorities deem fanciful or obsolete? If he adheres to the guidelines of his own culture, do we still believe that he knows what he is doing? How would our answer be affected by the outcome of his therapy?

Such questions have particular relevance to our understanding of the body in the ancient Greco-Roman medical tradition. On the one hand, in the nineteenth century, physicians and patients began to lose trust in the humoral medicine whose authority had been guaranteed for centuries by the names of Hippocrates and Galen. From a modern vantage point, the explanation of human nature in the late fifth-century B.C. treatise *On the Nature of a Human Being*, the first extant text to advocate a model of four humors, seems to be describing another species. We might respond to the text's alienating aspects by pointing out that the bodies of classical Greeks were undoubtedly much like our own and attributing the author's views to his ignorance about the body whose nature he purports to describe. If his therapies had any success, we could argue, it was because they were actually time-honored folk remedies, recently endowed with (spurious) theoretical pedigrees; or we could attribute the results to the placebo effect.[1] None of this would be to say, however, that the physician knew what he was doing.

Yet, on the other hand, the very premise that the physician requires specialized knowledge of the physical body appears self-evident to us precisely because Greco-Roman medicine exercises such a strong influence on the Western medical tradition. It is because of the Greeks, in other words, that medicine seeks to know the nature of the body—indeed, it is because of the Greeks that we believe there is a concept of a physical body whose nature (*physis*) must be known by the physician (*iatros*). If, then, we are asking ourselves how medical knowledge and technology impact the understanding of the body in Greco-Roman antiquity, it is useful to distinguish between the body as a physical object with relatively stable characteristics over the last three millennia and the body that emerges in the classical period as an object of expert knowledge and care.[2] In fact, having examined the historical contingency of the body qua medical object, we may be in a position to rethink our privileging of medical views of the human body in the contemporary Western world.[3]

In this chapter, I give a brief overview of the body that takes shape in the different medical paradigms that, while ranging widely, participate in the Greco-Roman tradition of attributing disease to physical causes and excluding gods or demonic agents.[4] I leave aside, for the most part, the question of whether ancient medical knowledge about the body is factually correct. I concentrate instead on the ideas about the body that recur in learned medical texts, although I do take a brief look at the interaction between learned medicine and the cult of Asclepius, which first takes root in mainland Greece in the fifth century B.C. and flourishes in the Hellenistic and Roman imperial periods. Such an approach necessarily privileges an urban dweller's experience of medical knowledge, and I cannot do justice here to the diversity of what Vivian Nutton has called the "medical marketplace" in Mediterranean antiquity.[5] Nevertheless, I hope that by adopting the perspective that I have just outlined, I can shed some light on the close relationship between Greco-Roman learned medicine and Western concepts of the physical body. The article follows a roughly diachronic arc. For, from our earliest medical writings, it is clear that elite physicians, however much they sparred with one another about the nature of the body and the best way to care for it, saw themselves as participating in a continuous tradition.

It is worth saying a few words about the nature of our evidence. We are fortunate to have in the Hippocratic Corpus sixty-odd treatises, mostly dating from the classical period (late fifth and early fourth centuries B.C.) and probably organized into a corpus by scholars in the Alexandrian period (third through first centuries B.C.).[6] Although these treatises have come down to us under the name of Hippocrates, we have no way of knowing whether any of them were authored by the historical Hippocrates; it is certain, moreover, that the texts were written by a number of different physicians.[7] Our evidence for the lively period between the fourth century B.C. and the writings of Galen in the second century A.D. is fragmentary and distorted by the prejudices and agendas of our source texts.[8] With Galen our evidence is again rich (his texts represent over 10 percent of extant Greek literature from before 300 A.D.).[9] On the basis of this material, I sketch an overview of some of the most provocative and influential models of the body generated by those who would have identified themselves as working within medicine (*iatrikê, medicina*). In the interests of providing some background, however, I begin with the Homeric and early archaic evidence.

THE KNOWLEDGE OF THE ARCHAIC HEALER

We cannot assume that the body as a whole is the healer's natural object of knowledge and care in the early archaic period. For, insofar as we can tell from our limited evidence, the concept of the body may not have been useful in this period. In Homeric epic (ca. eighth century B.C.), the person splits at the moment of death into an ethereal *psychê* bound for Hades and a corpse. Before this moment, however, the living person does not fracture along dualist lines. There is nothing (mind, soul, person), then, against which "the body" stands out as a singular entity.[10] Rather, the person can be described in various ways, each of which involves what we would call body. His emotions and thoughts are realized through a collection of substance-forces that scholars have tried, with limited success, to map anatomically.[11] He is at once a visible form (*eidos*), a built structure (*demas*), and a coordinated, agile group of limbs (*melea*); he is covered with skin (*chrôs*) that can be easily broken. But *sôma*, the word that regularly gets translated as body in the later Greek evidence, is used only a handful of times in the Homeric epics. In each case it is applied either to the corpse of an animal or a human corpse that has not been given proper burial.[12] The *sôma*, in short, is neither "the" body nor the concern of the *iatros*.

The expertise of the healer is, in fact, rather limited in the Homeric epics. In the *Iliad*, the healer exhibits special knowledge of drugs (*pharmaka*) and wound care. He is even less visible in the *Odyssey*, showing up once among other traveling craftsmen who might be welcomed into a royal household.[13] Moreover, that neither wound care nor pharmacology is the exclusive province of the *iatros*: at one point in the *Odyssey*, for example, a wound is treated with an incantation, and women in the *Iliad* prepare remedies.[14] Nevertheless, it seems clear that the healer's epistemic advantage lay primarily in his skilled use of the knife and his pharmacological knowledge, which may have enabled him to treat internal ailments.[15]

What is important to recognize in the present context is what the epic *iatros* apparently does not know. When Apollo sends a plague against the Achaean troops in the first book of the *Iliad*, Achilles asks for a seer, a dream-interpreter, or a priest to come forward to interpret not the cause of the plague—everyone agrees that Apollo is responsible—but the *reason* for the god's anger.[16] His request prefigures the medical interest in creating broad etiological frameworks through which to understand disease. Yet, no one suggests that the most prominent healers among the Achaeans, Machaon and Podalirius, have any insight to offer, as Celsus, a Roman encyclopedist who wrote a history of Greek medicine up until his own time in the first century A.D., did not fail to recognize.[17] It is the seer Calchas, rather, who steps forward to shed light on Apollo's anger.[18]

Calchas offers the Achaean army a form of expertise grounded in divination, not in the mechanics of the physical body. The close relationship between divinatory knowledge and healing can be understood in light of the fact that in the archaic period—and indeed throughout Greco-Roman antiquity—many diseases that lacked an obvious cause were attributed to divine or demonic agency, as was also the case, for example, in ancient Babylonia and many other cultures.[19] It is true that there are ailments that a hero suffers as a result of forces or stuffs within him: Achilles's anger in the *Iliad* can be seen as a kind of escalating disease for which the implacable hero is himself responsible. Here again, however, the *iatros* is no expert. It is the other heroes who try to temper Achilles's anger with healing words, albeit to no avail.

In short, then, despite that *we* might be willing to say the object, or one of the objects, of the archaic healer's knowledge is the physical body, our evidence suggests that it is unlikely that contemporary sources would have concurred, for the simple reason that "the" body is not yet available as a discrete object of expert knowledge. Nor is it clear that "the disease" is the object of his knowledge, since in the scattered references we have from early Greek poetry, diseases are allied with nebulous demons who may have been engaged by a diviner or purifier.[20] The healer acts on the patient by means of *pharmaka* and the surgical knife, tools that will remain fundamental to the medical art. Yet, in so doing, he does not presumably explain why the disease occurred and why his drugs are effective. It is the seer or the dream-interpreter who deals with the unseen causes of suffering by locating suffering in a broad, mortal-immortal social context.

In a fragment of the lost epic *The Sack of Ilion*, however, we find evidence of another perspective on the knowledge of the archaic healer. The fragment, which may date from the sixth or even the fifth century B.C., assigns the two major healers of the *Iliad*, Machaon and Podalirius, different skills.[21] The former has "defter hands," giving him the edge in removing missiles, cutting, and healing wounds. Podalirius, who is virtually invisible in the Homeric epics, is given a more prestigious skill, namely, the precise knowledge to know and to cure mysterious diseases.²² He is thus the first to recognize the signs of Ajax's madness—his flashing eyes and his disordered thought—after Ajax is denied the arms of Achilles. Unfortunately, we do not know how he responds to these signs in the epic.

If Podalirius's knowledge seems to diverge from the knowledge of the healers we have discussed thus far, this may be the result of our limited evidence from the archaic period. What is so intriguing about the *Sack of Ilion* fragment, however, is that its description of Podalirius neatly anticipates the persona who will come to dominate our earliest corpus of extant medical texts: the physician as the decipherer of corporeal signs. At the same time, whereas Podalirius in some way has a superior understanding of the changes to Ajax's eyes and thought, what distinguishes the physician in the later texts is his grasp of the nature of bodies and diseases. Let us turn to these texts now in order to see how the physical body (*sôma*) emerges as an object of knowledge and a site of technical intervention.

SEEING THE PHYSICAL BODY

"The nature of the body is the starting point of the medical *logos*," observes the author of the classical-era medical treatise *On Places in a Human Being*.²³ His observation builds on a number of assumptions that are crucial to the conceptualization of medicine in fifth- and fourth-century B.C. Greece: there is a thing called the body; it has a nature; and medicine should be based on an account (*logos*) of that nature. Our understanding of how these ideas take shape is unfortunately hampered by the absence of evidence for the early phases of naturalizing medicine. As a result, the rupture between the archaic *iatros* and his classical counterpart is exaggerated for us. Nevertheless, it is difficult to deny that the extant medical texts from the classical period represent a historical shift in how disease was viewed and the domain of the *iatros*. These texts offer explanations of disease from which the gods are systematically excluded as causes. Several of them launch attacks, sometimes virulent, on magico-religious etiologies, as in the opening pages of the treatise *On the Sacred Disease*.²⁴ In the place of the gods we find a roster of "natural causes," which, in many treatises, must be identified if the physician is to intervene successfully in the disease.

Rather than say that the medical writers "discovered" a timeless truth about human disease, namely that it is due to external (environmental, dietetic) factors and pathological processes inside the body, we might say that they gradually shape an object capable of assuming the causal power once ascribed to the gods: the *sôma*, in which and through which disease is now believed to take shape. Physicians and defenders of the medical *technê* are active participants in the public debates about human nature and the intellectual salons that flourish in Athens and throughout the Greek world in the latter half of the fifth century.[25] By the fourth century B.C., many medical treatises are circulating, and the well-educated layperson is expected to be acquainted with the basic principles of medical accounts of the body, disease, and human nature.[26] The cultural saturation achieved by medicine in this period suggests that the body at the heart of the new medicine may have begun to influence, at least in some quarters of the Greek-speaking world, how embodiment was not only conceptualized but also experienced.

The emergence of the physical body owes much to the broader "inquiry into nature" in this period.[27] From Aristotle to present-day histories of philosophy, a triad of thinkers in sixth-century B.C. Asia Minor—Thales, Anaximenes, and Anaximander—have been celebrated for offering explanations of phenomena that were traditionally ascribed to Zeus (e.g., lightning) in terms of natural causes (water, air, the hot, and so on). Fifth-century "physicists," such as Empedocles and Anaxagoras, sought to explain phenomenal reality in terms of microscopic elements. These elements interact in an ongoing process of composition and dissolution driven by chance, necessity, and the nature of each element and compound, rather than by the intentions and emotions so crucial to the circulation of power in the divine world.[28]

One way of understanding the new medicine, then, is to see it as describing how persons are necessarily implicated in the impersonal cosmos being described by the physicists.[29] Most of the interactions humans have with this world-in-flux exist below the threshold of what we can sense of ourselves. Moreover, they happen automatically: although we may control the kinds of encounters we have with the external world by paying attention to our diet and our activities, our physical interaction with that world happens without our conscious participation. *Sôma* thus comes to designate the part of the person that participates mechanically in the physicists' cosmos.

The *sôma* does not simply participate in the larger physical world: it is also a microcosm of that world and subject to similar processes and exchanges. In the early fifth century B.C., Alcmaeon of Croton, said by a later doxographer to have written on "medical things," gives us our first extant naturalizing account of disease: he claims that it is caused by the excessive power of a single force among the many different forces (the hot, the cold, the bitter, the sweet, etc.) that together constitute the human body.[30] The idea that the body naturally comprises a number of potentially dangerous impersonal stuffs or humors—their number and characterization vary from author to author—proves a powerful one for the medical writers. Thus, while the disease process is catalyzed by a powerful influx of force from the environment, such as a change of temperature or a sharply bitter food, the chain reaction that eventually produces the symptom takes place within and by means of the body itself. The author of the late fifth-century treatise *On Ancient Medicine* nicely sums up what he takes to be the discovery of the first physicians: "they saw that these things [i.e., the humors] are inside a human being and that they hurt him."[31]

Medicine responds to the specific vulnerability represented by the body qua unstable compound caught in an indifferent world by developing technologies to manage the fragility of human nature. As the passage just cited suggests, these technologies are founded first and foremost on the ability of the physician to look past the surface of the body to "see" into its physical substratum. For, as we have seen, much of what counts as *sôma* cannot immediately be felt, let alone understood by the embodied person. Nor does the physician have direct access to this hidden world, given what one author calls the "density" of the body.[32] Insofar as the body is a kind of "black box," whose interior is beyond investigation, the physician relies heavily on the symptom: "The vision of unseen things is through the phenomena," to quote a tag that is ascribed to Anaxagoras and could serve as the motto of the early medical writers.[33]

In magico-religious paradigms, too, corporeal phenomena provoke inferences about things unseen. No one—except, of course, Homer's audience—*sees* Apollo string his bow against the Achaean troops or the arrival of his arrows. His presence, rather, is inferred from the sudden onset of plague. And it is not just the god's presence that is inferred but also his anger and the reasons behind it. In a similar way, the medical writers understand a symptom as giving them access, first, to a hidden space populated not by anthropomorphic beings but by fluids and channels, hollows and bones.[34] Symptoms yield information, too, not about the reasons behind an intentional act of harming but about the forces that act in this space—humors, as well as the forces of life and death.

In the Hippocratic treatise *On Regimen in Acute Diseases*, for example, the author provides an explanation of why patients sometimes die with a mark on their flanks "as if a blow had been received."[35] Rather than blame the mark on demonic anger, the author claims that it is caused by a massed humor present in the body at the time of death. Although the patient had felt the humor as pain, it had not been seen—until now. Its materialization on the surface of the body represents the end of a series of events inside the body, catalyzed by the physician's ill-advised decision to administer gruel: "one bad thing is added to another," writes the author, until that "bad thing" has become powerful enough to cause the patient's death. Without concerning ourselves with the finer points of the pathological process envisioned here, let us use this example to highlight three ways in which a nascent medical semiotics helps to generate a concept of the physical body.

It is clear, first, that the writer believes that a symptom—first pain, then the mark—can bring a humor to the attention of both the patient and the physician. Pain, in other words, brings to light the presence of something that is usually part of us without us being aware of it.[36] The author of the treatise *On the Nature of a Human Being*, who is arguing, presumably in a public arena, against those who say that a human being is made up of only one thing, points out that pain can be felt only by a body whose nature is not uniform: here, pain becomes a proof of our composite nature.[37] Elsewhere we find medical writers making more specific claims about our bodies (e.g., that winter fills the body with phlegm, that bodies are harmed not by gods but by diseases) that are also deemed proven through phenomenal signs.[38] Thus, it would seem that by the latter half of the fifth century B.C., physicians were inviting patients to "see" themselves first and foremost as composite, labile, and often disordered physical bodies, rather than as social agents capable of attracting the anger of gods and demons. Such a mode of "seeing" takes place primarily through symptoms.

Second, symptoms can provide more specific information about what is going on inside the body, information that allows the physician to intervene in the disease process. In the passage we just saw, the author interprets the visible mark on the body as the final moment in a chain of events inside the body that he reconstructs from the patient's symptoms (pain, rapid and heavy breathing, and so on). On other occasions, the physician uses symptoms in order to act. The surgical treatises are full of signs (*sêmeia*) that allow the healer to recognize the nature and location of a fracture or dislocation.[39] There are signs to know when a disease is "settled" and the patient can be purged.[40] The author of the treatise *On Regimen* boasts that he has developed a system of "pre-diagnosis" for identifying the onset of disease in the body far before "the healthy is conquered by the diseased."[41] In these instances, we see the physician trying to gain access to the underlying state of the body or events that have occurred there. The body is imagined as a space hiding the disease. Symptoms, correctly interpreted, can give the physician the edge in the race against the trouble developing inside the body.[42]

Finally, symptoms are used to predict the outcome of a patient's disease. Thus a corporeal sign may be good or bad depending on whether it forecasts recovery or death (or relapse). There are whole treatises (*On Prognostic, Prorrhetic* I and II) devoted to teaching the physician how to identify good and bad signs and to make his predictions accordingly. Often we are given very little indication as to why a sign is believed to be positive or negative, an omission that once led scholars to believe that the prognostic treatises represented the accumulation of collective empirical observation. No doubt some of the prognostic signs have a basis in experience: the *facies Hippocratica* is still held to be a reliable predictor of death. Yet careful work in recent decades has demonstrated that many of the signs deemed critical—that is, signs that occur at specific points in the disease and indicate in which direction it will turn—also have a basis in the theoretical and pretheoretical assumptions of humoral pathology and contemporary number theory.[43] How does prognostic interpretation imagine the body? Whereas the signs we have seen thus far are used to prove a point about the body's nature, to shed light on its present condition, or to indicate a specific event inside the body, prognostic signs are held to be signs "of the whole body."[44] A "good" sign indicates the body's strength; its life force; or, as it is sometimes represented, its inner heat.[45] A bad sign, conversely, signifies the ascendancy of the disease. Prognostic symptoms thus help to represent the body as governed by a kind of life force that maintains the cohesion of the composite body and defends it against external forces.

Here, then, are three things that symptoms in the medical treatises enable physicians to "see," thereby encouraging the crystallization of the concept of the physical body: the person's composite nature; his hidden inner depths, where disease forms; and a nonconscious, automatic principle of life. This picture is necessarily schematic. Yet it does give us the basic blocks with which to build an overview of how concepts of the physical body morph and multiply in the following centuries under the influence of competing ideas about what the physician can and should know and do. I take up each of these aspects now in turn. I begin with the idea that bodies are unstable compounds requiring management. I then trace briefly how the idea of a principle of life develops, before considering how the desire to *see* the hidden life of the body gives rise to systematic human dissection; I examine, too, the powerful challenges to the idea that medicine requires knowledge of the unseen, which begin to appear in the Hellenistic period. After revisiting the concept of the body as an object of care in the imperial period, I close with a look at Galen and the formation of Galenism.

CARE OF THE BODY AND CARE OF THE SELF

A distinguishing feature of medicine in the late fifth and early fourth centuries B.C. is its prominence in the broader cultural and intellectual sphere. Physicians are not engaged in disinterested research on the nature of the body. Rather, they are actively debating its characteristics and trying to persuade general audiences and patients to interpret symptoms—both their own and those of others—through the lens of specific medical explanations: rhetoric is an indispensable tool to the rise of the new medicine.[46] Laypersons are being encouraged to educate themselves in the basic principles of medicine.[47] After all, if the body is a source of vulnerability, it is in the interests of those most vested in maintaining control over themselves and others—namely adult freeborn men—to care for their bodies.[48]

It is against this backdrop that we should imagine the rise of dietetics, as well as a particular kind of "technical" gymnastic training—both practices designed to ward off disorder in the body and to shape its external appearance.[49] The most extensive extant treatise on these subjects, *On Regimen*, is typically dated to 400 B.C., but its author refers to many earlier writings on the subject.[50]

The treatise sheds light on the growing sophistication of a kind of quotidian, prophylactic care that the author presents not only to an elite audience but also to an audience of people who cannot afford to take care of their health "at the expense of everything else."[51] Its premise is the necessity of protecting the composite body—for this author a mixture of fire and water—against seasonal changes and other possible catalysts of disease. Its strategy turns on the patient's constant adjustment of the body's economy of forces through diet and exercise, as well as on his monitoring of the body for the first signs of trouble. Lack of care, *ameleia*, is playing with high stakes. For, it is on the physical composition of the body and the soul that not only health but also intelligence and character depend.[52] Despite the fact that our evidence for the middle and later fourth century B.C. is fragmentary, it appears that the management of the body within a broadly humoral context becomes increasingly precise in these decades. An extensive excerpt from the works on regimen by one of the leading physicians of this period, Diocles of Carystus, outlines an almost hour-by-hour regimen for patients interested in maximizing health.[53]

At the same time, we also have evidence of resistance to the "excessive care of the body." In the third book of Plato's *Republic*, Socrates attacks modern medicine for enabling people to cling to worthless lives. Not only does medicine preserve those who, having turned their bodies into "stagnant swamps," would be better off dead,[54] Socrates alleges; it also ruins perfectly happy lives. For "the excessive care of the body . . . makes any sort of learning, thought, or private meditation difficult, by forever causing imaginary headaches or dizziness and accusing philosophy of causing them . . . it is constantly making you imagine that you are ill and never lets you stop agonizing about your body."[55]

Socrates's complaint here is part of a challenge to medicine's therapeutic authority by those advocating what begins to be called, already in the fifth century B.C., a "care of the soul." The therapies of the soul are established both on analogy with the care of the body and in competition with it.[56] For Plato's Socrates, it is because medicine does not treat the soul, the font of untrammeled appetites and false beliefs, that it can only prop up unlivable lives instead of fostering true human flourishing through the cultivation of reason and the disciplining of emotions. It is likely that the idea of a soul somehow separate from and yet analogous to the body helped to crystallize a concept of the body as not-person. In Plato's early Socratic dialogues, as well as in later dialogues such as the *Phaedo*, he approaches the body as a mere instrument or something alien to our true nature.[57] At the same time, contrary to what the cliché of Platonic dualism would lead one to expect, it is possible to see the physical body becoming more and more entangled in Plato's concept of the soul. By late dialogues such as the *Timaeus*, the *Laws*, and the *Philebus*, the body is sufficiently implicated in the soul and critical enough to our ethical health that it is the province of philosophy, not medicine, to manage the turbulence of embodiment.[58]

Aristotle sets out to establish a strong division between ethics, on the one hand, and biological and physiological investigation, on the other. Nevertheless, he, too, is fascinated with the gray zone of the psycho-physiological (e.g., dreams, perception, melancholy), and the medical analogy is a staple of the ethical treatises.[59] The analogy, which makes the soul the object of ethical therapy, grows extremely popular in the Hellenistic philosophical schools.[60] Yet, as we will see further in Galen, neither the separation of ethics and medicine nor the supremacy of the former is ever fully secured. Medicine's ambitions to describe the nature of a *human being* (and not just a body) never disappear. Indeed, its resources for doing so grow increasingly sophisticated in the Hellenistic period.

HELLENISTIC REVELATIONS: THE LIFE OF THE BODY AND THE TRUTH OF THE CORPSE

The growth of dietetics and the concomitant development of ethics qua therapy of the soul indicate popular fascination with the models of embodied selfhood generated by medical inquiry. They also point to a burgeoning interest in the functions and the structure of the healthy body in the fourth century B.C. For example, Aristotle and Diocles of Carystus dissect animals in the interests of creating an analogical model of the human body.[61] In so doing, they are building on their predecessors' interest in the body's inner structures (*schemata*),[62] as well as its skeletal architecture. (The surgical works in the Hippocratic Corpus suggest that practitioners had considerable success treating dislocations and fractures, and their procedures remained standard for centuries.) But for Aristotle, and perhaps for Diocles, too, describing the human body was not simply about description. A model of the body should also capture the relationship between the structure of each part and its function vis-à-vis the organism as a whole. It should, that is, be formulated from a teleological perspective.[63]

Both Aristotle and Diocles also appear curious (if somewhat mystified) about a basic principle of life, which seems to take the form of an inner, vital heat that enables the body to triumph over external threats (e.g., in digestion) and, in Aristotle at least, over the formlessness of matter.[64] Of equally hazy importance to life is *pneuma*, "breath" or "air," which will become a major player in Hellenistic accounts of the soul and the body. In Aristotle, *pneuma* may communicate sensory information to the "ruling part" (*hêgemonikon*) of the soul, which he consequentially located in the heart; *pneuma* also apparently translates the desires of this part into voluntary motion.[65] The evidence for the role of *pneuma* in Aristotle, however, is hardly overwhelming; we have scarcely more evidence for *pneuma* in Diocles.[66] Nevertheless, what we have provides a glimpse of the significant role that *pneuma* will play in Hellenistic theories of sensation and voluntary motion.

The interests of fourth-century B.C. writers in the traffic between body and soul and the movement of life through the body were encouraged by their anatomical investigations. Animal dissection enabled investigators to see a complexly networked body where the earlier medical writers had imagined a vague system of channels for the circulation of humors. Praxagoras of Cos, working in the latter part of the fourth century B.C., is often credited with first differentiating arteries and veins, which had been previously lumped together under the term *phlebes*.[67] For Praxagoras, the difference concerned the substance transmitted: nutrient-rich blood for veins, *pneuma* for arteries. He also seems to have made the *neura*, "sinews" from Homeric times, outgrowths of the arteries, thereby creating a connection between the *pneuma* in the arteries and voluntary motion.[68] *Pneuma*, it would seem, thus comes to play a pivotal role in Praxagoras's understanding of the dynamics of human life. His conceptualization of the pneumatic body will prove particularly influential for the Stoics.[69]

Praxagoras's other major contribution to the Greco-Roman medical tradition is his claim that the regular contraction and dilation of the *pneuma*-bearing arteries can be sensed in the pulse. Although hitherto noticed only as a pathological sign (tremors, palpitations), the pulse becomes for Praxagoras a powerful index of health and disease in the body as a whole. He thus lays the groundwork for "sphygmology," a science of the pulse, which takes hold as a highly subtle diagnostic tool that puts the physician in direct, that is, haptic, contact with the vital forces of the patient.[70]

The advent of systematic human dissection in third-century B.C. Alexandria takes Praxagoras's inquiries into the networked body to a new level of complexity. The manipulation of the human corpse had been, until this point, taboo throughout the Greek world. Why the practice became possible at the court of the Ptolemies in Egypt is a matter of speculation, although one important factor was no doubt the royal sanction of the practice (the Ptolemaic court supplied the bodies of condemned criminals).[71] More certain is the powerful impact of anatomical investigation on the concept of the physical body. So powerful has this impact been that, whereas the humoral body appears a relic of the past, the anatomical body, that is, the *seen* body, often simply *is* "the" body for us, even today. Yet, as Shigehisa Kuriyama has argued, the Greek "anatomical urge" appears less inevitable in cross-cultural perspective. He attributes this urge to a cultural tradition that equates knowing with seeing, a tradition that, as we have seen, is integral to establishing medicine in the classical period as a privileged kind of insight into the inner body.[72] Rebecca Flemming and Phillipa Lang have further implicated anatomical inquiry, which would have been coded as a specifically Greek practice in the bicultural world of Hellenistic Alexandria, in Ptolemaic cultural imperialism.[73]

Dissection discloses a new world. Herophilus of Chalcedon, one of the most celebrated anatomists working at Alexandria, painstakingly details the inner landscape of the physical body, pointing out, in the process, structures that still bear the illustrative names he gave them.[74] Through repeated dissections, he charts the paths of the arteries and veins and offers further support to the perceived diagnostic usefulness of the pulse by relating it to the natural dilation and contraction of the arteries in all living beings.[75] Herophilus also isolates the nerves and traces their origin to the brain, in which he locates the ruling part of the soul; he seems to have further distinguished "voluntary" nerves (for motion) from sensory nerves.[76] There is strong evidence that the desire to see this intricate system encouraged Herophilus and other Alexandrian anatomists to vivisect not only animals but also criminals furnished by the Ptolemaic court.[77] The practice generated debates well into the imperial period about the costs of medical knowledge and the very feasibility of seeing life through the exposure of "the parts that nature had previously hidden" to the light.[78]

The challenge posed by anatomical investigation to a humoral, primarily fluid model of the body, which is dynamic first and foremost in the shifting nature of its composition, was first articulated in detail by Herophilus's controversial contemporary, Erasistratus of Ceos. Unlike Praxagoras and Herophilus, both of whom seem to have retained the basic tenets of humoral pathology, Erasistratus explains disease as the malfunctioning of those conduits that had come to dominate the physician's field of vision: arteries, veins, and nerves. Rather than describe disease in terms of the delicate humoral economy of forces, he attributes fevers and inflammations to a single event: the leakage of blood from the veins into the arteries, either through a cut or through an excess of blood in the veins.[79] The systematization of these new theories of pathology works in tandem with Erasistratus's formulation of a physiology in which health depends on the integrity of "elastico-fluid" systems.

To speak in terms of "elastico-fluid" systems, as the historian of medicine Mario Vegetti does, gives some indication of the influence of contemporary mechanical models on the Erasistratean conceptualization of the body.[80] According to Erasistratus, what is happening when blood from a vein enters a severed artery, for example, is the natural rush of fluids toward a vacuum, in this case created by the escape of *pneuma* from the artery. Erasistratus relies heavily on this biophysical principle in his explanation of normal physiological processes as well.[81] The fragments are full of analogies between machines and body parts, bodily processes and the mechanisms of Hellenistic pneumatics, hydraulics, and hydrostatics.[82] It is true that technological analogies can be found in our earliest biological and physiological writings.[83] With Erasistratus's analogies, however, the mechanical principles have grown more sophisticated (with a correspondingly complex explanation of bodily processes). Moreover, such analogies may have become provocative, if we understand the greater attention being paid in the fourth century B.C. to the body's innate tendency toward life as creating a framework in which mechanistic explanation is seen as incompatible with vitalist principles: Galen, at any rate, will oppose approaches that he sees as strictly mechanical to his own teleological vitalism.[84] Erasistratus's own position, however, may have been more nuanced. He seems to have held not only the idea that life processes can be elucidated on analogy with machines but also the idea that the body is naturally equipped for the functions required for its flourishing.[85] Whether he saw a contradiction between these ideas or how he might have dealt with it are not questions the fragments allow us to answer.

CHALLENGES TO "THE VISION OF UNSEEN THINGS"

Erasistratus's own version of the Hellenistic fascination with the various (vascular, sensory, motor) webbed systems in the body prominently featured something he called the "*triplokia*," a braided bundle of arteries, veins, and nerves that constitutes the fabric of the body.[86] A distinguishing feature of the *triplokia* is that it is invisible to the naked eye. It is thus representative of a new class of unseen things produced by anatomical inquiry.[87] For, the more of the body that dissection reveals, the more is thought to lie below the surface of the visible. The evidence for Herophilus suggests a cautious thinker who drew back from speculation about the unseen substratum of the anatomical surface, whether in the form of microscopic structures or causal mechanisms and forces.[88] Yet, Erasistratus, for example, and, in the second century B.C., the physician Asclepiades of Bithynia, were more than willing to speculate about a new class of "things seen by reason." For Erasistratus, these include not only the *triplokia* but also the minuscule valves that allow blood to spill over into the arteries in cases of plethora; for Asclepiades, *onkoi*, tiny, featureless particles that circulate in tiny channels through the body.[89]

Speculation about the unseen strata of the physical body through inferential and analogical reasoning had been, as we have seen, a venerable practice in naturalizing medicine. In the Hellenistic period, however, resistance to the very idea that the physician can know such unseen things (conditions, mechanisms, causes) springs up, perhaps in response to the new set of unseen causes and structures created by anatomical investigation, perhaps in response to the practice of anatomy itself. Physicians and medical writers had always been a contentious group. Yet the challenge to medicine's quest for hidden causes appears to have created at some point in the third century B.C. a formal school of "Empiricists," headed by a renegade student of Herophilus, Philinus of Cos.[90] The emergence of the Empiricists initiates a period in which the learned medical tradition is fragmented into different sects.

The Empiricists hold that the knowledge relevant to clinical success requires no training in the principles of the body's organization or functioning. Rather, the physician can glean everything he needs to know from his own experience, from reading case histories written up by others, and by making educated guesses on the basis of similar cases in the past. Those physicians against whom the Empiricists define themselves never form a coherent group. Nevertheless, by the Roman imperial period they are identified collectively as Dogmatists or Rationalists. In the early decades of the first century A.D. in Rome, another school opposed to the search for hidden causes takes shape: the Methodists. The Methodists hold that any pathological condition of the body falls into one of three general categories—looseness, tightness, or a mixture of the two.[91] The physician need only identify the condition (or "community") to know the proper therapy: one of Methodism's major exponents, Thessalus of Thralles, declared that anyone could learn medicine in a mere six months.[92]

The debates among the sects bear primarily on epistemological questions, and a detailed account of their differences would take us too far afield. What we can observe, however, is how these new sects redraw the lines of the medical body.[93] For both the Empiricist and the Methodist, the symptom no longer indicates hidden events, forces, or conditions inside the body. The Empiricist draws a direct line, rather, between the corporeal phenomenon and a therapeutic response (rather than an internal event or condition) on the basis of previous successes and failures. For the Methodist, inferential reasoning is unnecessary insofar as each of the three possible pathological conditions of the body can be apprehended directly by the senses. Given the immediate transparency of the body, the Methodists had little use for anatomy. Empiricists were equally dismissive of the anatomist's expanded field of vision, arguing that the corpse can tell us nothing useful about the living body.[94] Both sects, then, challenge medicine's claim to arcane knowledge by putting the relevant facts about disease in the public domain, as it were, and radically curtailing the data relevant to diagnosis and treatment. The medical body is recast primarily as a body of surface phenomena and a body encountered at the bedside.

MANAGING THE BODY IN THE EARLY IMPERIAL PERIOD

The limits of our evidence make it difficult to know how anatomical inquiry and the sects' debates affected public perceptions of the body, particularly the body vulnerable to disease. The Alexandrian physicians are part of a larger intellectual community under the patronage of the Ptolemies, and Hellenistic poetry bears evidence of the poets' exposure to their colleagues' investigations. Apollonius, for example, incorporates corporeal objects generated by dissection into his description of the effects of Eros's arrow on Medea.[95] In the Roman period, laypersons are evidently aware, and wary, of dissection practices (although systematic human dissection had been limited to Alexandria, where it ceased after just a few generations).[96] But laypersons can also be extremely knowledgeable about and interested in medicine: the Roman encyclopedist Celsus, for example, was deeply familiar with medical theories and practices, although he was almost certainly not a physician.

The fascination with all things medical, coupled with a deep distrust toward physicians, is characteristic of the reception of Greek medicine in the Roman world.[97] The doctoring of both family and slaves had been the task of the *paterfamilias*, and the old guard of the Roman Republic was not always amenable to the professionalization of healing at the hands of foreigners. According to Roman tradition, Archagathus, the first Greek *medicus* (and for the Romans, all *medici* were Greek), was driven out of Rome after being nicknamed "the executioner" for his harsh treatments.[98] Nevertheless, although we know little about the specific circumstances of the arrival of Greek medicine in Rome, it was undoubtedly prominent among the intellectual and cultural imports that flowed into Rome in the last centuries B.C. as Rome conquered the Greek-speaking world. The epistemological debates that dominated Alexandrian medicine were an integral part of the Hellenistic philosophical tradition to which Roman elites were exposed. Medical treatises were written for the general public, and Greek medical terminology and ideas made their way into a wide range of literary and other nonspecialist texts written in Latin from as early as the third century B.C. Greek doctors practiced both as slaves and as citizens in the highest echelons of Roman society. In spite of Roman resistance to Greek physicians, then, these doctors were in popular demand, and educated Romans commanded a sophisticated knowledge of medical theories and practices.

One of the consequences of the dissemination of medical ideas in everyday life was a preoccupation with the "care of the self" in the early centuries A.D., much as we saw was the case in the fourth century B.C. Despite radical changes to earlier concepts of the body in the Hellenistic period, the body taken as the object of the care, that is, as a fragile and labile composite organism in need of constant surveillance, remains familiar. Moreover, despite the debates between the medical sects about the limits of causal explanation, the relationship between behavior (diet, exercise) and health, so central to early Greek medicine (and of continued importance in the Hellenistic world),[99] remains robust enough in the Roman context to keep medicine intertwined with ethics. The body, in other words, continues to function as a tableau of one's way of life.[100]

For leisured Romans and Greeks in the first centuries A.D., then, the care of the body, informed by general medical ideas about healthy living (what and when to eat, when to bathe and how, and so on) was "a moral obligation."[101]

Texts such as Plutarch's treatise *Advice about Keeping Well*, written toward the end of the first century A.D., the letters of Seneca, and the correspondence between Marcus Aurelius and Fronto open a window onto a culture that scrutinized the body and was preoccupied with diet, bodily habits and practices, and corporeal phenomena.[102] The focus on the body parallels the emphasis in the Hellenistic and Roman philosophical schools on caring for the soul, and indeed body-care remains the model for soul-care.[103] There has been much speculation about the social and historical reasons for the "medicalization" of daily life and the sense of the body as something vulnerable and unstable. What is clear is the heightened attention to the physical body. Indeed, this awareness of the body's fragility may have impacted early Christian notions of the body as corrupt and terrestrial.[104]

The concept of the body as something highly sensitive to food, drink, and other influences appears, too, in the scattered testimonia we have from the supplicants of the healing god Asclepius. Our first evidence of the Asclepius cult dates from fifth-century B.C. Greece, although there is archaeological evidence of healing cults dating much further back. In early testimonia, Asclepius cures a wide range of ailments, from extended pregnancies to blindness to paralysis, for those who come to spend the night in his shrines.[105] Although the god sometimes cures with *pharmaka* in these accounts, he also heals with the touch of his hand (or the touch of his serpent or his sacred dogs) or by more violent and fantastical means: in one dream, for example, a mother dreams that the god cuts off her daughter's head and hangs her upside down, upon which a large quantity of fluid drains out of her body.[106] There is, however, increasing overlap between the care of the body advocated by human physicians and that recommended by the god as time goes on.[107] By the first centuries A.D., Asclepius often appears as a uniquely gifted personal physician. The Asclepieia in the large urban centers, particularly in the Greek East, were places where people gathered to compare symptoms and discuss current affairs, much as they did in nineteenth-century sanitoria.[108]

Our best source for this world is the rhetorician Aelius Aristides, who wrote of the god's benefaction in six books of *Sacred Tales* (of which five are extant). The *Sacred Tales* are full of detailed accounts of Aristides's maladies and his dreams.[109] Although he comes to Asclepius because the best physicians have failed to diagnose his illness, and although he exhibits a particular concern for bodily purity that challenges humoral notions of "relative health," the control that Aristides exercises over what goes in and out of the body (in consultation with his dreams) is consistent with the practices of body care in contemporary elite culture.[110] Of course, not every supplicant was a man of leisure, and Asclepius could intervene in a more pragmatic way when necessary.

We have followed over six hundred years the elaboration and transformation of the three ideas that I identified as critical to the crystallization of the physical body in early medical writing: the concept of the physical body as an unstable compound; the concept of a substratum of the body wherein lie its inner truths; and the concept of a nonconscious principle of life. All these concerns (and many more) are taken up in the immense corpus of Galen of Pergamum (129–ca. 200/216 A.D.), whose own vision of medicine prevails for over a millennium. I turn now to a brief overview of Galen's perspectives on the body before closing with a brief look at medicine in the centuries after his death.

GALEN AND THE BODY

What makes Galen's legacy so daunting is both the range of his interests and his sometimes convoluted attempts to make his multiple perspectives converge on ostensibly unified objects. The son of a wealthy architect, Galen studied philosophy under representatives of the major schools of the day—Platonic, Peripatetic, Stoic, and Epicurean—and medicine at the great school in Alexandria, an education that no doubt played a role in his dual commitment to theory and experience. His first patients were gladiators. As his fame at Rome grew, he became the personal physician of emperors. From 162 A.D., he was active in the public medical demonstrations and debates popular in Rome, and he wrote prodigiously (we know of over 350 works). A product of a deeply competitive society, he was harsh toward critics and rivals and less than forthright about his affinities and his debts. He revered Hippocrates, whose true heir he believed himself to be.

It is probably Galen who codified the four humors of the body (blood, yellow bile, black bile, phlegm) that become canonical in later medicine. For, although he himself framed the quaternary schema as a Hippocratic legacy, only one extant treatise, *On the Nature of a Human Being*, works with such a model.[111] Yet Galen also recognizes, under the influence of Aristotle, four basic elements (fire, air, earth, water), which are represented in the body by humors, and four qualities (the hot, the cold, the dry, and the moist), which combine to form temperaments.[112] The humoral pathology transmitted under the name of Galen, then, takes on a particularly Aristotelian hue, which helps to ensure its viability in the Arabic, Byzantine, and Western Christian worlds, where Aristotelianism is deeply rooted. As in the earliest medical writings, health for Galen is a balance between the various constituent parts of the body. While ideal health is something of an illusion, a state according to nature (*kata physin*) is both possible and desirable.[113] The most perfect body was assumed to be male: females were by nature weaker, colder, and formed for the very tasks society held them to—above all childbearing.[114]

Galen's desire to recover a Hippocratic ideal—a desire that has its historical origins in the tradition of Hippocratic exegesis and lexicography that dates from Hellenistic Alexandria and undergoes a resurgence in the second century

A.D.[115]—should not, however, lead us to believe that he was retrograde in his interests or perspective. Well versed in anatomy as a result of his studies at Alexandria, he was engaged publicly and privately throughout his career in dissecting and vivisecting animals—he preferred Barbary apes and rhesus monkeys, since they were closest to human beings—in the interest of defending his claims about the body and seeking new information about its inner parts.[116] Galen championed a tripartite physiological system that he credited to another of his idealized ancestors, Plato, but that in truth owed much to the veins, arteries, and nerves of the Alexandrian body.[117] And anatomical knowledge facilitated Galen's surgical operations. These operations, in which Galen drew on a tradition that was at least half a millennium old and had flourished in the wake of anatomical investigation at Alexandria,[118] could be highly complex.

Galen's commitment to anatomical investigation, together with his fondness for philosophical logic, located him among the "Dogmatists," who held that knowledge of the body's inner workings was necessary for the proper interpretation of symptoms. Indeed, Galen's extant writings abound in examples of Galen as "medical detective," capable of discerning the hidden reality behind the morass of corporeal phenomena; attentive to every detail; practiced in pulse taking (he wrote thousands of pages on variations in the pulse).[119] Yet he was sympathetic, too, to the Empiricist program, as well as to empirical investigation; he was wary of generalization and deeply committed to clinical work. Knowledge of the body, on his view, originates with the experience of the body, and bodies themselves are required to confirm or refute conjectures.[120] Galen thus proudly refused membership in both the Dogmatist and the Empiricist sects—he was scornful of the Methodists, with their disregard for medical knowledge and training—and believed that he was in a position to profit from the advantages of each. His sustained emphasis on the regularities of the body and nature, on the one hand, and the importance of empirical inquiry, epistemic flexibility, and contingency, on the other, conjure up a body that both upholds an ordered vision of the world and challenges overly schematic models.

The idea of the ordered body was, in fact, one of Galen's deepest-held beliefs. Whereas scholars have shown that the teleology of Aristotle and later Peripatetics is sometimes nuanced or hesitant,[121] Galen is an enthusiastic and polemical teleologist. He is eager to demonstrate that every part of the human body is designed for a given purpose by a beneficent creator.[122] He is allergic to explanations that he sees as mechanistic—that is, as based on biophysical principles, rather than on innate, end-directed faculties—and quick to criticize those who fail to uphold a view of nature as purposive: Erasistratus and the Erasistrateans regularly come in for attack on these grounds. Galen's trust in a kindly demiurge commits him to the view that the body itself represents the best possible ordering of matter, which, as in Plato, preexists and at times escapes the formal work of the creator. As a result, he in no way demonizes the body, as do the neo-Platonists and the early Christians, and he refuses to accept the idea of a god who can overrule the laws of nature with his mere intention, such as the god of the Judaic and Christian traditions.[123]

Despite his Platonism, Galen could never bring himself to understand the soul as something incorporeal, as contemporary Platonists argued, and he remained throughout his life professedly agnostic about its nature. Evidence abounds, however, for his belief in the interaction between the mind and the body. Galen held that one's physiological constitution determined character, thereby laying the groundwork for medieval ideas about character in both the Arab world and the Latin West, and he advanced explanations of psychic disease in bodily terms.[124] Galen was also a firm advocate of regimen and dietetics as enterprises that could strengthen character. What Galen did believe about the soul, then, was in a materialist vein.

In a few pages, one can give only a few indications of Galen's major positions and his relationship to his contemporaries. Yet such a sketch is useful not only because Galen was clearly a giant in his own age but also because he became the gatekeeper to the Greek medical tradition for later centuries. I close with a brief look at the fate of his writings and their impact on ideas about the body in subsequent generations.

LATE ANTIQUITY

We are ill informed about medicine in the century following Galen's death.[125] When our evidence thickens midway through the next century, it becomes clear that the longevity of Greek medical views on the body in general, and of Galen's views in particular, is due to the flourishing of the medical school at Alexandria from the fourth through the seventh centuries A.D. For in Alexandria lay the roots of the Byzantine and Arabic medical traditions, which championed Galen long before he was taken up as a master in the Latin West in the eleventh and twelfth centuries. One of the distinguishing features of the Alexandrian school was the rigid division between theory and practice.[126] I adopt it here to discuss first the learned tradition, then the more pragmatic legacy of Greek medicine.

Alexandria was, in late antiquity, a center of medical learning. Yet the tradition of anatomical inquiry had yielded to a primarily philosophical and nonexperimental approach to understanding the body, an approach that took to heart Galen's emphasis on the logical and theoretical foundations of medicine.[127] The body envisioned was very much the Galenic body, comprising four humors—themselves based on the four Aristotelian elements—and four basic qualities, as well as a tripartite system of life pegged to the brain, the heart, and the liver.

Health was seen as a successful equilibrium achieved within the inner mixture and sustained through its relations with the world around it. Galen was placed on the same footing as Plato and Aristotle, a position that reflects the integration of medicine into a broad program of natural philosophy. His work was the subject of commentaries and lectures that, at least during the tenure of one of Alexandria's most famous iatrosophists ("medical rhetoricians"), Magnus of Nisibis, people flocked from overseas to hear.[128] Galen's writings, which the Alexandrians judiciously narrowed down to sixteen core tracts, formed the basis of the medical curriculum and the prism through which the earlier Hippocratic writings were viewed.[129]

Galen's views on the body were disseminated, too, through the encyclopedias of medical theory that began to be compiled from earlier literature in the fourth century A.D. for a general educated public. Not all the extracts in these encyclopedias, such as those by Aëtius of Amida (first half of the sixth century A.D.) and Paul of Aegina (first half of the seventh century A.D.), are from Galen. Yet it is Galen who dominates the first of these compilations, written by Oribasius of Pergamum, a fourth-century A.D. product of the Alexandrian medical school who worked under the patronage of the Emperor Julian. And it is primarily Galen's word, decontextualized and repurposed, that prevails in later centuries.[130]

The gradual process by which Galen was winnowed down, abridged, and codified eventually produces, in place of an author, a medical philosophy: Galenism. After the Arab conquest of Alexandria in 642 A.D., Galenism, coupled with the Aristotelianism that had thrived alongside neo-Platonism in Alexandria, finds its natural home, first, among Christian Syrian physicians, who, from the sixth century A.D., had begun to translate the Alexandrian medical syllabus into Syriac; then among the educated elite of the Arab world, who prized Greek medicine alongside Greek philosophy, mathematics, and astronomy, placing it front and center of the gentleman's education, as had been the case in both classical Greece and imperial Rome. Galen was thus translated yet again, this time into Arabic. As a result, ideas about the importance of diet or the environment on health, for example, and the principles of humoral pathology lived on. The strong interest of Arabic culture in physiognomy, the practice of inferring character from physical traits, gave even greater weight to Galen's attempts to correlate psychological and ethical characteristics with physiology.[131] At the same time, Galen himself occasionally came under attack by independent-minded thinkers, such as Muhammad ibn Zakariya' ar-Razi (known in the Latin West as Rhazes), who taught medicine and had an active clinical practice in ninth- and early tenth-century A.D. Baghdad, and Aristotelians defending their master in matters where Galen had disagreed with him.[132] Commentaries on the core Galenic texts of the Alexandrian medical school were also produced in Latin in sixth- and seventh-century A.D. Ravenna, a Byzantine center. Nevertheless, it is through the Arabic translations that Galenism comes to have a significant impact on the Latin West in the Middle Ages.

By the fourth century A.D., Christianity had succeeded in becoming institutionalized in the Roman Empire, and there was ample room for tension between medical views of the body and those fostered by the state religion. Galen's materialist views on the soul were a source of dismay to his later Christian admirers.[133] Prominent physicians were among the last public intellectuals to recant paganism, and they could get into trouble for their commitment to natural causality, which always threatened to encroach on divine territory.[134] More benignly, the dismissal of demonic causality by some physicians—others were quite willing to accept that demons may be at fault for an illness[135]—could simply be viewed as misinformed.[136] The debasement of bodily life in Christian ascetic traditions appears strikingly at odds with the practices of body-care underwritten by the broadly accessible medical ideas that we have seen.[137] Yet, in many respects, Christian authorities and intellectuals seem to have accommodated the practice of Greek medicine, whose knowledge and *materia medica* could be attributed to God's grace. They could find, too, strategic points of overlap, between Galen's divine demiurge, for example, and their own creator. In any event, apart from ascetic communities, most people, presumably, would have been open to a variety of practitioners and approaches in the pragmatic interest of a cure.

There was no doubt a good deal of continuity between the body concepts (medical or magico-religious) that had been the norm in the pagan Empire and those of the Christian world. In fact, given the limits of our knowledge of folk medicine and the physicians working outside of, or on the margins of, learned traditions,[138] it is difficult to know for certain how deep an impact the naturalized body had made in the centuries of Greco-Roman antiquity. In many quarters, a body defined in terms of physical elements or material life forces would have always coexisted with an embodied person open to malign intentions and demonic agents.[139] Moreover, the body being described by learned medicine was not always the body assumed by local physicians: Aulus Gellius reports some educated Romans discovering a physician in Attica who does not know the distinction between veins and arteries.[140]

Nevertheless, it does appear that in the wake of the social unrest of the third century A.D. and the split within the Empire in 364 A.D., there are fewer physicians treating suffering within the framework of the physical body, at least in the Western part of the empire, and people were left increasingly to their own devices.[141] After the seventh century A.D., the situation in the Byzantine world may have been similar. It is unsurprising, then, that the compendia of previous medical writings mentioned previously encompassed not only the genre of carefully organized theoretical positions, as represented by authors like Oribasius, but also useful collections of recipes. Herein lies much of the pragmatic heritage of the learned tradition, which had itself always borrowed freely from older and parallel healing traditions. From the Hippocratic medical writings to the magisterial five-book *On Materia medica*, written in the first century A.D. by Pedanius Dioscorides of Anazarbus, to Galen's extensive research on animal and plant substances, pharmacology had been a theoretical enterprise founded on beliefs about the unseen powers of plants and the natural world as a whole.[142] Dioscorides, for example, organized his work according to a complex system of *dynameis*, "powers." Yet when his work was copied in later centuries, the theoretical framework was dispensed with—a similar fate befell Galen's pharmacological work, later conflated with Dioscorides's—and the material was reorganized alphabetically (and often with illustrations) for practical application: what mattered were results, rather than causes.

It is the body caught in a dynamic and complex web of natural causes that perhaps best captures the object of medical knowledge in Greco-Roman antiquity and the early medieval period, despite the challenges of the Empiricists and the Methodists. The complementary figure to this body is the knowing technical agent, who is capable of intervening in the body (and hence, in the person). He commands this power as a result of his knowledge, which is very often based on the claim to see beyond the corporeal surface to what is unseen. Knowledge of the physical body is thus imperative for survival and well-being, given the threats to the body from outside and its own inherent instability—a sentiment that will sound familiar in an age obsessed with expert advice about health. Indeed, it may be the familiarity of the physical body that keeps us from recognizing how it was not simply the notorious enemies of the body in the Western tradition, the Platos and the Descartes, who shaped the concepts of "the" body that have proved most tenacious. The very premise that the body is a physical thing, encompassing all that is somehow estranged from *the person* by virtue of its participation in nature and requiring expert medical care, has its roots in the learned Greco-Roman medical tradition.

CHAPTER FIVE

Popular Beliefs about the Human Body in Antiquity

PAGE DUBOIS

In a portrait of a "character," a type, that of the squalid man, written by the fourth-century B.C.E. author Theophrastus, a successor of Aristotle, the modern reader cannot miss the distaste of an elite, philosophical author for a lower-class person who is unkempt and uncivilized, distinguished from the citizen who knows how to conduct himself properly:

> The squalid man is the sort who goes around in a leprous and encrusted state, with long fingernails, and says these are all inherited illnesses; he has them like his father and grandfather before him, so it won't be easy to smuggle an illegitimate child into *their* family! You can be sure he is apt to have sores on his shins, whitlows on his fingers, which he doesn't treat but lets fester. His armpits might belong to an animal, with hair extending most of the way down his sides. His teeth are black and decayed.
> And things like this: he wipes his nose while eating, scratches himself while sacrificing, shoots spittle from his mouth while talking, belches while drinking. He sleeps in bed with his wife without washing. Because he uses rancid oil in the baths, he smells.[1]

From such a portrait we learn not only what Theophrastus himself considers to be the appropriate maintenance of the body for a citizen and a civilized person, but also how the policing of the body fits into notions of class distinction. This portrait exhibits an intense commitment to cleanliness, and to smelling good, to the use of aromatic oils that both the Greeks and Romans prized. They clearly, at least in some classes, saw the natural body as in need of much care and attention, of the sort that Michel Foucault alludes to in his volumes on Greek and Roman antiquity in his *History of Sexuality*.[2]

Classicists have found fault with many of the details of Michel Foucault's two volumes about antiquity, the *Use of Pleasures*, about classical Greek civilization, and the *Care of the Self*, about Rome, in their broad outlines of a historical development from the fifth century B.C.E. into the early centuries of the common era. Yet in their ambitious survey of developments over almost a millennium of ancient culture, these volumes offer a paradigm for thinking about sexuality, and the human body, that is unparalleled. One of the inherited characteristics of classical studies, especially in the Anglo-Teutonic strain, is a reluctance to generalize, a desire to stick close to the evidence, to avoid lyricism or speculation concerning centuries of change in antiquity. As a heuristic device, embracing a Gallic lyrical speculation, I want to consider the issue of popular beliefs about the human body in antiquity through the lens of Foucault's arguments.

The first volume of *History of Sexuality* focuses on modern ideas of self and sexuality, arguing for production rather than repression of discourses about these matters. The psychoanalytic session replaces the confessional; medical cataloguing of sexual pathology, including the naming and identifying of homosexuality as an exclusive practice, comes to manage sexual life in the nineteenth and twentieth centuries, overcoming such notions as the random and incidental practice of sodomy in earlier periods. After setting out this history, Foucault decided to go to antiquity, conceiving that he needed to discuss the production of the desiring self as it emerged in the course of Greek and Roman antiquity, as a prequel to his discussion of self and sexuality in the twentieth century. The second volume of the *History of Sexuality*, then, focuses on Greek ideas of selfhood and the emergence of the self as an agent of mastery. Foucault located the project of self-fashioning in relation to household, wife, diet, and boys and derived much of his evidence for the history he tells in prescriptive texts, works of medical advice, and even in the philosophical dialogues of Plato. The emphasis here, in this description of the Greeks, lies on a kind of *askesis,* that is, practice, a practice at controlling, mastering oneself in relation to diet, but also in relation to others. Foucault describes the growing attention the thoughtful man pays to the self in Roman society and how care of the self entails care of the body. Before describing the changes that occur from the period of Greek classical civilization to the so-called golden age of the Roman empire, Foucault in *The Use of Pleasure* gave an account of the philosopher's management of his diet, for example, an essential feature of the mastery of the self that included not just feeding the body, but also his management of his relations with his household, with his wife, and with boys, in erotic relationships. In the Roman period: "The increasing medical involvement in the cultivation of the self appears to have been expressed through a particular and intense form of attention to the body."[3] He points out that this is not the same attention that the Greeks and Romans paid to the body in an athletic or military context, part of a citizen's education. Rather, the new attention is focused on the intersection of body and soul, a point of possible contamination. The body is a vulnerable thing and must receive the concern of its master to preserve it as the site of the soul.

Much of the material adduced by Foucault to demonstrate these new features of Roman life comes from medical writers, devoted to care of the body, and from philosophers, concerned with the health of the soul. While Foucault is especially concerned with aphrodisiac matters in this third volume of his history of sexuality, and he devotes most of his attention to questions of sexual activity, pleasure, and the management of the sexual life, he also points to a new anxiety about the body that must have been generalized, applying not just to the elite, the leisured philosophical class, but also to the myriad urbanites not just of Rome itself who frequented the baths that were built in every municipality across the whole of the empire, and who shared concerns about the cultivation of the body's health, the body that was newly seen as needy. Epictetus, the philosopher who had himself been a slave, writes of similar preoccupations, and Foucault cites views on domestic architecture that reveal a fear about the fragility of the body:

> One can cite ... the analysis submitted by Antyllus of the different medical "variables" of a house, its architecture, its orientation, and its interior design. Each element is assigned a dietetic or therapeutic value; a house is a series of compartments that will be harmful or beneficial as regards possible illnesses. Rooms on the ground floor are good for acute illnesses, hemoptyses, and headaches; upper-floor rooms are favorable in cases of pituitary illnesses; [and so on].[4]

Antyllus, a physician writing in the second century C.E., advises fellow physicians and patients on how best to manage a suffering or recalcitrant body. Other writers such as Athenaeus focus on diet, while still others advise hot baths; the first-century Celsus, who seems to have gathered the information from medical sources, including Hellenistic sources, to benefit himself and other laymen, suggests moderate exercise and then: "The proper sequel to exercise is: at times an anointing, whether in the sun or before a brazier; at times

FIGURE 5.1: Greek open air shower baths for men. Wellcome Library, London.

a bath, which should be in a chamber as lofty, well lighted and spacious as possible."[5] Plutarch devotes an entire treatise to advice about keeping well.

In the period of the Roman empire, the emphasis changes, to a care for the self, care that imagines others also caring for themselves. In an interview with the Americans Paul Rabinow and Herbert Dreyfus, while he was working on the ancient material, Foucault presented his sense of the shift he was concerned with:

> Well, the *substance éthique* [ethical substance] for the Greeks was the *aphrodisia* [sexuality]; the *mode d'assujettissement* [mode of subjectivation] was a politico-aesthetic choice; the *form (sic) d'ascèse* [form of asceticism] was the *techne* [practice] which was used—and there we find, for example, the *techne* about the body, or economics as the rules by which you define your role as husband, or the erotic as a kind of asceticism toward oneself in loving boys, and so on—and the *téléologie* [teleology, goal] was the mastery of oneself. . . .
>
> Then there is a shift within this ethics. The reason for the shift is the change of the role of men within society, both in their homes toward their wives and also in the political field, since the city disappears.[6]

I would add to Foucault's arguments some further attempt at explaining why this shift occurs, although he seems reluctant to do so. In fact, there are vast

Fig. 542. Femme grecque au bain.

FIGURE 5.2: Greek woman bathing. Wellcome Library, London.

social, economic, and political differences in the shift from Greek city-state to Roman empire and a vast bibliography describing these differences and discussing the degree of change in everyday life over these centuries. Some scholars argue that in fact the city did not disappear, and that Athens, for example, continued its political existence for many centuries after the fifth and fourth centuries, in which the independent democracy flourished. But it seems important to note that in fact Athens became part of a Macedonian empire, then part of the Hellenistic world of kingdoms and minor empires after the death of Alexander, and that the Romans' conquest of Greece led to the domination of all localities in the formerly independent Hellas. Everyday life, even though it may have involved some administration of the city itself on the part of citizens, was lived at the pleasure of the Romans and their emperor in Rome. So one might argue that ideas of the human body, like all other aspects of life and consciousness over these centuries, were transformed. If the free citizens of the democratic city of Athens had beliefs about human bodies, these were inevitably changed as those bodies came under the domination of foreign, Roman masters. And the retreat inward, to focus on one's own body, and care for it, makes sense in light of these larger political mutations. A free citizen of an ancient Greek city-state, especially in the democracies, did have powers of mastery—over households, slaves, wives, boy-lovers—and the focus on the human body in this period rests on that mastery. In the Roman period, especially after the ascent of the first emperor Augustus to absolute power, political choices changed, and the mastery performed in the democratic assembly of Athens, the power invested in the republican senate of Rome, gave way to another focus, on the individual, who in some sense was powerless in relation to imperial domination. Thus the individual body, the self, becomes the arena of care and attention.

One salient problem for the present survey is that Foucault is most often discussing philosophical ideas about the self, the body, and sexuality in these volumes, pointing to the vanguard thinkers of the day as he maps changes from the classical period of Greece to the developments in ancient Rome that he sees as preceding and indeed determining the prescriptive modes of early Christianity. But in fact it seems possible to me to discern, within the texts of the elite thinkers cited by Foucault, tendencies in the broader cultures of Greece and Rome that were shared by the populace at large, and that are visible in material practices such as the treatment of slaves and the performance of magical rites, to take just two examples.

So I will argue in this brief survey, which can only sketch in the most desperately abbreviated way a vast history of class, geographical, ethnic, and historical differences in popular beliefs about the human body, that we see, in the broadest possible terms, a change from a focus on mastery to one of anxiety concerning the human body. My treatment here will be somewhat arbitrary, because of the unmanageable extent of the evidence, and also because of an inevitably limited knowledge of it all, lying as it does within many different subfields of classical studies—Hellenic versus Roman; epigraphy, or the study of inscriptions; art history; archaeology; literary studies; the study of magic; popular culture in general. Some topics I am forced by the brevity of this chapter to neglect are questions of the body in war, in sports, in gladiatorial combats. Though my overarching argument concerns developments and change from Greece to Rome, these two societies, one of which comes to dominate the other, share a great deal, and I will focus to a great extent on the details of their common experiences of the body.

HUMAN BODIES AND ANIMAL BODIES

In a treatise called *Physiognomics,* attributed to the ancient Greek philosopher Aristotle, we find evidence of ancient Mediterranean beliefs, in popular culture and even in elite culture, concerning the human body.[7] For example, the author states:

> The signs of the coward are soft hair, a body of sedentary habit, not energetic; calves of the legs broad above; pallor about the face; eyes weak and blinking, the extremities of the body weak, small legs and long thin hands; thigh small and weak; the figure is constrained in movement; he is not eager but supine and nervous; the expression on his face is liable to rapid change and is cowed.[8]

Such a guide to deciphering the signs and meaning of characteristics of the human body forms part of the encyclopedic corpus of works of the Aristotelian school, many of them probably composed not by Aristotle himself but by students and followers bent on creating a mastery of the natural world through description and catalogue. This part of the treatise follows on a section that connects human bodily signs with their resemblances to animals: "Soft hair shows timidity and stiff hair courage. This is based on observation of all the animal kingdom."[9] We find other such conclusions: "Those who have large upper parts and are vulture-like and hot are somnolent."[10] Furthermore, "[b]ulging eyes mean stupidity; this is appropriate and applies to the ass."[11] Bodily features can be analyzed to draw further conclusions about disposition and character, which seem likely to be part of a vast landscape of popular beliefs:

> Those who have a bright-red complexion are apt to be insane, for it is an excessive heating of the parts of the body which produces a bright-red skin; those who are excessively heated would naturally be insane.[12]

Later in the *Physiognomics,* the author (or authors) offers an interpretation of why differing body types produce different characters:

> Excessively small men are quick; for as the blood travels over a small area, impulses arrive very quickly at the seat of the intelligence. But the excessively large are slow; for as the blood travels over a large area the impulses arrive slowly at the seat of the intelligence.[13]

We see how a learned treatise, medical knowledge and diagnostics, and popular culture can come together to produce an elaborate mapping of kinds of human beings, carefully distinguished one from another even as they exhibit resemblances to various animal species. The writer continues to return to one of the themes of the Peripatetic school, the emphasis on the mean, on moderation, as he values especially "the man of moderate size."[14]

Although such observation is attributed to the Peripatetics, philosopher-scientists before there was a division between these types of thinkers, it may draw its insights from beliefs in popular culture, in which remarkable connections are often made between human beings and animals. Some magical texts assert the possibility of metamorphosis, of change from the human body to an animal's; Apuleius's novel *The Golden Ass* recounts the tale of Lucius, a man who was transformed by magic into a donkey, and who lives many adventures before he succeeds, with the help of the goddess Isis, in returning to his human form. There are also accounts of transformations of human beings into wolves; the *Satyricon*, or *Satyrica* of Petronius describes such a metamorphosis, set in a cemetery, a tale told by one of the guests, another former slave, at the feast of the freedman Trimalchio:

> When I look for my buddy I see he'd stripped and piled his clothes by the roadside. My heart was in my mouth—I just stood there like a corpse. He pees in a circle round his clothes and then, just like that, turns into a wolf![15]

As a wolf, the friend attacked some flocks and was speared in the neck; the speaker later finds him at home, a human being again, being treated by a doctor for a wound in the neck. Having a human body in these popular contexts is a flexible matter. And there are many tales of metamorphosis from human bodies into animal and plant shapes in the great mythological work of the Roman poet Ovid. The faithful husband and wife Baucis and Philemon, who hospitably receive the gods into their humble home, become an oak and a linden tree growing from the same trunk at the end of their human lives.[16]

The work of Theophrastus, *Characters*, also attempts to create a mapping of human kinds, in this case portraying various kinds of human beings and at times giving evidence about what Greeks seem to have thought about issues of decorum and cleanliness of the body, required especially when entering sacred areas, precincts of the divinities and temples. But even in everyday life, when the sacred was not an issue, the Greeks seem to have condemned those who did not keep their bodies clean. For example, in describing the "obnoxious man," character number 12, Theophrasus writes that he "is the sort who, when he meets respectable women, raises his cloak and exposes his genitals. . . . In the theater . . . [w]hen the audience is silent he rears back and belches, to make the spectators turn around."[17] Obviously, the man violates the appropriate decorum, the proper management of his body, in a way that is associated with vulgarity and perhaps with a lack of respectability, even of upper-class status.

The nature of the evidence for popular beliefs about the human body in antiquity presents numerous problems in itself. First of all, the literary and textual material we have from this period is for the most part from elite sources, those unusual persons in the ancient world who were literate and wrote for an audience, and whose works were preserved by the tradition. Each of these elements—the rarity of real literacy, the few of the literate who wrote to be heard or read, and the vulnerability of all ancient writing to destruction through various means, accidental or deliberate—contributes to a problematic picture in which it is difficult to uncover popular beliefs, difficult to know whether they are indeed popular or rather simply attributed to the populace at large by elite authors. There is some archeological evidence, material cultural remains, that offer sometimes enigmatic information. The conclusions to be drawn from burial practices, for example, or from evidence in magical texts or objects need to be interpreted carefully and assigned, often tentatively, to classes other than the elites who, as mentioned previously, leave behind written records of their views. So this account of popular beliefs about the human body must be taken with a grain of salt; how can we know for sure who believed in what in a world so distant from ours? If we take the views of the author of the Peripatetic school's *Physiognomics* to have some relationship to popular ideas about the human body, we must also take into account the fact that it forms part of the massive encyclopedic projects of this school, its attempts to categorize, interpret, and account for all the phenomena of the universe, from metaphysics to botany.

GENDERS

The author of the *Physiognomics* draws his customary parallels between human beings and animals when distinguishing between genders:

> [T]he female sex has a more evil disposition than the male, is more forward and less courageous. Women and the female animals bred by us are evidently so; and all shepherds and hunters admit that they are such as we have already described in their natural state. . . . in each class each female has a smaller head, a narrower face and a more slender neck than the male . . . the female has knock-knees and spindly calves, neater feet, and the whole shape of the body built for charm rather than for nobility, with less strong sinews and with softer, moister flesh.[18]

He continues with analogies drawn between the world of animals and that of human beings, pointing to the deficiencies of the female in both cases:

FIGURE 5.3: Votive offering: pregnant woman wearing a binder, Greek. Wellcome Library, London.

Those who have well-made, large feet, well-jointed and sinewy, are strong in character; witness the male sex. Those who have small, narrow, poorly-jointed feet, are rather attractive to look at than strong, being weak in character; witness the female sex.[19]

Although at times reluctantly acknowledging that the female body may be more attractive to look at, the author finds its physical appearance signifies ethical inferiority in both the human and the animal domains. "Those who walk with feet and legs turned out are effeminate; this applies to women."[20] "Ill-proportioned men are scoundrels; this applies to the affection and to the female sex."[21] The author concludes the treatise with a summary that damns the female body: "The male sex has been shown to be juster, braver, and, speaking generally, superior to the female." Thus the mapping of the human body, analysis of its various distinctions, and analogies with animal bodies all lead to the confirmation of a cultural conviction, or a constant ideological labor, demonstrating the inferiority of the female sex.

FIGURE 5.4: A seated Greek woman on an obstetrical stool being held in position by her husband whilst giving birth aided by a midwife, another attendant dresses the first baby. Line engraving by A. Tardieu after Maréchal. Wellcome Library, London.

The classicist Giulia Sissa has written about ideas of the female body in ancient Greece in her book *Greek Virginity*, and her arguments confirm this sense that the female body houses an inferior creature.[22] The female body, for example, was seen as having two mouths, the vagina and the upper mouth; one cure for hysteria, the wandering of the womb identified with various afflictions of women, involved burning aromatics at the lower mouth, so that the womb would come to this opening and settle there, attracted by the pleasing odors, abandoning its travels through the body. The Greeks believed too that the womb could be settled down by frequent heterosexual intercourse and prescribed in law a certain frequency of marital coitus.

COMEDY

One of the richest sources for knowledge about popular ideas of the human body is comedy, especially in the theater. Although the comic writers were sometimes, not always, from elite culture, the jokes and ridicule of mythic and historical persons that formed the fabric of comic performances led to laughter in

FIGURE 5.5: Woman with a tumor of the breast. Greek votive offering. Collection of Prof. Meyer-Steineg. Wellcome Library, London.

the audience, obviously, since Aristophanes's plays won prizes in the Athenian dramatic contests and the plays of Plautus and Terence were performed and preserved as well.

Some of Aristophanes's comedies are named after animal choruses, and much play is made of the bodies of the human actors who sang and danced in these roles, imitating animals and sharing certain characteristics that excite laughter in the crowd. *Birds, Frogs, Wasps*; these men's bodies are costumed to resemble animals. And *Wasps* in particular is an especially rich source for understanding popular ideas about the human body and the features they share with other animals. We see slave bodies here as well as citizen bodies, old and young bodies, male roles embodied, and female bodies characterized in the course of the comedy.

Slave Bodies

Slave bodies are seen as much beaten, scruffy, and marked by the blows of their masters. The first scene of the comedy shows two slaves guarding the house of their master, and they discuss privately the possibility of a beating if they fail at their task. Xanthias's ribs will complain if the master is displeased, and this vulnerability of the slave body recurs throughout the play. The chorus leader, with his battalion of man-wasps, also threatens the slaves: "Now let the man go. If you don't, I do declare you'll envy turtles their shells!"[23] The old man whom they have been guarding, prisoner of his son, reminds them of his past care for them in language that also stresses the comic vulnerability of slave bodies, recalling an occasion when he "did a right manly job flaying you raw."[24] Later in the play, the slave Xanthias complains of the beating he has taken:

> Ah tortoises, I envy you your shells! It was good and brainy of you to roof your backs with tile and so cover your sides. Me, I've been bruised within an inch of my life by a walking stick![25]

In this case, rather than comparing the human body to an animal's, the joke calls attention to its vulnerability, its lack of protection in the face of aggression from a hostile and punitive master.

Obscenity: Greek

Comic texts abound with metaphors and analogies between the human body and animal bodies, and with inanimate objects that resemble in some way human body parts. Much of Aristophanic obscenity relies on such analogies. For example, the old man of *Wasps* leaves a symposium, an upper-class dinner, in a most indecorous, drunk, and disorderly state and dragging with him a female flute player, an entertainer at the party. He first compares her to an insect, "my little blonde cockchafer," and then offers her his phallus:

> Grab hold of this rope here with your hand. Hang on, but be careful, the rope's worn out; all the same, it doesn't mind being rubbed.[26]

The old man's phallus is seen as a rotten rope, and this metaphor shares the Aristophanic stage with many such references, especially to male and female genitalia. In the same scene in *Wasps*, the old man denies that the flute-girl is a woman, insisting rather that she is a torch burning in the marketplace, that her genitalia are pitch oozing out from the torch, her anus a "knothole."[27] Much of the comic action of *Wasps* depends on the analogy between jurors of the Athenian law courts and the insects for which the comedy is named. Throughout, the chorus is dressed as wasps, and they and others refer frequently to their waspish character. And one feature of their costumes seems to have been a stinger. Although scholars debate whether this stinger was the phallus or another protuberance, from the back of the costume, it is often associated with a violent, aggressive phallicism on the part of its possessors, qualities consistent with other representations of the male body in ancient Greek society. The old man's son describes the chorus with dismay:

> You sorry fool, whoever riles that tribe of oldsters riles a wasps' nest. They've even got stingers, extremely sharp, sticking out from their rumps, that they stab with, and they leap and attack, crackling like sparks.[28]

One can almost characterize such a description as a wish for phallic potency, for aggressive capacities that the old men, with their rotten ropes, have relinquished. All this contributes to a popular view of the male body dominated by a lecherous and aggressive penis. And one could catalogue a vast array of such allusions, as Jeffrey Henderson does in his very useful study of obscenity in comedy.[29] He lists many figures for the phallus, for example, including words normally signifying neck; finger; skin, for foreskin; sinew; thing; equipment; fig, with many variations on the various stages of development of the fruit; acorn; chickpea; barleycorn; spear; oar; goad; shaft; boat-pole; handle; bolt; sword; spit; peg; top; seal; drill; and so on. These comic metaphors focus on the erect or flaccid penis and spring up everywhere in Aristophanes's dialogue. As Eva Keuls pointed out in her book *The Reign of the Phallus*, ancient Athens was a site defined by phallic imagery; she cites many Greek vases, the herms, boundary markers with the head of Hermes, and an erect phallus protruding from a pillar that dotted the urban landscape and marked off sacred space and private property.[30] Popular beliefs about the human body in ancient Athens centered on the male sexual organ, the erect phallus that connoted power, protection, and aggressivity.

Like Anglo-Saxon, the Greek comic language, vulgar and inventive, produced a huge number of figures for the sexual organs. And the female genitalia too are often named and mocked in this popular genre. Jeffrey Henderson, in his valuable catalogue of obscene language in Attic comedy, lists a great number of metaphors relying on comic allusions to the female body, and especially its sexual organs: box; hole; piggie, as well as pig and suckling pig; and sow. Here the talent for obscenity meets the predilection for analogies between animals and human beings; Greek women practiced depilation of the pubic area, and popular beliefs saw a resemblance between the naked female body and that of a pig. Other jokes about the female body refer to the dog, the bull, and animals of the sea, while plants were also called into play: the fig, the pomegranate seed, roasted barley, myrtle berry, rose, and the whole garden, as well as meadow, thicket, grove, plain, and many other terms that refer to the female sexual organs as gates and passageways, rings and circles, holes, cooking implements, and the smoke from cooking that was suggested by undepilated pubic hair. Comic lines sometimes refer to the female organs as hot coals, as various foods, and the list, carefully assembled by Henderson in an exhaustive account, goes on and on.³¹ It is remarkable that comic interest in the female body focuses almost exclusively on the labia and vagina. The Greeks seem, at least in what we can discern from the comedies of Aristophanes, little interested in the female breast. Some have argued that the preference for depilation of mature female bodies is a result of the practice of pederasty, and that female bodies that were boyish were preferable to more voluptuous forms, and this preference may be visible in Greek statuary of the female form, which at least in the classical period does present a modest, less curvaceous female body than that preferred in some other cultures.

Interestingly, although comedy featured oversized, padded phalluses and buttocks for its actors, in artistic depictions actual penises are often shown as presumably smaller than life size, decorously boyish. In the statues and figures representing naked athletes, for example, a small penis is shown tied to the thigh, as it was in athletic contests.

The stingers of the wasps of the comedy are not just sexual equipment, however; they are also part of the armory used against the enemy by the citizen body, in a depiction of the community of male human bodies that relies on analogies with an insect swarm. The chorus recalls how it responded long ago, when the city of Athens was attacked by the barbarian Persians, and likening their waspish bodies to those of warriors:

> We who sport this kind of rump are the only truly indigenous native Athenians, a most virile breed and one that very substantially aided this city in battle . . .
> [W]e charged forth with spear, with shield. . . .
> Then we pursued them, harpooning their baggy pants and they kept running, stung in the jaws and the eyebrows.³²

The male warrior body is seen in popular culture not as vulnerable, like the roofless slave, but as equipped with weapons to chase off the city's invaders. Even in their more pacific activities, engaged in jury service in the Athenian courts, the poet continues the analogy with wasps, suggesting that all the bodies of the jurors cluster together in a swarm:

> [W]e gather together in swarms as if into nests, some of us judging in the archon's court, some before the Eleven, and some in the Odeum, packed in tight against the walls like this, hunched toward the ground and hardly moving, like grubs in their cells.³³

In order to make the audience laugh, Aristophanes must be alluding to some shared vision of the male citizen body here as he caricatures the bodily movements of his fellow inhabitants of Athens.

Obscenity: Roman

The record of Roman popular beliefs about the body is somewhat fuller than that of the Greeks, in part because of the survival of graffiti inscribed by people who were often barely literate. Amy Richlin, in her important study of Roman humor, lists many examples of graffiti that exhibit Roman attitudes toward the human body.[34] She cites one graffito:

> Here I have now fuckid a gril beautiful too see,
>   Praised by many, but there was muck inside.[35]

Her translation reproduces the grammatical errors in the Latin original[36] and reveals the misogyny and perhaps fear of women's bodies in at least one Roman. There are graffiti that address others by name, calling them "pricks," or "prike," as Richlin translates *metula,* for the common obscene term *mentula.*[37] The most common accusations circulating against enemies were that men were effeminate and women unchaste. And there is a strong tendency, greater even than in Greek culture, that sometimes mocked older women's flabby bodies, to express disgust and even horror at the bodies of old women. The great poet Horace, for example, addresses a woman in his eighth epode, here again translated by Amy Richlin:

> You, foul by your long century, ask
> what unmans my strength,
> when you've a black tooth, and old age
> plows your brow with wrinkles,
> and between your dried-out cheeks gape filthy
> an asshole like a dyspeptic cow's?[38]

Horace's twelfth epode continues the theme of disgust and horror at aging female bodies: "What a sweat on her shriveled limbs, and what a bad smell/ grows everywhere, when (after my cock is limp)/she hurries to quiet her unconquered lust; nor does/her wet powder stick now, her blush/painted on with crocodile dung."[39] Such poems stress the corruption and decay associated with these female bodies. Although Aristophanes mocks the lust and bibulousness of old women in such comedies as *Ecclesiazousae* and *Thesmophoriazousae,* he rarely dwells so intensely in the invective of the Romans directed at older women's sexuality.

Roman poets, especially Catullus and Martial, seem to draw on popular ideas about the human body in their poetry of invective. The Romans considered the mouth to be a site of possible contamination and impurity, and much sexual humor or aggression relies on the mouth. A virile man could see himself as violently thrusting his penis into the mouth of another; he would impose dominance on his enemy, who would be soiled by such contact. So Catullus offers various threats in his poetry, threats often evaded by translators but that receive their full popular force in Richlin's translation of Catullus 16:

> I will bugger you and I will fuck your mouths,
> Aurelius, you pathic, and you queer, Furius,
> who have thought me, from my little verses,
> because they are a little delicate, to be not quite straight. . . .
> You, because you have read "many thousands
> of kisses," think me not quite a man?
> I will bugger you and I will fuck your mouths.[40]

Catullus here refers to his own poetry, to a celebration of kissing between lovers, and then ferociously aggresses Aurelius and Furius, establishing his dominance through a threat of anal and oral rape. The mouth of the person raped is made impure, stained by a kind of filth associated with oral intercourse, and the whole person is contaminated and made filthy through this means of degradation.

The figure of Priapus presides over Richlin's work, Priapus the god with the immense and erect penis who stands as a guardian at Roman doorways, in Roman gardens, threatening any intruder with penetration and rape. We have poems dedicated to or concerning this god, the *Priapeia*, written by both Greeks and Romans, the Roman especially focusing on his predilection for anal rape. Richlin points out that the Roman poets frequently address their penises, and the god Priapus, sometimes in situations of impotence:

> Are you pleased [by my impotence],
> Priapus, who under the tresses of a tree
> like to sit, red, with your reddening phallus,
> your holy head bound about with vine tendrils?[41]

These lines, translated by Richlin, are from the *Virgilian Appendix* and end with a farewell to the god: "Goodbye, unspeakable deserter of my loins,/goodbye Priapus." Anxiety about potency and the omnipresence of representations of the hyperpotent divinity are part of the landscape of Roman popular beliefs about the human body.

In Rome, the poets similarly refer, in what we must often assume is popular language, to differences between male and female bodies and represent the sexual organs. The Romans seem to have been especially sensitive to questions of the mouth, understanding it to be vulnerable to contamination in various ways. Much of the obscene poetry of Catullus and Martial, for example, alludes to sexual practices that affect the mouth.

OTHERS

The author of the *Physiognomics* extends his reflections on the body beyond male and female, human and animal, to include those the Greeks sometimes called "barbarians," that is, those of neighboring lands, around the Mediterranean, who eventually became part of the Roman Empire. He argues, drawing analogies again with the animal kingdom, that:

> Those who are too swarthy are cowardly; this applies to Egyptians and Ethiopians. But the excessively fair are also cowardly; witness women.[42]

The mean is best; "the complexion that tends to courage is in between these two,"[43] he notes, and praises the tawny-colored who are like lions, in contrast to the "reddish," who resemble foxes.

Although it has been claimed that the ancient Greeks did not harbor racist ideas concerning Africans, for example, the author of the *Physiognomics* associates the Ethiopians with other generalizations he makes concerning human and animal bodies: "Stiff hair on the head betokens cowardice; this refers to the affection, for when men are frightened the hair stands on end. Those with very woolly hair are cowardly; this applies to the Ethiopians."[44] Bodily differences mark the barbarian; Herodotus refers to the peculiar bodily practices of the Egyptians, for example, in his *Histories*, noting that "almost all Egyptian customs and practices are the opposite of those of everywhere else":[45]

> Everywhere else in the world, priests have long hair, but in Egypt they shave their heads. In times of mourning, it is the norm elsewhere for those most affected by the bereavement to crop their hair; in Egypt, however, in the period following a death, they let both their hair and their beards grow, when they had previously been shaved. . . . They knead dough with their feet and clay with their hands, and they pick up dung with their hands too. Other people, unless they have been influenced by the Egyptians, leave their genitals in their natural state, but the Egyptians practise circumcision.[46]

We learn not only what the Greek traveler believed about Mediterranean neighbors, but also the norms of the Greeks themselves. The author of the Hippocratic treatise *On Airs, Waters, and Places* may be expressing popular opinion when he describes how various climatic and geographical features of peoples' homelands produce different kinds of bodies; the people who live near the river Phasis, which leads into the Black Sea, are affected by the humidity of their environment:

They are big and stout and their joints and veins are obscured by flesh. Their skin is yellowish as if they had jaundice and their voices, because they breathe the air which is moist and damp and not clean, are the deepest known. They have little stamina but become quickly tired.[47]

The author attributes these and other defects of the Scythians to their climate and to their habits. The Scythians "are the most effeminate race of all mankind" because they wear trousers and because they "spend so much of their time on horseback so that they do not handle their private parts, and, through cold and exhaustion, never have even the desire for sexual intercourse."[48] Although these are the views of a learned author, a physician, perhaps, a member of the Hippocratic school, or tradition, his opinions may reflect popular ideas concerning the differences among various peoples who lived around the Mediterranean Sea in antiquity.

Some of the Roman authors who benefited from the great expansion of the Roman empire and the wealth that poured into the city of Rome itself from its conquests expressed anxiety about contamination from the many new peoples who entered the city, and whose bodies expressed their differences from the ideal Roman body, pure, clean, well-tended, and exercised. Strangers from the east seemed especially alarming; the rhetorical writers connect the voluptuous prose of the Asiatic style with the dangerous bodies of Greek and Eastern immigrants.

MAGIC

One last area of popular culture that reveals popular beliefs about the human body in antiquity is magic. Although magic was once relegated to superstition and neglected as a phenomenon of ancient societies unworthy of scholarly attention, it has in recent years received much study, study that offers a window into popular ideas. Earlier scholars nominated the worship of the Olympian gods as religion and the practice of magic as something lesser, but we learn a great deal about the people from considering the rituals, beliefs, and material culture of magic in classical and Hellenistic Greece and in Rome.

The human body was thought by some to survive death, in some form, and to haunt the living. The ghost of the dead Patroclus appears to Achilles, seeking burial so that he can enter the land of the dead, in the twenty-third book of the *Iliad*. Although the Homeric poems seem to represent the soul after death as a transparent, airy creature, needing blood and sustenance even to speak, as in the scene of Odysseus in the underworld, in book eleven of the *Odyssey*, other texts present some kind of survival and persistence of the dead. Even Plato, who condemns superstition and popular beliefs, in his *Phaedo,* written about the impending execution of Socrates, discusses the possibility of corporeal

FIGURE 5.6: Asklepios/Aesculapius: The ancient Greek deity of healing. Wellcome Library, London.

survival since he seems to condemn the body, elsewhere citing the Greek phrase *soma sema*, "body = tomb," the body that weighs down the soul:

> [W]e must suppose, my dear fellow, that the corporeal is heavy, oppressive, earthly, and visible. So the soul which is tainted by its presence is weighed down and dragged back into the visible world, through fear, as they say, of Hades or the invisible, and hovers about tombs and graveyards. The shadowy apparitions which have actually been seen there are the ghosts of those souls which have not got clear away, but still retain some portion of the visible, which is why they can be seen.[49]

Such surprising views, coming from the mouth of the philosopher, must reflect widely held views about the persistence of the body after death, and such ideas are shared by other ancient writers. Plautus's *Mostellaria*, a Roman comic play, uses the tale of a haunted house, told by a wily slave, to keep a young man's father from discovering his escapades. Pausanias, a travel writer of the second century C.E. who visited the site of the battle of Marathon, fought between Greeks and Persians in the fifth century B.C.E., records that: "All night long there one can hear the sound of horses neighing and men at war."[50]

Daniel Ogden, in *Magic, Witchcraft, and Ghosts in the Greek and Roman Worlds*, a valuable anthology of texts concerning these aspects of popular culture in antiquity, cites the late text, the *Suda*, a compilation of ancient material, in an entry that tells us more about ancient people's conceptions of the body:

> *maschalisthenai*: Being armpitted: Men usually wipe their swords on the heads of the slain to avert the pollution of the killing. Or, to purify themselves, they would cut off the slain man's extremities and hang them around his armpit, so that, it is said, he would be too weak to avenge the murder.[51]

Even the imperial family of Rome was subject to fears of the dead. Nero, who had arranged for the murder of his mother, Agrippina, is said by Suetonius to have been haunted by his mother's ghost and to have summoned her in order to appease her rage.[52] There are many episodes of exorcism, attempts to rid the living of the persistent presence of the bodies of vengeful, restless dead.

The practices of magic intersect not only with those of ancient religion, which required rituals of purification before entering sacred ground, for example, but also with those of medicine. Physicians were consulted in cases of epilepsy, as were sorcerers. The Greek poet Pindar recalls the healing powers of the god Asklepios in a way that links incantations with physical manipulations:

> And those who came suffering from the sores of nature, or with the limbs wounded . . . he delivered them from pains, tending some of them with kindly incantations, giving to others a soothing potion, or swathing their limbs with drugs, or restoring others by the knife.[53]

Another revealing aspect of ancient magic was the practice of uttering curses, often directed against the bodies of others. Curses could "bind," that is, constrain, the tongues of opposing speakers in courts, could inhibit the work of rival shopkeepers and potters, and there are reports concerning the efficacy of such curses or spells on the body of the victims. Ogden cites a victim of sorcery whose "body felt drawn tight as if by purse strings, with his limbs being crushed together."[54]

Another popular belief about the body concerns the "evil eye," a look, or a withering ray that was sent out from the eye of an envious person and was consciously directed by its possessors, especially women. The curse of the eye could be deflected, however, by the phallus, and of course such beliefs persist in the Mediterranean region into the present, as babies are adorned with

FIGURE 5.7: Engraving of a bust of the Greek physician Hippocrates. Wellcome Library, London.

FIGURE 5.8: Greek physician and patient, plaster cast in W.H.M.M. Wellcome Library, London.

phalluses made of coral or other materials to protect them from such ill will. Plutarch discusses such popular practices in his *Moralia*:

> Democritus says that the envious emit images, not altogether without their own realization and impulse, and that they are full of the wickedness and evil-eying that derives from their projectors. Together with this wickedness and evil-eying these images mold themselves to, remain with, and take up residence with the evil-eye victims and damage and blight their body and mind.[55]

Possessors of the evil eye can injure themselves with a kind of boomerang effect if they look into water or a reflecting surface, sending the harm back upon themselves.

Erotic magic controlled lovers' bodies, attracting some, repelling rivals. The people of antiquity also used dolls and amulets, often in the shape of body parts, to attempt to control others' behaviors and bodies. One of the most compelling pieces of evidence concerning popular beliefs about the body in classical Greece is the so-called Mnesimachos doll, laid in a tiny coffin and unearthed from the city cemetery, the Ceramicus, in Athens.[56] The doll was buried about 400 B.C.E. and represents a tiny human body with an erect phallus; the case contains a list of names, with the addendum "and any other advocate or witness they have on their side."[57] The doll, made of lead, has its arms in

FIGURE 5.9: Detail of the decoration of a red-figured Greek vessel in the Louvre showing a satyr with a crippled leg. Watercolour, 1780/1820. Wellcome Library, London.

a bound position and lies inside the case, which may once have had lead nails piercing it. This seems to be a curse binding participants in a law case and is aimed particularly at the man Mnesimachos, whose name is listed and also written on one of the doll's legs. The Roman orator refers to such binding in one of his rhetorical treatises: "When I had rounded off the case for the defense, . . . [his opponent] all of a sudden forgot his entire case and said that this had been caused by Titinia's spells/poisons (*veneficia*) and incantations."[58]

CHANGES

The Greeks and Romans shared many of their popular beliefs concerning the human body, including notions about differences between male and female bodies, between their own and barbarian bodies, about the efficacy of magic, religion, and medicine to curse or heal. But if we might return to Michel Foucault's description of the changes in the culture of antiquity as a long duration, from classical Greece to the time of the Roman empire, we can perhaps discern

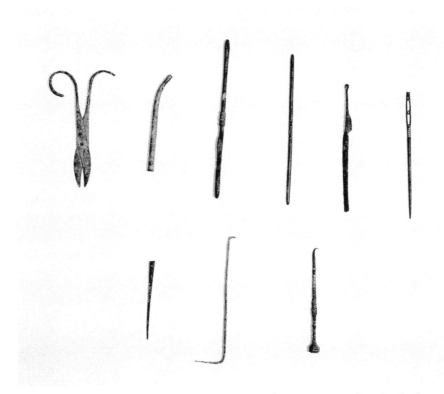

FIGURE 5.10: Ancient Roman and Greek surgical instruments: Blunt hook, bronze, encrusted. Roman; Surgical dissector. Bronze, encrusted, right angle hook on either end; No information; Probe or bodkin. Bronze, encrusted, double eye (may have been used surgically). 5-¼" long; Graeco-Roman, bronze surgical instruments; Double ended probe. Museum No. unreadable; Roman bronze probe; Surgical tube, bronze, encrusted, curved. 4" long. Found in Rome during excavations, 1932; Bronze scissors with curved pointed blades and hook handles crossed and joined by a rivet; part of one handle missing. Wellcome Library, London.

a shift in popular culture as well as in the world of philosophers, a change from questions of mastery of oneself and others, to a concern for care of the self, both reflected in bodily practices and even in the treatment of slave bodies. We might compare Aristophanes's old man, in *Wasps*, who has problems mastering his own addiction to the law courts but confidently administers the bodies of his slaves, boasting to them as he remembers how he "did a right manly job flaying you raw."[59] The master of the classical Greek world, whether a philosopher or not, needed to control his own body; keep it from bestial excess, self-indulgence, and uncleanliness; and control the bodies of his household, those of his wife, his children, and his slaves. We see a different attitude in the Roman Republican author Cicero, who shares with his fellow Romans anxieties about purity, health, and contamination and extends his concern even to his well-educated secretary and beloved former slave Tiro, to whom he addresses these words in a letter, expressing concern for his freedman's health:

> One thing, my dear Tiro, I do beg of you: don't consider money at all when the needs of your health are concerned. I have told Curius to advance whatever you say, . . . Your services to me are beyond count—in my home and out of it, in Rome and abroad, in private affairs and public, in my studies and literary work. You will cap them all if I see you your own man again, as I hope I shall. (Leucas, November 7, 50 B.C.E.)[60]

Such a tender missive is difficult to imagine coming from the pen of an ancient Greek writer and may testify not only to the possibility of humane and generous relations between some Romans and their slaves, but also to new popular beliefs about the body, ideas that persist into later antiquity and into the Christian middle ages.

CHAPTER SIX

Reflections on Erotic Desire in Archaic and Classical Greece

FROMA I. ZEITLIN

Eros is a god.[1] So is golden Aphrodite. Myths of their origins vary, as do the differing concepts of kinship between them. But the fact of divinity acknowledges the overwhelming power of passion; it tells us also that, like all divine forces, carnal desire is something that is ageless and deathless forever. Eros, as Greek culture tells us again and again, is an emotion, an image, a figure, an idea, a supernatural force. It affects both mind and body, clouding the intellect, loosening the limbs. Radiant with a beauty that stuns the senses, Eros brings the promise of pleasure, but more often it is perceived as a mixed blessing, if not an affliction. It burns the soul, sweeps over the lover like a mountain storm, lays him low with sickness, even a madness of spirit, and pierces the bone to the marrow. With a few notable exceptions in the world of the gods, no one is immune to erotic influence, neither beasts nor humans, neither mortals nor immortals. To ignore Eros or to try to escape his effects is to invite disaster. Gods are vengeful and punitive more often than providential. They enforce the

---

This chapter is reproduced with the kind permission of The University of Michigan Press © 1999. To preserve the integrity of the original text, for this chapter the notes and references appear at the end of the chapter.

universal laws of human existence; they claim the rights of nature and necessity. In the case of Eros and Aphrodite the connection with nature is stronger still because sexual desire is deemed essential for creation, for the reproduction of all living things. Dynamic, mobile, elusive, ever renewed, erotic energy circulates in all zones of existence. Whether in celebration of or submission to this compulsion, Eros is called a *tyrannos*. His *imperium*, like that of Aphrodite, is the entire cosmos—land, sea, and air. On more than one occasion the cry goes up that his power is insufficiently acknowledged in comparison to the cults and prestige of the ruling Olympian gods. This is a trope. From the archaic period through the classical, and Hellenistic ages and down to the end of antiquity, Eros is a continuing obsession that permeates the cultural fabric in different registers and in different modalities of expression. In myth and in cult, in art as in literature, in drama as in philosophy, in the earliest epics as well as in the late romances of the Greco-Roman period, the Greeks never ceased their explorations into the physiology and psychology of desire. Recording endless and varying encounters with the power of eros through storytelling, dramatic enactments, personal lyrics, visual imagery, and theoretical speculation, this culture has exercised its own power over us, its own tyranny over the Western imagination.

The insistent preoccupation and the sheer quantity of cultural data might be sufficient in itself to generate and sustain this tyranny. But in its general capacity to represent experience as a convincing form of reality, in its reservoir of powerful archetypal images that lay claim to some privileged kind of truth about human nature, the Greek legacy invokes the prestigious authority of classical antiquity to persuade those who followed of the universal potency (delights along with the dangers) of instinctual life. As models for imitation and elaboration in art and literature, as attestations of deeper psychic impulses or conveyers of mental categories, Greek representations of Eros promote (and project) a superior power to express the universal regime, eternal reality, and irreducible mystery of sexual desire. Precisely because this same cultural prestige established Greece over the centuries as the touchstone of some profound and original wisdom, the Hellenic preoccupation with the identity and operations of Eros has also served at times as the counterweight to the repressive effects of Western culture and its differently driven systems of religious and political beliefs. In more than one era the land of the pagan past becomes the site of an imagined libidinal freedom and truth, whether as a montage of satyrs, maenads, and goat gods playing blissfully in a pastoral setting or, on the other hand, as a repertory of emotional responses that fuels the discovery and validation of a universal scenario for the development of unconscious psychic life, embodied, for example, in such familiar mythic figures as Oedipus and Narcissus.

PROBLEMS AND METHODS

To return to the Greeks, then, is never a neutral undertaking, a study of just one culture or period among many others. This principle holds true even today, when, in our contemporary climate of research, mystery has yielded to demystification, universals to historical specifics, and the claims of nature are ascribed, rather, to notions of cultural constructs, social institutions, and discursive practices that change and diversify over time. Indeed, in the current use of anthropological approaches to the general study of the ancient Greek world, no area has been subjected to a greater intensity of focus than the investigation of eros as a historically and culturally conditioned set of rules, ideas, images, fantasies, norms, and practices. The emphasis now falls on difference rather than on perceived affinities or continuity of influence. The Greeks are not "us"; they are truly "others," and Eros, above all, can serve as the ultimate proof of the conviction that every culture determines its own variety of meanings for "nature," its own shifting representations of the human body, its own perceptions of sex and gender, and, above all, its own historical "psychology."

Let me amplify the significance of these observations. As Jeffrey Henderson puts it: "That the same two sexes occur in every society is a matter of biology; that there is always sexuality, however, is a cultural matter. Sexuality is that complex of reactions, interpretations, definitions, prohibitions, and norms that is created and maintained by a given culture in response to the fact of the two biological sexes."[2] Furthermore, once the categories of male and female are taken beyond the simplest anatomical and physiological facts, once it is perceived that these categories are not synonymous with the behaviors associated with each—that is, that masculine and feminine are defined by socially prescribed traits, roles, and obligations and not by biology—then the way is clearer to perceiving the range and extent to which sexual ideology pervades and sustains the larger areas of society. Sexuality is not just an affair between individuals in the private sector. Beyond direct social regulations that govern modes of courtship, sanctioned or unsanctioned sexual unions, rules of marriage, reproduction, and inheritance, initiations into adulthood, or prescribed communal rites that celebrate the body and sexuality, the erotic domain is available as both model, symbol, and metaphor for other disparate forms of social interaction. Through extensions, analogies, and mirrorings of erotic experience, sexuality offers a multivoiced "language" that can be deployed along an entire continuum in all areas of human relations and concerns. More accurately, it is "one of the languages for defining, describing, interpreting, and hence transacting all manner of other business."[3] Erotic expression is therefore deeply implicated in the social, economic, political, and religious institutions of any society. This is certainly the case in ancient Greek culture, with its gendered and sexually active pantheon of divinities, its highly developed structures of symbols and myths, and its general tendency to produce highly codified systems of thought and values.

If the study of sexuality occupies a special place in historical inquiry as the point of intersection between public and private, the social and the psychic, the institutional and the subjective, this same set of conditions must also apply in turn to the conduct of the study itself. Hence, when it comes to mapping the erotic territory of ancient Greece, the newfound emphasis on cultural difference may not be as objective as it might seem at first. Every age inevitably poses the questions appropriate to its own needs and interests. Indeed, if sex, sexuality, and sexual behavior have recently become legitimate objects of inquiry across all historical periods, the reasons are not far to seek in the social climate of our own times. In our concern for individual freedom of choice and a tolerance for a wide variety of life styles in a more permissive society, the bans of censorship have fallen. Liberated from the taboos, both verbal and visual, that used to discourage open publicity about such intimate matters, the topic of sexuality has become a prominent part of daily discourse. The search to identify, define, and understand the workings of eros in a given culture is therefore inevitably conditioned by individual experience and outlook to a much greater extent than any other field of inquiry. Our own status as subjects must come into play, whether consciously or unconsciously. The same kinds of data may be collected and analyzed, but an observer's eye may well be led to highlight one feature more than another, may be alert to different aspects and details of erotic behavior, and, above all, may adopt, modify, or resist a prevailing interpretative grid or style of discourse, depending on one's own position (social, intellectual, sexual, political) in the current heated debates about sex and gender, nature, the body, and society. Hence, if we turn back to Greek culture, now in the spirit of difference rather than identification, we may find there is more at stake than increasing our fund of knowledge about the ancient world or even contributing to the larger project of composing a general history of sexuality that no longer takes nature as its indisputable guide and its touchstone of universal "truth."

Who is speaking for whom and in what voice? An anthropological approach that would report on sexuality within the solidly androcentric bias of ancient society without taking serious account of its hierarchical implications promotes the risk of a discourse that replicates the very terms of the society it aims to explore. Or, in the interest of "objectivity," the anthropologist may fall into binary thinking to impose a misleading symmetry on the categories of male and female experience (e.g., marriage for women, war for men). On the other hand, it is no accident that some of the major advances in theorizing questions of sexuality and in charting its historical vicissitudes have been due to the less dominant voices in our society—those, in fact, who have been most *defined* by the ruling norms according to their "sexuality" (women and male homosexuals) and those, therefore, most concerned with sexual identity and difference. The gains in our understanding from these points of view are substantial; so are the misapprehensions that may result if the focus is too narrow or the ideological agenda too pervasive. For example, a certain "orthodox" line that stresses the categories of "active" and "passive" as central to ancient ideas about sexuality, regardless of gender, encourages certain feminist theorists to press the grievance of female victimhood and oppression over a more complex view of relations between the sexes. That same line inspires gay theorists to underplay differences between the sexes in order to promote a parity between heterosexual and same-sex relations at the cost of undervaluing the role and importance of the family, procreation, and to a large extent, I would add, the extensive repertory of male imaginative fantasies about femininity. If some feminists may be tempted to locate inferior status in women's ancient Greece along the spectrum of other patriarchal cultures as a sign of "sameness," gay theorists will, for their part, stress "otherness." These remarks are, of course, far too reductive. They are meant only to set the boundary markers of a field that is notable, as might be expected, for the richness of its continuing debates and the wide variety of opinions and approaches.[4]

In larger terms, however, current discussions about sexuality in the Greek world must confront three major lines of inquiry: Freud, Foucault, and Feminism(s). While Freud and Foucault may be said to join hands in their male-centered views of the world and of the human subject, the divergences are noteworthy and significant: universal psychology vs. contingent historicity, techniques of repression, projection, and sublimation vs. techniques of power, mastery, and knowledge. A feminist outlook intrudes as the third (and perhaps unwelcome) partner to raise the stakes of the game: to destabilize claims made about that erotic being—"woman," the "Other"—to probe further into some of the possible explanations for male views of female sexuality, and, finally, to explore the tensions, contradictions, and slippages of boundaries that undermine confidence in the comfortable pronouncements of official doctrines, both ancient and modern.

If, when it comes to antiquity, Freud (and his followers, especially Lacan) seem too bounded by modern cultural assumptions about the person and his or her horizons of expectation in emphasizing individual choice and emotional fulfillment, the problem of sexuality limited solely to institutional structures and sociological determinants of age, class, gender, and status seems no less unrealistic a set of coordinates through which to consider Greek ideas about Eros. There is more to Freud (and psychoanalytic thinking in general) than the Oedipus complex as there is more to Greek definitions of feminine functions than the famous statement of the orator Demosthenes: "We have courtesans to give us pleasure, concubines to attend to our needs, and wives to give us legitimate children" (*Orations* 59.122). Without recourse to a psychological point of reference, how can we do justice to the complexities of Greek representations of Eros, the conflicts and anxieties that attend its expression, the proliferation of sexual misadventures in myth and literature?

Foucault, for his part, is certainly not immune to criticism. But his development of the idea that the "uses of pleasure" become central to the project of male self-fashioning already in the classical period and his emphasis on the need to integrate erotics into the broader fields of regimen, economics, and philosophy (and vice versa) are both astute and meritorious.[5] True, he restricts himself to too limited a repertory of prescriptive texts (e.g., Xenophon, Hippocratic Corpus, Plato, Aristotle), and, although he refers in passing to different and even dissident voices, his single-minded focus, conducted from a purely male point of view, leads at times to some significant errors and distortions. Page duBois rightly objects that Foucault "remains fixed inside the universe defined by Plato, where the single, masculine, rational, and disciplined self is identical with the subject of history. He takes for granted and thus 'authorizes' exactly what needs to be explained. Hence his work is a superior example of [those] very prejudices, historical and sexual," that may and should be exposed to inquiry.[6] Yet, by comparison to other cultures, anxiety about the status and definition of the masculine self along with a marked concern for strategies of acquiring (and maintaining) power—over others as over oneself—are among the most particular (and insistent) themes of Greek concepts about Eros that lead out far beyond Foucault's particular set of concerns. This is a society that from Homer on promotes competition between men in war, athletic exploits, political debates, and even artistic production, with the aim of meeting standards of male achievement in a contest that must always be renewed (and hence never satisfied). The male self must put himself at risk because that self is always at risk. He is judged by the degree to which he maintains autonomy and bodily integrity and defends himself against threats of ridicule, shame, and, most of all, effeminization. The assaults of Eros always compromise the boundaries of that self, who is—both willing and unwilling, both subject and object of a desire that is conceived as emanating from a coercive external source, whether in the personified form of the god Eros or in the irresistible attraction exercised by the beloved. Indeed, a certain paradox for men inheres in these representations of the psychology/physiology of desire: if phallic aggression also means submission to another's control, the lover is both active and passive at the same time, in a position of both strength and weakness. According to this norm, desire is not a reciprocal affair, nor, it would seem, are the terms of its fulfillment.

The myth of Narcissus, despite its relatively late appearance in text and art, is characteristically Greek in its implications. Refusing the constraints of desire, Narcissus is punished by falling in love with his own reflection. The offense and the retribution are warnings to young men that self-love is an impossible dream—but that it is dreamed at all suggest its temptations to the male self, who can both experience the intensity of desire and yet remain self-sufficient unto himself. The motif is extended in Socratic teaching in Plato's dialogues (especially *Symposium* and *Phaedrus*), in which the goal of a lover is to find in the beloved a reflection wherein he might see himself and transcend that duality, not only in the interests of achieving the true love of Beauty but also in the fantasy of becoming pregnant and giving birth to himself.

SOURCES AND THEIR LIMITATIONS

Despite the abundance of data across the expanse of Greek culture from archaic to late antiquity, the kinds of evidence we possess place certain inevitable restrictions on our knowledge. We have virtually no secure means of assessing gaps between theory and practice, between what people said in public and what they did (or thought) in private. We know little beyond the textual and pictorial evidence, most often produced for and by a cultural elite. Even information about courtesans and prostitutes is sadly deficient. Although such evidence as comic obscenity, dream books, folklore, magic love charms, and medical works may give us a glimpse into sexual attitudes and ideas about the desiring body that include other classes of society, the virtual absence of women's own voices is notable. That absence indeed verifies their muted social roles in a culture of marked asymmetry between male and female roles and respective levels of participation in cultural life, if we "hear" and "see" women, therefore, in the literary and pictorial traditions, we must keep in mind that they are figures, not persons. Their images are filtered through conventions of artistic representation and norms for social behavior. This obvious fact does not mean that we cannot form some idea of the modes of acculturation for women or that we thereby exclude as invalid any perceptions men may have regarding female sexuality. But without access to evidence about women's private lives, their habits, their attitudes, their own cultures of expression, we cannot speak with any confidence about how women might have understood their erotic experiences in ways different from and apart from men.

Sappho is, of course, a notable exception. In a sense this is an unfortunate historical accident. We neither have access to the whole body of her poetry nor to the precise social conditions of seventh-century Lesbos. The tantalizing fragments that remain suggest an approach to Eros in keeping with essentialist and stereotyped notions about femininity in its more romantic, more personalized intensity, reinforced by the idea that the passions of Eros are the source of female creativity. These fragments have also given rise to an entire industry that aims to determine once and for all the nature of her relations with other women—ranging from schoolmistress to full homoerotic partner, with many variants in between. Even an anthropological perspective that takes musical training of girls as a preparation for their future lives as wives and mothers[7] cannot overcome the virtual uniqueness of her position in our sources and the consequent tendency for modern critics to exaggerate or, at times, invent her place in both erotic and literary history.[8]

Aside from Sappho (and fragments from a few other women poets), then, we mostly "know what certain men publicly articulated and took for granted." Women remain, as the same critic observes, both "a public myth and a potent social reality."[9] More than one text suggests that husbands were suspicious of other women's potentially corrupting influence over their wives, when they are in one another's company apart from men. Numerous references from the archaic period onward attest that husbands imagine feminine conversation in the privacy of the house as occasions for gossip and unwholesome talking about sex. Since women are credited in any case by their menfolk with a dangerous excess of sexuality that requires strict protocols to contain its potential misuse, the loss of women's unheard voices on the subject is all the more profound. But this absence from our sources may be precisely the point in a system that tended to equate chastity with silence but also counted on women's conspiratorial secrecy to veil an unspoken and "unspeakable" truth. This secrecy, in fact, is culturally codified in fertility rituals reserved for women alone in which obscene ribaldries form an obligatory part of the ceremonies, as in the Thesmophoria, an important festival in honor of Demeter.[10] We see this complex of ideas expressed comically in Aristophanes' *Thesmophoriazousae*, which is staged as taking place at that festival, in which ritual obscenity is reformulated in the plot that has women (i.e., male actors dressed as women) bring charges against the tragic poet Euripides for having spilled their secrets in his representations of sexually outrageous heroines on the public stage. Euripides slanders respectable women, they claim, with these lurid tales of female misconduct, but their protests are undermined by the comic setting itself, which here as elsewhere delights in its typical stereotypes of women's libidinal excesses.

My intention is not to seek some imaginative reconstruction of women's lost voices as a way of redressing the imbalance in our sources. Quite the opposite. Rather, I would like to use my position as "other" with respect to the dominant male subject to enlarge the perspective by which we view and evaluate Greek sexual attitudes in the formative periods of the archaic age (eighth to sixth centuries B.C.E.) and the classical era of the fifth and fourth centuries B.C.E. (largely represented by Athens). If we have grasped the essential otherness of this culture in its uses of eros to acculturate and maintain social identities in a civic context, then it may be salutary to resist the idea that the only outcome of imaginative scenarios and erotic discourse is to reinforce prescriptive norms and hierarchical realities.[11] Teleology has its uses, but the end is not the whole story. The route to getting there also counts in assessing the continuing threats to the security of male dominance as well as dreams of increasing it. The distance between social unequals does not only set the rules for their interaction; it also conjures up ideas of resistance, competition, reversal, and subversion by both partners.

NORMS AND COUNTERNORMS

The Greek social system was built on the principle of a divided world, which enforced strong distinctions between the genders, assigning to male and female their defined roles, attributes, spaces, and spheres of influence. Institutionally, eros had its uses in acculturating the young of both sexes to perform their approved functions: boys as future active citizens in public space, girls to be brought under the yoke of matrimony as wives and mothers in the private domain of the house. As a rule, our sources of these periods do not emphasize what we would call conjugal love. Aphrodite is a dangerous commodity for women: before marriage the girl is a vulnerable virgin who fears unwanted advances and prizes bodily integrity; afterward, as a wife (and, later, mother), sexuality is considered a dangerous temptation.

This insistence on the nature of female sexuality in an ancient Mediterranean culture is one that is not unfamiliar to us in the long history of patriarchal, husband-oriented societies that transfer a woman from her father's house to that of her spouse and enjoin a strict chastity upon her (with its accompanying suspicion of her hypersexuality) in order to be certain of her loyalty and to ensure the legitimacy of the paternal line. Her professed inferiority only increases his insecurity, his conviction that women are all too adept in the uses of guile and deception, especially when it comes to sexual betrayal. The logic that attributes to women less control than men over their carnal appetites as justification for placing them under masculine control also, paradoxically perhaps, gives women a greater control over men, especially in a culture conditioned by the ethics of honor and shame. The threat that a woman's sexual misconduct will bring shame upon her husband's family is an ideological constraint on her behavior. But it gives her leverage in readjusting the balance of power and encourages her husband to be mindful of her presence and of her rights in fear of her retaliation. Medea may be an anomalous figure—a foreigner, adept in the uses of magic and also of divine ancestry—but she claims to speak for the general situation of all Greek wives (to which the female chorus gives its assent), and Jason pays a terrible price for his betrayal of her.

The assertions of male dominance, therefore, are far from secure, susceptible to being undermined not only from within by female deceptiveness and capacity for revenge but by an open confrontation between male and female interests. More than one myth (and, certainly, tragedy) stages a battle between the sexes, and, if men inevitably win at the end, they do so not by brute force but by negotiation and compromise. The law of the talion is a highly operative principle in Greek thought and practice. It extends to vengeance in the execution of justice, but it also applies to the principle of compensation (or compromise), which levels out to some extent the potential for unlimited power of the stronger over the weaker and certainly challenges any facile reduction of sexual difference to neat categories of active and passive roles.

What seems to me to be more distinctive about this Mediterranean culture, however, is the institution of pederasty: the highly stylized erotic relations between men and boys and the emotional investment in the public affirmation of the value of such relationships. While it is true that the most substantial data comes to us from a very particular period (late archaic/classical) in the city of Athens,[12] the persistence of these practices into later periods (although with differing evaluations of its significance and even desirability) suggests a very different attitude about same-sex relations between men and, consequently, a different axis of sexual definition than one based solely on the anatomical differences between male and female.

Whatever uses the institution of pederasty may have had in educating pubescent boys through stated aims of leading them to emulate the manly virtues of their more mature lovers, the existence of pederasty has a profound effect not only on relations between male and female but also on the evaluation of feminine sexuality. It is no accident, I think, that the high anxiety one may perceive in Greek attitudes toward female sexuality and, indeed, toward the female body, along with a striking degree of fear of effeminization, can be linked to, is conditioned by, the alternative (at least, temporarily) of pederasty. It seems logical, therefore, that only with the waning of this publicly approved system as an integral part of a citizen's life do we perceive a concomitant rise in the valuation of conjugal sexuality and of the significance of affective relations between men and women.

These brief remarks are too brief, of course, to give any satisfactory account of the range of Greek attitudes and behaviors in anthropological terms. Such accounts may be found elsewhere.[13] But with the aim of liberating ourselves from the tyranny of a functionalist perspective that interprets our cultural data as aimed solely at producing social conformity, let me propose the following mixture of items as a point of departure:

If Eros is called a *tyrannos*—
If erotic desire is experienced as an assault on the self and its boundaries, whatever its sensual delights, and its outcome may be figured as fusion or fragmentation—
If Eros is a madness, a wound, a disease, a fire—

If erotic protocols favor a relationship between partners unequal in status and age, regardless of the gender of the beloved—

If courtship is most often represented in the form of pursuit and flight—

If love is a sport that takes on the language of the hunt and of horse taming in images of net, snare, and trap or those of bit, bridle, and reins—

If, consequently, erotic discourses elide expressions of reciprocal desire and mutual satisfaction—

If the idea of "erotic justice" is extortionate, demanding that the beloved return the lover's desire or else be condemned to suffer the same rejection when it is his or her turn—

If female sexuality is perceived as excessive and threatening, roasting a man with a fire that brings him to a premature old age—

If woman is created for man as a punishment, a thievish creature in return for the theft of fire, the instrument too of definitively dividing men from gods—

If the blood sacrifice of virgins (Polyxena, Iphigenia) is an important theme—

If, on the contrary, outraged virgins (Danaids) slay their bridegrooms on their wedding night, contrary to the dictum that men shed blood but women bleed—

If the love of boys is a significant alternative in the city-state—

If Zeus takes Ganymede to Olympus to be his cupbearer where he remains eternally young, pouring out endless streams of precious liquid—

If, conversely, when the goddess Dawn (Eos) takes the boy Tithonus to her bed, she forgets to procure his exemption from the desiccating effects of age so that he dwindles into a feeble child—

If Socrates' innovative myth of the birth of Eros makes him the child of Penia (poverty) and Poros (resources), brought into being at the feast in honor of Aphrodite's birth by his mother's theft of his father's substance—

If this myth is yet another variant on the widespread, but not exclusive, notion that the female body is figured on the model of excess and lack—

If in Aristotle's biological theories the female provides the matter, the male the form and spirit—

If the goddess Aphrodite in Hesiod's account of how the world came to be is born from the semen of the severed phallus belonging to the primordial male god, Ouranos (Heaven), consort of Gaia (Earth)—

If his drops of blood fall to the earth and produce the Erinyes, agents of retribution—

If the mythic imagination invents many other ingenious strategies of reproduction and birth, including Zeus' well-known obstetrical adventures, which produce Athena from his head and Dionysus from his thigh—

If in retaliation for the theft of maternal functions Zeus' consort, Hera, invokes her female capacity for parthenogenesis and brings forth monsters—

If Ixion can assault Hera but impregnate, instead, her cloudlike substitute (Nephele)—

If this union produces the race of lustful hybrid centaurs and earns for Ixion the penalty of being fastened to a fiery wheel forever—

If the appropriate punishment is to know no respite from the torments of desire—

If erotic pursuit finds only a phantom, an empty dream, an illusion—

If clouds can be impregnated—

If, in some versions, the real Helen never went to Troy—

If, finally, Eros and Thanatos are not strangers of each other—

Admittedly, this allusive list is both highly selective and purposefully mixed. I chose this assortment of items to focus our questioning on a complex of erotic attitudes and fantasies that exceed a purely social explanation and immeasurably darken the pleasures of desire. As Patricia Storace wryly observes, "the source of sexual tension in ancient Greek myth is not so much the drive to the ecstasy of consummation, as uncertainty whether either or both partners will survive the sexual act."[14] Yet, in recalling the bitter side of Eros, do we forget the sweet? True, desire is a tempestuous storm that sweeps over the helpless lover; true, Boreas, the storm wind from the north, snatches off maidens he desires. But where are the Zephyrs, the gentle breezes? Where are the gardens, the meadows, the perfumed scents of fruit and flowers? Where is the golden sheen of sexual allure, the persuasive charms of *charis* (grace) that emanate from the object of desire? Where are the companions of Aphrodite—Pothos (longing), Himeros (desire), Oaristus (love talk), and Eros to compensate for her negative associations with Apate (deceit) and Dolos (trickery)? Above all, where is beauty? Socrates, in the myth just mentioned, invokes the occasion of Eros' conception at the celebration of Aphrodite's birth to claim this as the reason why Eros is "a lover of beauty, because Aphrodite herself is especially beautiful." More generally, where are the aesthetics of Eros that strikes a subject with amazement *(thauma)* at the sight of a beautiful face and body and links this enchantment *(theixis)* to the charms of poetry, music, and art?

All these elements are indeed present in many representations of and allusions to the gentler pleasures of Eros in both literary and iconographic sources. The lover may speak wittily of a playful Eros in lyric poetry; vase paintings may show an Aphrodite flanked by the personified figures of Himeros and Pothos, while choral song may portray her as a tender and gentle influence in the company of the Muses. But, like Pandora, created by Hephaestus at Zeus' behest in retaliation for Prometheus' theft of fire in Hesiod's two texts, *Theogony* and *Works and Days,* the beauty of sexual allure also proves more often to be a snare and delusion rather than an unmixed joy. The lovely shape that is the maiden who resembles the immortal goddesses and is adorned with all the shining garments and crown of a bride is designed to be a "bane *(pêma)* to mortal men." Aphrodite is bidden to pour grace *(charis)* over her head but also adds "cruel longing and cares that weary the limbs." To complete her person Hermes endows her with a shameless (bitchy) mind and a thievish nature. So, while the Charites (the Graces) and Peitho (Persuasion) give her golden necklaces and the Horae (the Hours) crown her with spring flowers, Pandora (and through her the pleasures of eros) is named from the outset as a treacherous gift given by the gods to men's sorrow and ruin (*WD* 59–82), a seductive creature who will steal man's substance and dry him up before his time (705).

If I have, however, stressed the negative aspects of eros as threatening to the boundaries of the self, both in sex and reproduction, was it always this way? Or should we accept a commonly held view that ascribes to Homer a relatively unproblematic view of heterosexuality ("it is the way of the world for a man to enjoy sex with a woman" [*Iliad* 9.276]) in keeping with its aristocratic formations by contrast to the "sexually repressive society of late archaic and early democratic Athens, with its segregation of women and strict citizenship laws," as Jeffrey Henderson sensibly suggests?[15] The answer is yes and no, depending on the context and the attitude in question. Certainly, pederastic relations are not mentioned in the Homeric epic, and it is only later authors who ascribe such a bond to the comradely love between Achilles and Patroclus on the battlefield at Troy. The idea that a warrior, whether lover or beloved, would fight more bravely in the presence of the other and suffer increased shame if he did not do so, as particularly in the Spartan military system, for all its apparently ancient, even tribal aspects, is not openly promoted and extolled until the later period.

On the other hand, there is more to say about the subsequent emphasis on eros than what Henderson perceives as the compensatory "exaltation and exaggeration of sexuality" on public festive occasions, such as celebrations of Dionysus with expressions of comic obscenity and men costumed as phallic satyrs. The same repressive constraints, he argues, also "made sexuality a central topic of ethical debate, notably in the Socratic 'schools'; and prized dramas that portrayed individuals ruined by (tragedy) or evading (comedy) ideal norms of kinship, citizenship, the complementary natures of polis and *oikos,* and the behaviors assigned to men and women."[16] To these cultural products we may also add the explicit sexual iconography of vase paintings in the late archaic period and the later proliferation of little Erotes and the figure of Aphrodite in romanticized scenes depicting private life. But, if there is no doubt of an increasing preoccupation about the nature of eros—its physics and metaphysics—and with modes of sexual expression in the city-state by comparison to the archaic age, Henderson's hypothesis also seems far too limited to account for the differences he notes. His model of inhibition and restraint is more suited to festive comedy than to the rest of the data in which eros figures but is put at the service of different causes such as conflict in the family, unwanted sexual advances, or uneasy passage to adulthood. Chief among these larger issues seems to be a marked concern with the status of the individual in relation to society that manifests itself in many spheres of life, including the conduct of eros, which contributes in no small part to the project and problems of self-definition in a newer era.

In any case, despite the substantial differences between these later developments in a civic context and what we may infer of the aristocratic society that forms the background of Homeric epic, we should not overlook the marked continuities in erotic attitudes. Indeed, if we consider the matter more closely, we will see that typical male anxieties about sexuality and its effects on the self are already in place in both epics, both of whose plots are based on the fact (Helen) or possibility (Penelope) of a wife's infidelity. For all the conjugal ideals promoted by the *Odyssey*, the specter of Clytemnestra, who killed her husband on his return, casts a long shadow over all women, including Penelope, and, even more, female eroticism poses obstacles to Odysseus's return with bodily risks of detention, unmanning, or engulfment (Circe, Calypso). A second source of continuity with the archaic age is one that engages sacred and not just social concerns in a system of beliefs that ascribes erotic power to divinity. Hence, throughout antiquity, in mythic, mystic, and psychological terms, questions about an individual's status as a creature of erotic desire also entail the larger metaphysical questions of the lines that divide mortals from immortals and nowhere more evident than in the fact that members of both categories are sexual beings, who may even couple with one another.

It is to this aspect of Greek ideas about eros that I now turn in order to focus closer attention on a specific text. And what better example than to turn to the figure of Aphrodite herself and to the hymn in her honor, probably composed in the late archaic period, a poem that both playfully and profoundly offers reflections on the conundrum of relations between mortals and immortals in ontological terms as well as those between the sexes in the context of social relations and relative hierarchies of status.[17]

THE HOMERIC HYMN TO APHRODITE

Eros and Aphrodite may be used interchangeably to refer to sexuality, sexual desire, and sexual pleasure, either singly or together. If Eros, however, may be more accurately limited to the instinct of amorous desire, Aphrodite is engaged in all the activities that lead from sexual appeal to the performance of sexual acts. Erotic allure is a gift (*dôron*) of Aphrodite; her works (*erga*) are the consummation of desire. *Ta aphrodisia* (things of Aphrodite) cover the entire erotic spectrum of sexual behaviors. While the status of Eros fluctuates according to theogonic, philosophical, and mystic traditions about whether he is the oldest or youngest of the gods or both at the same time, he is more often depicted as subordinate to the goddess. He is sometimes her attendant, sometimes her child. In a gendered system of values it matters that one is female and the other male; it matters, too, that the female is dominant over the male. As an acknowledged member of the Greek pantheon of gods, Aphrodite is entitled to the full range of privileges accorded to divine rank; like other gods, she wields dangerous power, requiring honor and worship, and punishing affronts or challenges to her influence (e.g., Hippolytus). Unlike them, however, she also operates in the world of the gods, where sexuality represents male potency but also, paradoxically, a limitation in having to yield to the constraints of erotic desire. Aphrodite's influence over all domains of existence attests to the cosmic and universal force of eros. But, if desire is configured as an irresistible force that acts on the body and soul of the lover, its embodiment in a female figure intensifies the ambivalence that surrounds sexuality from the devious paths of seduction to the dynamics of actual performance.

This ambivalence that, at its limit, may be configured as a political struggle between the sexes is best represented in the example of Zeus himself, the chief divinity, "father of gods and men," for whom susceptibility to erotic desire offers a challenge to his authority and his claims to dominance. On the one hand, his notorious sexual appetite is proof of a super virility, which allies the sign of the phallus with political and cosmic power, symbolized by the scepter and thunderbolt, and his pursuit and conquest of female partners may be included among his other, nonerotic exploits that enforce his mastery over others. Yet, as an easy prey to erotic seduction, Zeus in turn can be mastered by the imperious demands of the flesh that "take away his wits" through deceptive feminine wiles. The most notorious case is the scene between Zeus and Hera in the *Iliad* (14.152–360) in which under a series of false pretexts to cover her own political designs that would turn, at least temporarily, the tide of battle at Troy, Hera first mobilizes Aphrodite's magical resources to increase her allure and distracts her spouse's vigilant eye through sex and its enervating aftermath in the unconsciousness of sleep. Everything that transpires in this scene accords well with Greek ideas about the dynamics of eros between a man and a woman, fulfilling all the suspicions men may have about the seductive tricks of their wives, who, it is presumed, not only use these to entice their husbands into sex but, even worse, may also use sex as an instrument for furthering other ends. Aphrodite is an innocent party, as it were, to Hera's deception of Zeus, having been duped into believing Hera's tale about using it to reconcile an estranged couple in the divine pantheon. But the scene demonstrates nonetheless the subversive potential of Aphrodite's irresistible power to challenge the assumptions of masculine preeminence and to undermine the order of hierarchical status in the universe.

The Homeric *Hymn to Aphrodite* confronts these dilemmas in explicit form. The hymn recounts Zeus' plan to assert his superior power over the goddess by subjecting her to the same desire she has inflicted on others but from which, until now, she herself has been exempt. As a poem ostensibly designed to honor Aphrodite and, like other hymns to the individual gods, to celebrate and define her spheres of power and modes of her intervention in the world, the curious ambiguity that results from the initial premise of the plot subverts the confident opposition between praise and blame, between triumph and defeat.[18] In the process of both exalting and diminishing the power of Aphrodite, the hymn offers us the most brilliant and most subtle exploration of the quandaries and conflicts of Greek erotic psychology.

The most common term for sexual intercourse is *mixis,* the mingling of two partners (or their seeds) in the act of love. This is not a neutral concept in a culture that puts such a high value on the integrity of the individual body and is so preoccupied with defining and maintaining boundaries between different domains. One such boundary is the line that divides male from female. A more pervasive principle that organizes the entire cosmos is the line that divides beast from humans and humans from gods. Sexual desire not only induces *mixis* between male and female within these three separate species. It also operates as a force that crosses boundaries in allowing gods to mate with mortals. While in other contexts the mixing of human and divine seed accounts for aristocratic pedigrees and claims to political entitlement, in this hymn that capacity is construed by Zeus as a threatening transgression—of hierarchy, propriety, and male self-esteem. The purpose of the hymn may be to reassert these boundaries at the end by dividing the world of eternity from the world of aging mortality. But, in questioning the puzzling nature of the relationship between gods and mortals by means of the sexuality they share, the poem also reveals the degree to which the divinization of erotic desire inevitably mixes with the humanizing of the divine. Ontological certainty in an anthropomorphic concept of divinity is never quite as stable as official belief would claim.

The opening of the hymn declares the universal imperium of the goddess over all living creatures in all domains of existence—sky, land, and sea. It will culminate with the dictum that no one, "neither blessed god nor earthly mortal, has ever escaped Aphrodite" (34–35), with the exception of three virgin goddesses (Artemis, Athena, and Hestia) who have been granted special exemption.[19] The triumphant proof of this principle, the pinnacle of praise for the goddess's power, is her power over Zeus:

> She even led astray the mind of Zeus who delights in thunder
> and who is the greatest and has the highest honor.
> Even his wise mind she tricks when she wills it
> and easily mates him with mortal women,
> making him forget Hera, his revered wife.
>
> (36–40)

The sexuality of the gods may later become a theological problem, subject to rejection of the idea or its allegorization. Here it is framed as an internal struggle for power in the divine pantheon. Aphrodite has challenged Zeus' hegemony, and her victim retaliates in kind. He will "place sweet longing in Aphrodite's heart to mate with a mortal man" (45–46). Aphrodite had flouted all the social rules: she not only promotes erotic mismatches between partners of unequal rank, but, what is worse, she encourages adultery, which in the hymn is now a source of moral opprobrium that arouses the same indignation in the society of the gods that it would in the human world. While it was Zeus himself in the scene in the *Iliad* who confidently (and tactlessly) invoked his previous amours, listing them each by name, as the way to demonstrate how much more he desired his wife now than on any previous occasion of lovemaking, here the terms have dramatically changed. Zeus' sexual infidelities are an offense to the cherished wife, to whom a husband owes respect, and Aphrodite's influence is a source not of pleasure but of reproach (*oneidos*). Moreover, Aphrodite has added insult to injury by advertising these shameful indiscretions in order to boast of her exploits, "how she had joined in love gods to mortal women, who bore mortal sons to the deathless gods and of how she had paired goddesses with mortal men" (48–52). The phrasing implies that it is only the gods and not the goddesses who, in mating with mortals, leave behind tangible (and permanent) signs of their impropriety in the offspring they sire, leaving the rest of the hymn to repair this imbalance in the retribution that Zeus' plan will shortly visit upon Aphrodite herself.

Now it is Aphrodite's turn to undergo the same indignity she had formerly visited upon the gods and to be compelled to acknowledge that mortifying fact. Rising from Anchises's bed, she tells him that she has incurred shame among the gods because of him. Until now the gods had feared her scheming wiles and the arrogant disclosure of her successful exploits. Now others will know that she too was driven out of her mind with desire for a mortal, and, if this were not enough, she is pregnant too. She might at least be able to keep this last embarrassing matter secret—unless, that is, Anchises is less than discreet. "If you reveal the name of the child's mother," she declares, and "boast with foolish heart to have mingled in sex with Aphrodite, an angry Zeus will smite you with a smoking thunderbolt" (281–89). In this reversal of the balance of power males now have the upper hand—from both sides. Her human lover may betray her secret; she must look to Zeus' power to protect her honor.

It has been suggested that Aphrodite's defeat spells the definitive end of these disgraceful liaisons between gods and mortals. Henceforth, so the argument goes, she will continue to instigate desire, but only among those of the same kind.[20] The text makes no such overt claims. The full range of sexuality still belongs to Aphrodite; it can and might erupt at any time and in another place. But, even if the gods might continue to mate with mortals, at least the scandal of publicity will be avoided. What the goddess has lost, it seems, now that she too has a secret to hide, is the right to self-praise, the pride she has shown in celebrating the success of her wiles. Yet, ironically, the secret will be, must be, told. This is the task of the poet-worshiper, who, in relating the scandal of Aphrodite's humiliating defeat, rejects the demand for secrecy by the very requirements of his task to voice her praise.

If the work is structured as a hymn in acclaim of the goddess's erotic power, which Zeus both appropriates and turns against her on the divine level, the meeting between Aphrodite and her mortal lover exemplifies her power over men. The poem, as Ann Bergren points out, "attempts to resolve the tension between a cosmos controlled by Aphrodite and one controlled by Zeus into a stable hierarchy in which immortal males 'tame' the principle of sexuality as an immortal female, who herself 'tames' the mortal male."[21] But, even so, matters are not so simple. There is a further tension in Aphrodite's status as both female and goddess that casts the problem of mortal and immortal into another light. Male gods may ravish their mortal women in disguise, a cloud, a shower of golden rain, the form of a satyr, a swan, or even in person. A female divinity, however, cannot simply enforce her desire. The protocols of courtship require that the other, the male, desire her. For a female to be desirable means to be a young and beautiful maiden on the threshold of marriage, a girl not yet experienced in sex. She is desired because at that moment she partakes in the "gifts of Aphrodite," bestowed upon her by the goddess.

In the rhetoric of sexual encounter the glamour that surrounds a lovely mortal derives from the momentary radiance of a divine élan that makes her "like to the gods." This is the standard compliment to human beauty, which strikes the beholder with the force of a divine epiphany. After lovemaking Aphrodite will indeed reveal her identity to Anchises as a true manifestation of divinity. But this fact only acknowledges the further deceptiveness of her seductive wiles, since she deceives him at first by denying her divinity in order to allay his fears ("I am surely no goddess: why do you liken me to the immortals?" [109]) and represents herself as a nubile maiden who is looking for matrimony and a respectable status in the world of human social relations, based on her fine lineage and her family's resources (110–42). The confusion, however, implies that, if beautiful mortals may look like goddesses, beautiful goddesses may look like mortals. In Bergren's reading this hesitation between the *two* alternatives is epistemological in nature: "How to tell the truth from an imitation of the truth." The answer is never clear, she continues, since we know from Hesiod's *Theogony* that the first woman, Pandora, was fashioned as a virgin bride, with "a face like to the immortal goddesses," created at Zeus' orders as a deceptive gift to mortals in return for Prometheus' theft of fire.[22] Yet the question of imitation, as the text itself tells us, is not limited to the female form. When Zeus places desire for Anchises in Aphrodite's heart, this young man, we are told, "resembles an immortal in his body" (54). The goddess sees him and instantly falls in love, ensnared both by Zeus' will and by Anchises' godlike form. Physical beauty is an attribute that in the case of both sexes owes its ultimate referent to models of permanence and perfection. The result is that in this situation of planned mésalliance between gods and mortals, the literal and figurative properties of language both coincide and diverge to reflect on the more general rhetoric of courtship and the divine nature of eros.

These brief remarks do not do justice to the intricate circularity of this text and its paradoxical contradictions, particularly with regard to the unexplained "logic" that allows Zeus to instigate sexual desire in the first place. I limit myself, however, to two further reflections. By his tactics Zeus restores the balance of power between the sexes in the cosmic hierarchy. He does so by placing Aphrodite in the same position as she had previously put the male gods. Yet the retaliation cannot be exact. Males inseminate; females procreate. Male gods leave their future progeny behind, when they have consummated the union; so does Aphrodite, when she leaves the child (Aeneas), a mortal, for others, the nymphs, to rear. But the fact that she is subject to pregnancy and childbearing, like any woman, already attests to a more fundamental, irreducible difference between the genders that links goddesses more closely than gods to matters of the body and the vicissitudes of the life cycle. We may discern these differences in assessing the significance of the two myths about other Trojan youths that Aphrodite tells Anchises, once she has revealed her divinity. The first is the myth of Ganymede and Zeus, the second is that of Eos (Dawn) and Tithonus.

> Wise Zeus abducted fair-haired Ganymede
> for his beauty, to be among the immortals
> and pour wine for the gods in the house of Zeus,
> a marvel to look upon, honored by all the gods.
>
> (202–5)

The following one relates to Anchises how "Eos abducted Tithonus, one of your own race, who resembled the immortals." Eos had gone to Zeus to ask him to make her lover immortal and live forever with her in bliss. He granted her wish, but in her foolishness she forgot to ask the favor of eternal youth for her lover. Tithonus "took his delight in the goddess" until gradually he started to age, and when she ceased to desire him she still kept him, fed him, and put him away in a closed chamber, where "his voice flows endlessly and there is no strength" (218–38).

The male god has the power to bestow immortality; the female does not. She is also not as clever as he is, nor as theologically adept. She neglected to consider that divinity consists of two prerequisites: to be both "deathless *and* ageless forever." A second difference between the two stories is a spatial one, pertaining both to geographical location and to dimensions of public and private zones. When Zeus abducts the boy, he takes him to the heights of Olympus, seat of the blessed gods, to serve them and to enjoy their public admiration amid the joys of the banquet. She, on the other hand, takes her lover into the privacy of her bedroom, which itself is situated in a distant place, by the stream of Oceanus at the ends of the earth.

What may be implied by these different outcomes? In the first place we might ask why the beloved in each case is a male? The simple answer to this question is Aphrodite's motive for telling these stories to Anchises as the reason why she cannot promise *him* the same immortality. But the difference in status behind the fates of Ganymede and Tithonus depends on the distinction between heterosexual and homosexual eros. The latter, by necessity, cannot entail the further process of procreation and hence the body is exempt from the mortal sequence of birth, aging, and death. The contrast may be framed, as here, as one between the wet and the dry. Ganymede, the cupbearer, pours out endless streams of liquid, an inexhaustible supply of male substance.[23] By contrast, Tithonus shrivels as he ages, possessing nothing but the "endless flowing of a shrill, clear voice" and reduced finally, as other versions tell us, to a chirping dry cicada.[24] All Aphrodite can promise Anchises is that he will not undergo the unhappy fate of poor Tithonus, who lives forever but in a diminished state of eternal dependency on a female nurturer, an outcome whose logic perhaps accords well with his initial role as the sexual plaything of a woman. The point of both tales is to reassure Anchises that sex with a goddess will not unman him and that his line will continue. Unlike the barren union of Eos and Tithonus, Aphrodite can promise her mortal lover: "You shall have a dear son who will rule among the Trojans, and to his offspring children shall always be born" (196–97). She knows because she now carries his child. Sexuality and fertility go together, despite the incongruous mixing of female eroticism and maternity, along with the shame that now covers her in the eyes of the immortal gods.

Aphrodite's reassurance that his sexual contact with her will not unman him, however, remains only a conditional promise. Anchises's manhood depends on his keeping her secret. Later tradition suggests that he did not (and indeed, the very recital of the poem already reveals it), relating that he was indeed struck by Zeus' thunderbolt or subjected either to paralysis or blindness.[25] The evidence is vague, but the threat that Aphrodite has the power to deprive a male of his virility recalls not only the generalized fear that feminine sexuality is dangerous to the male but also the best-known version of how she was born. Elsewhere, Zeus' efforts to tame the power of Aphrodite and to bring her under his control results in a genealogy that makes her the daughter of Zeus and Dione (whose name is simply a derived adjective of Zeus' name [Dios]). The other myth, known to us from Hesiod's *Theogony*, records her birth from the semen of the severed phallus of Ouranos, when Cronus castrated him, in obedience to his mother's, Gaia's, wishes (*Th.* 187–206). On the one hand, Ouranos (or, at least, his generative part) initiates the first challenge to female fecundity and her capacity for parthenogenetic birth. On the other hand, the price is the loss of his virility: he can no longer enjoy sexual relations and his procreational function is at an end. Aphrodite's origins from male semen may be interpreted in a number of ways. The separation of the primordial couple, Earth and Sky, was essential if their children, still hidden in the depths of the mother, were to see the light of day. It was Ouranos' refusal of this necessity that occasioned Gaia's intrigue against him. Aphrodite is born at this moment of disjunction as the principle of desire that henceforth is needed to rejoin male and female in the act of sex. But that this principle of sexual union is female and born only from a prior act of male castration is matched by Aphrodite's power, as Anchises fears, to duplicate the circumstances of her birth in other males, reversing upon them the very action that resulted in her creation.

The plot of the Homeric *Hymn to Aphrodite* turns on the enigmatic paradox of a sexuality that instigates desire in others without itself succumbing to that same desire. If sexuality is embodied in a beautiful female, as in this instance, this riddle is framed as a threat to male dominance, and the problem is solved by a simple application of the law of reversal: "subdue" the one who "subdues" you; use the same means she has used, and produce the same reaction in her. A more logical alignment of woman and goddess portrays Aphrodite as herself a partner in an adulterous triangle. Married off to the ugly, lame artisan god, Hephaestus, she consorts instead secretly with that virile specimen of manhood who is Ares, the god of war. This triangle is conceptually profound. Love and War are opposites. When they join they create a daughter, Harmonia, as the reification of the very principle of a union of contraries. They also produce Fear and Terror, a result that maintains their fundamental incompatibility—or perhaps is another way of depicting the awesome two-sidedness of sexual mixing. At the same time, Hephaestus matches with Aphrodite as a pair of contrivers, each endowed with cunning skill and both associated with fire: the fire of the forge, the fire of sexuality. Aphrodite indeed is rightfully associated with both males, but, in accounting for her own desire, the relationship can only be configured as a triangle of a cheating wife, a handsome, if stupid, lover, and an ugly cuckolded husband.

In the celebrated song of the bard in the *Odyssey* (8.266–366), Hephaestus catches the errant couple in bed together, and, by the finely woven golden net he has suspended invisibly above them in advance, he entraps them and exhibits them in the shame of their locked embrace to all the gods for their amused laughter. Whether Aphrodite consorts with mortals, as in the hymn, or with gods, as in the *Odyssey,* the final outcome is the same: humiliation of the goddess in public exposure and the charge of fomenting or embodying the disruptive threat of an illicit sexuality that shames the male partners in turn. At the same time, the two accounts differ in representing Aphrodite herself as impersonating a virgin bride in the hymn, while in the epic she plays true to her role as an adulterous wife. Hera, we may recall, in her deception of Zeus in the *Iliad* engages the magical breastband of Aphrodite to arouse a desire, whose intensity can only be measured by comparison to the thrill of the first time. Hera's feigned reticence that is designed to arouse Zeus all the more is the proper demeanor assigned to the uninitiated bride to be brought under the yoke of matrimony. Once initiated, however, the wife's sexuality continually falls under suspicion as leading her to desire another man, not her own. Aphrodite intervenes at the start of a marriage in endowing the bride with her own attributes of sexual allure; when she intervenes again, once the bride has become a mother-wife, it is to promote a dangerous and illicit compulsion that, in the world of mortals, leads to destructive, tragic results. This is the theme, in fact, that enters tragic plots in the fifth-century theater and that through dramatic enactment both verifies and transcends the archaic models of eros and conceptualization of desire to turn its catalytic effect on the more general dynamics of social relations in a contemporary setting.

The significance of the theater in this regard holds true particularly in the case of Euripides, the third and last of the great tragic poets, who was notorious, even in his own day for his interest in erotically charged plots. The scandal and censure that in some instances followed his productions suggest a response to a serious breach of propriety in public discourse.[26] But resistance to the poet's disruptive candor also attests to a changing of the times, and his emphasis on eros and feminine psychology in the framework of traditional myths indicates a heightened availability of eros as a subject for deeper reflection and a newer validation of the quality of a private, individual life. These shifts in outlook expand in the next period, the Hellenistic age and beyond, to the idea of an eros that is closer to our own in its romantic aspects of two mutually love-smitten partners and in the profound and long-lasting emotional impact on the person, an eros that includes but goes beyond the basic physical act of sex. But that is another story.

NOTES

1. This essay is a somewhat revised and much abridged version of a longer piece, entitled "Eros," that appeared in Italian translation in Settis 1996, 369–430. The notes have been kept to a minimum as well as citations of secondary work. For some examples, however, see nn. 4 and 13. My thanks to James Porter for helpful discussion.
2. Henderson 1988, 2:1250.
3. Halperin, Winkler, and Zeitlin 1990, 4.
4. The bibliography on issues of eros, the body, and sexuality in antiquity has vastly increased in recent years. It includes not only general discussions but also particular treatments on topics as diverse as literature, historiography, law, medicine, rhetoric, and iconography.
5. See especially Foucault 1985 and 1986. Foucault himself, it should be noted, is indebted to the work of many classical scholars.
6. duBois 1988, 2.
7. See the pioneering work of Calame 1977.
8. It is useful, for example, to recall that Sappho's poetry served as a model for young men in love as well and was sung at all-male symposia. In one source Solon himself is pictured as becoming enraptured by a song of Sappho as sung by his own nephew at a symposium (*Aelian via Stobaeus* 3.29.58). More generally, see Plutarch *Sympotic Questions* 711d: "Even when Sappho's songs are sung, or Anacreon's, I feel like putting down my drinking cup out of respect." See also 622c: "There was a debate at the house of Sossios, after the songs of Sappho had been sung." See further Rösler 1980; J. Herington 1985. I do not mean, of course, to disparage the large and ever-growing bibliography on Sappho, some of which is excellent, but only to suggest that in the absence of other evidence her figure and poetry are all too often taken as a norm.
9. Henderson 1988, 1249.
10. See, e.g., Zeitlin 1982, 29–57.
11. David Cohen's remarks are salutary: "A culture is not a homogeneous unity . . . The widely differing attitudes and conflicting norms and practices represent the disagreements, contradictions, and anxieties which make up the patterned chaos of a complex culture. They should not be rationalized away" (1991, 201–2).
12. Sparta is another, even stronger case in point. As a militaristic state, it required pederasty as an integral part of training young warriors. But in its emphasis on collective life, which dictated a different balance between public and private life, including different marriage customs and roles for women, Sparta is sufficiently atypical to warrant a separate discussion.
13. I offer a few samples. See Henderson 1988; Calame 1977 and 1992; Carson 1985; Zeitlin 1986; the various essays in Halperin, Winkler, and Zeitlin 1990; Katz 1989; Dover 1974 and 1989; Detienne 1994 and 1979; Vernant 1991; Borgeaud 1988; Loraux 1993 and 1995; Halperin 1990; Winkler 1990; and the various essays collected in Zeitlin 1995.
14. Storace 1996, 10. "This thread," as she continues, "runs through all the stories of the young men and women who are killed after lovemaking with each other or the gods." From this perspective it could even be said that "the body of myths as assembled, invented, reinvented, and anthologized from a variety of sources by the ancient Greeks is one of the *least* erotic of the world's mythologies."
15. Henderson 1988, 1256.

16. Henderson 1988, 1256.
17. The original essay also contained analyses of the "Deception of Zeus" scene in the *Iliad,* briefly alluded to in subsequent pages, and also Euripides' *Hippolytus* and Plato's *Symposium.*
18. For further details and discussion, see Bergren 1989, 1–41.
19. The reasons for these exemptions neatly divide the ideological roles of virgins and women: war and crafts for Athena; the wild spaces external to the city but also the "city of just men" for Artemis; and, for Hestia, as Jenny Clay puts it, "the chaste focus of the domestic hearth" (1989, 161).
20. See Clay 1989, 166–70, with citations of previous bibliography, and also 192.
21. Bergren 1989, 7.
22. Bergren 1989 2; cf. 11–13.
23. On the equation between the "flowing stream of wine" and "the flood of passion" that Zeus, as the lover of Ganymede, pours in upon the lover, as Plato says, see the *Phaedrus* (25 Sc). The name Ganymede can be interpreted to mean "glistening genitals."
24. Now that Tithonus' vital juices have diminished, he is no longer useful for "irrigating" a female partner in intercourse. In Greek medical theory the male is generally associated with the hot and the dry, the female with the cold and the wet, but seminal fluid is, of course, an essential masculine property. For further discussion of these distinctions and their contexts in medical theory, see, e.g., the excellent discussion by Hanson (1992, 31–71), with further bibliography. My point in assigning the "wet" to Ganymede and the "dry" to Tithonus in order to indicate that in the case of these two male figures the distinction is one between eternal youth and aging, between the vital juices of wine (and semen) as the elixir of life and the desiccation of old age (and dry bones in death).
25. See Clay 1989, 199–200, for discussion of the meager evidence and relevant bibliography.
26. Aristophanes is our best contemporary witness (especially in *Frogs* and *Thesmophoriazousae*) to the scandal of Euripides' overly realistic portrayal of "bad" women.

REFERENCES

Bergren, A. 1989. "The Homeric Hymn to Aphrodite: Tradition and Rhetoric, Praise and Blame." *Classical Antiquity* 8:1–41.

Borgeaud, P. 1988. *The Cult of Pan in Ancient Greece.* Trans. Kathleen Atlass and James Redfield. Chicago.

Calame, C. 1977. *Les choeurs de jeunes flues en Grèce archaique.* 2 vols. Rome.

———. 1992. *1 Greci e l'eros: symboli, pratiche e luoghi.* Rome and Ban.

Carson, A. 1985. *Eros the Bittersweet.* Princeton.

Clay, J. 1989. *The Politics of Olympus: Form and Meaning in the Major Homeric Hymns.* Princeton.

Cohen, D. 1991. *Law, Sexuality and Society.* Cambridge.

Detienne, M. 1994. *The Gardens of Adonis: Spices in Greek Mythology.* 2d ed. Trans. Janet Lloyd. Princeton.

———. 1979. *Dionysos Slain*. Trans. Mireille Muellner and Leonard Muellner. Baltimore.
Dover, K. J. 1974. *Greek Popular Morality in the Time of Plato and Aristotle*. Oxford.
———. 1989. *Greek Homosexuality*. Cambridge, Mass.
duBois, P. 1988. *Sowing the Body: Psychoanalysis and Ancient Representations of Women*. Chicago.
Foucault, M. 1985. *The Use of Pleasure*. Trans. Robert Hurley. New York.
———. 1986. *The Care of the Self*. Trans. Robert Hurley. New York.
Halperin, D. 1990. *One Hundred Years of Homosexuality and Other Essays on Greek Love*. London.
Halperin, D., J. J. Winkler, and E. I. Zeitlin, eds. 1990. *Before Sexuality: The Construction of Erotic Experience in the Ancient Greek World*. Princeton.
Hanson, A. 1991. "Conception, Gestation, and the Origin of Female Nature." *Helios* 19:31–71.
Henderson, J. 1988. "Greek Attitudes toward Sex." In *Civilization of the Ancient Mediterranean*, ed. M. Grant and R. Kitzinger, 1249–63. New York.
Herington, J. 1985. *Poetry into Drama: Early Tragedy and the Greek Poetic Tradition*. Berkeley.
Katz, M. A. 1989. "Sexuality and the Body in Ancient Greece." *Mltis* 4:155–79.
Loraux, N. 1993. *The Children of Athena: Athenian Ideas about Citizenship and the Division between the Sexes*. 2d. ed. Trans. Caroline Levine. Princeton.
———. 1995. *The Experiences of Tiresias: The Feminine and the Greek Man*. Trans. Paula Wissing. Princeton.
Rösler, W. 1980. *Dichter und Gruppe: eine Untersuchung zu den Bedingungen und zur historischen Funktion fruher griechischer Lyrik am Beispiel Alkaios*. Munich.
Settis, 5. 1996. *I Greci: storia, cultura, arte, societa*. Torino.
Storace, P. 1996 (Oct. 3). "Marble Girls of Athens." In *The New York Review of Books*, vol. 43, no. 15, 7–11.
Vernant, J.-P. 1991. *Mortals and Immortals: Selected Essays*. Ed. F. I. Zeitlin. Princeton.
Winkler, J. J. 1990. *The Constraints of Desire: The Anthropology of Sex and Gender in Ancient Greece*. New York and London.
Zeitlin, F. I. 1982. "Cultic Models of the Female: Rites of Dionysus and Demeter." *Arethusa* 15:29–57.
———. 1986. "Configurations of Rape in Greek Myth." In *Rape: An Historical and Cultural Enquiry*. Ed. S. Tomaselli and R. Porter, 122–51. Oxford.
———. 1995. *Playing the Other: Gender and Society in Classical Greek Literature*. Chicago.

CHAPTER SEVEN

Marked Bodies

Gender, Race, Class, Age, Disability, and Disease

BROOKE HOLMES

The slipperiness of bodies is not always a laughing matter for those purporting to be their masters. Put the antics of love's victims or errant slaves onstage, however, and ancient Greek and Roman audiences could find considerable pleasure in insubordination. In one comic mime from Hellenistic Greece, a mistress, Bitinna, discovers her pet slave Gastron has been in another woman's bed. The punishment is to be corporal—a thousand lashes on the front, a thousand on the back. Gastron, begging for another chance, is stripped and bound. But at the last minute, Bitinna changes her mind and decides to discipline his body with a more lasting message, a tattoo that plays with the Socratic imperative at the heart of Greek philosophy: "Since although he is human, he doesn't *know himself*, he'll know as soon as he has this inscription on his forehead."[1]

The vignette recalls some of the most widespread and controversial ways of understanding marked bodies in recent years. On the one hand, Gastron's tattoo illustrates with uncommon vividness the inscription of the body, a popular idea in poststructuralist accounts of how subjectivity emerges from our embodied experience of sociopolitical regimes of power. These accounts have often emphasized less literal practices of inscription, such as the regulation of diet or sexual practices. Nevertheless, the image of the body as a passive surface subjected to a master discourse has proved to be a lightning rod for debates about the relationship between power, bodies, and selves. Gastron's tattoo may be seen as literalizing his subjugation to a system of power where his body is not his own. The mark of punishment is equally a sign of subjectivization.

On the other hand, marked bodies can be understood on analogy with marked terms in linguistics, that is, words or forms that depart from the default term. A classic example is gender in language: words like "lioness" or "poetess" or the feminine pronoun seem to flag our attention in a way that the corresponding terms—"lion," "poet," "he" or "his"—do not. Marked terms may expose, as feminist theorists of language have argued, the unspoken assumption that the universal, that is, unmarked, subject of language is not universal at all but masculine. If we entertain the possibility that a similar situation characterizes the classification of ancient bodies, Gastron's tattoo can be seen as signaling another kind of markedness, and, in turn, another problematic universal or norm. In Greco-Roman antiquity, the subjection to torture, whipping, mutilation, rape, and tattooing defines the slave body. In the words of the fourth-century B.C. Athenian orator Demosthenes: "If you wanted to contrast the slave and the freeman, you would find the most important distinction in the fact that slaves are responsible in person (i.e., in body) for all offenses, while freemen, even in the most unfortunate circumstances, can protect their persons."[2] At Rome, too, class and legal status determines the integrity of individual bodies. Thus, while the *lex Porcia* protected citizens from corporal punishment, slaves could be tattooed or beaten.[3] They were available to their masters as passive objects of sexual predation and abuse. Within this ideological framework, historical or mythic-historical outrages against the bodies of senators or free women play the role of exceptions that prove the perverse asymmetries of power under the Empire or outbreaks of political chaos.[4] Gastron's tattoo, then, might also be seen as materializing the slave body's vulnerability as precisely that which marks it vis-à-vis (unmarked) elite bodies.

The first of the two provisional readings I have just outlined turns on a pair of familiar binary oppositions, namely active-passive and inside-outside: the slave's body is marked by power imposed from outside. The second reading is structured by the tension between norms and deviance. If we bring these two readings together, we might be led to conflate marked bodies in the ancient world with forcibly inscribed bodies, understood either literally or as the stigmatized identities produced by systems of power. In making this connection we would have some support from the ancient evidence. For throughout Greco-Roman antiquity, the normative subject is defined, as we have already begun to see, by his corporeal integrity, embodied signs of self-mastery, and his exercise of mastery over others: all other bodies are defined in opposition to his. While Christianity, with its claim to break down oppositions between Jew and Greek, slave and freeman, male and female,[5] provides considerable resources for challenges to this axiom, the active male subject remains a remarkably stable model into late antiquity and beyond.

Such categories, however, are less clear-cut in practice, both as they are elaborated in texts and images and, it would seem, at the level of lived bodies. Marks travel across bodies. They appear and disappear in accordance with nature, habit, and time, thus blurring the line between classes of bodies and inviting us to consider the relationship between flesh, identity, and mutability. Not only deviant bodies, but also normative ones are marked by power, although the weight of necessity is unevenly distributed according to class, gender, health, and age. Marks are overdetermined, appropriated, reinterpreted.

Nevertheless, the oppositions that I just introduced (active/passive, inside/outside, normal/abject) structure the most pervasive representations and classifications of bodies that are used to discipline corporeal difference and naturalize systems of dominance in the ancient world. They contribute as well to the shape of modern theories of embodied subjects, many of which reach back to the ancient world for inspiration (e.g., Foucault's techniques of power or Bourdieu's *habitus*). The issues raised by Gastron's tattoo thus present a useful point of departure for thinking about the ways in which a concept of marking is apposite to our attempts to grasp how identity could be realized through the body in the ancient Western world.

Given the geographical and chronological scope of the material covered in this overview, together with the limits of space, I have adopted a synchronic approach in the hope that highlighting pervasive themes and problems will do more justice to the evidence than a partial catalogue organized by periodization, culture, or the categories of gender, age, ethnicity, disability, disease, and class. I begin by examining the relationship between normative bodies and the inversions they produce in the name of difference, before taking a closer look at how these social bodies are constructed, maintained, monitored, and destabilized. In the latter part of this overview, I examine in greater detail the fluidity of corporeal identity and how this fluidity affects practices of self-definition and the representation of others.

NORMS AND VIOLATIONS

To define marked bodies as deviations from bodies taken for granted, presumed inviolate, or extrapolated into universality, brings us face to face with a core set of concepts in the production of ancient identities within elite discourses. These discourses both require and create a normative subject: free, male, leisured, in the prime of life, healthy, and native to the geographical zones whose climates uniquely foster Greekness and Romanness. This normative body, while a statistical rarity, is the yardstick of everyone else.

Nowhere is this natural norm more productive of classes of deformity than in Aristotle. Given his role in developing ideas about nature that shore up cultural norms for later antiquity and the Western tradition, a brief examination of his taxonomy of bodies will be useful. Women are maimed men; children are dwarves; the elderly are near-corpses.[6] Barbarians are like slaves, slaves are like animals.[7] The association of corporeal difference with norms and abnormality is the legacy of Aristotelian taxonomy. The pseudo-Aristotelian *Problemata,* for example, explains the curly hair and "bandy legs" of Ethiopians on analogy with planks "warped" by the sun.[8] Another staunch defender of nature's will to perfection, the second-century A.D. physician Galen, put it like this. Achilles is beautiful absolutely; an ape can never compare (although he must be compared).[9] While an ape may, in fact, be beautiful qua ape, in the end he remains a grotesque double of Achilles. And yet, that ape is similar enough to Achilles to serve as his anatomical surrogate (Galen vivisected apes). Thus, while, on the sliding scale of Aristotle's biology, all difference is deterioration, bodies are surprisingly interchangeable.

For Aristotle, monstrosities such as slaves and women are natural in that they have a purpose, namely to deal with *all* bodies—including those of free males—and their reproduction. It is important, however, not to conflate the functions of these deviant types. "The female and the slave are distinguished in nature," Aristotle tells us, "for she . . . makes each thing for a single use."[10] Keeping bodies in their proper places is part of being a master: while women bear young, slaves are meant to labor on behalf of their masters, run their households, and attend to their physical needs, a muted symbiosis made uncomfortably clear in the case of one Domitius Tullus, a wealthy paralytic who complained of having to lick the fingers of his slaves every day when they brushed his teeth for him.[11]

Other monstrosities, however, such as deformed babies, are deemed by Aristotle to be useless to the community and would be exposed by law in an ideal state.[12] What we know of the fate of Roman slaves themselves in need of care indicates that their masters shared Aristotle's pragmatism. The Roman moralist Cato recommends selling off sick and old slaves along with worn-out tools and oxen, a practice that looks almost humane in light of the reality intimated by a first-century A.D. Roman law requiring masters to abandon, rather than kill, their infirm slaves.[13]

Thus, in dividing bodies into human and subhuman, Aristotle also ranges them according to functions, the highest function being the exercise of reason, the lower ones having to do with the labors of the body (and the lowest "function" being the lack of functionality altogether). As a result, as there are two ways of being beautiful, there are two ways of being healthy. Menstruation, for Aristotle, is the mark of a healthy female body. With other medical writers, he saw amenorrhea as a sign of a blocked uterus or constricted vessels—dangerous and potentially deadly conditions.[14] Indeed, most female ailments were blamed on their reproductive system until the Hellenistic and imperial periods, when physicians began to fold female bodies more snugly into the male model, without abandoning the importance of the uterus.[15] But while menstruation is healthy from one perspective, it also indicates to Aristotle an essential difference between male and female bodies. Unlike men, women are unable, for a variety of reasons, to concoct excess blood, a residue of nourishment, into seed, with the result that the excess is evacuated as menses.[16] For Aristotle, menstruation is also, then, a symptom of the weakness of the female body and its inability to contribute anything but matter—the seed being responsible for delivering form—to the embryo.[17] This weakness, associated with formlessness and the rule of the passions, expresses the natural inferiority of women. While Aristotle's views on menstruation and conception were not universally accepted, his naturalization of women's need to be ruled essentially was. "Women," writes the elder Seneca four centuries later in Rome, where, as in Athens, agency was phallic, "were born to be penetrated."[18]

On the one hand, then, all bodies should be *capable* of performing the functions deemed natural to them, with the most vulnerable (infants, slaves) becoming expendable should they fail to do so. Ideally, this capability should be evident. Soranus, a physician from the early imperial period, offers a checklist for parents to determine if a child is worth rearing: he should have a strong cry; he should be complete, and not sluggish, in all his parts; he should not have any orifices obstructed; he should have an appropriate size and shape; his joints should bend; and his entire body should be sensitive to the touch.[19] Another imperial-age physician, Rufus of Ephesus, wrote a book on buying slaves, with instructions on how to detect potential liabilities.[20] What matters is use value: the *Digest* of Justinian, a Roman legal compendium from late antiquity, declares that a slave with one eye or one jaw larger than the other is healthy, so long as he can use what he has properly.[21]

On the other hand, like the female body, even the healthy slave body should give evidence that confirms its (lowly) position on the scale of beings. One Greek treatise on physiognomy (see below) correlates corporeal signs, such as immobile or hunched shoulders, with a naturally slavish soul, and Aristotle himself believes that nature would like to distinguish the bodies of the free from the bodies of slaves.[22] Whether nature always gets what it wants is another story, to which we will return.

The corollary to naturally inferior bodies is the master's body. For one of the most important tasks of the normative male subject is to exercise control over those beneath him (women, slaves, children), just as the soul is entrusted with the control of the body.[23] This task requires a uniquely male bodily constitution characterized by a robust innate heat and a tendency toward dryness, qualities that produce signs, such as beards or hairy chests, that mark men out as natural masters actively fulfilling their function as masters. The paradox of unhealthy health represented by deviant bodies disappears. Instead, signs of weakness or passivity, such as a "womanly" knock-kneed walk,[24] indicate that a man is *failing* to fulfill his proper function. He thus becomes another pathological specimen, but a troubling one, given that his passivity is unnatural. The unmanly man forfeits his innate right to *transcend* his body in the exercise of reason and power. Consequently, Aristotle speaks of him as corrupted by body, that is, a man in whom the intentions of nature have come to naught.[25]

To return, then, to the opposition between marked and unmarked: Aristotle clearly takes the male body for granted as a universal and assumed norm, just as he takes for granted the masculine subject behind the ostensibly generic word *anthrôpos*. In this he was not alone in Greco-Roman antiquity. Ann Ellis Hanson has trenchantly suggested, for example, that early medical writers thought the womb wandered through a woman's body—causing a host of debilitating symptoms—because it had no home in a man's.[26] And we could ask for perhaps no clearer illustration of the Romans' social and economic investment in exempting the non-free from the human norm than a custom associated with the festival of the Compitalia. Households, hoping to slake the thirst of the underworld gods for real bodies, would represent their members with two classes of woolen effigy, corresponding to the two classes of family members: dolls for the free persons, balls for the slaves.[27]

We saw earlier that unmarked terms in language tend to suggest the idea of universals, which are challenged only when marked alternatives come to light. Yet it has also been suggested that the unmarked subject of language enjoys invisibility precisely because language may be disembodied: unencumbered by the body's needs, limitations, and particularities, the subject, especially of philosophical or scientific discourse, is free to pursue universal truths. While Aristotle is committed to the incorporeal nature of mind (*nous*), the ethical virtues and practices of reason are, for him, both grounded in and expressed through bodies. These bodies should *look* like rational agents, not like shapeless woolen masses. The body lacking in articulation is the marker of the barbarian, the effeminate, the child, the sick—those without the capacity to realize the human form fully.

There *is* no unmarked lived body in Greco-Roman antiquity, then, that is analogous to the male pronoun. However natural the right to mastery is, it must be recursively realized by free male bodies moving in a public field of vision. Moreover, one can never take it for granted in oneself—hence, the need for ethics and other techniques of self-mastery from the late fifth century onward. The normative body is not a state of being, but a collection of behaviors and visible signs. It is continually distinguished from its opposites and in danger of sliding into them.

Where do these signs come from? Are they truly symptoms of a hidden nature? Imposed from without? Forcibly inscribed or voluntarily assumed? Indelible or transitory? Fictional or lived? I would like to move beyond Aristotle at this point to take a closer look at the relationship between corporeal signs and character in other contexts before exploring challenges to the assumed isomorphism of bodies and identities.

LEGIBLE BODIES: ICONICITY AND SEMIOTICS

The idea that one's character, and hence one's nature, is realized in the public body recurs throughout archaic and classical Greek culture and well into later antiquity. Bodies in Homer's *Iliad* are ideally transparent, their appearance an index of the gods' affection. Agamemnon looks to Priam "like a kingly man," while the rabble-rousing Thersites is bowlegged and hunched.[28] In Book 13, the Cretan fighter Idomeneus uses the example of an ambush as an occasion where the coward and the brave man "show themselves clearly."[29] The heroic body proclaims control. The coward's skin, on the other hand, turns color; his pounding heart is mirrored in his shifty feet; his teeth chatter. The somatic type of the barbarian on the Attic stage is characterized by cowardice and lack of restraint, while those portrayed as fearful on vase paintings encompass the familiar range of Others—women, children, foreign peoples, and the elderly.[30]

Idomeneus's idea that certain situations are tests of character suggests, however, that the corporeal surface is less than transparent. Moreover, one of epic's greatest heroes, Odysseus, is known for violating the principle of iconicity with his talent for disguise and transformation.[31] Nonetheless, the commitment to the Iliadic ideal soldiers on in the aristocratic ethos as the principle of *kalokagathia*—the fusion of a beautiful body and a noble character—and its opposites.[32] Victoria Wohl has argued that in the fifth century, *kalokagathia* informs the idealized self that structures the collective identity of the Athenian citizenry as masters of empire.[33] To the extent that this aristocratic corporeal ideal succeeds as a point of psychic identification for the non-aristocratic Athenian citizen (e.g., the cobbler or the farmer), Wohl suggests it conceals the very real class differences that troubled the democracy. The idealized body of the citizen also gains definition through being opposed to others, such as the Persian or, as the citation from Demosthenes at the beginning of this chapter suggested, the slave.[34]

In the latter part of the fifth century, *kalokagathia* gains support as a contributor to ethnic self-identity from naturalizing medical theories that look to material causes, such as the hot and the cold, rather than divine favor or disfavor, to explain the alliance of character with appearance. The author of the treatise *Airs, Waters, Places* argues that both the characters and the bodies of whole races are determined by their common environment and transmitted through heredity, a theory that helps secure what Benjamin Isaac recently called proto-racism in the ancient world.[35] Unchanging and wet climates are especially damning. The medical author argues, for example, that people dwelling near the Phasis river have fat, lumpy bodies, just like the local fruit, which is stunted and "womanish" (i.e., porous and soggy).³⁶ The Phasians, like the Scythians described several chapters later, are mirror images of classical Greek sculpture's taut, muscled men.³⁷ Their ill-defined bodies allegedly signify laziness, cowardice, and thick-wittedness. Natives of less temperate climates exhibit, on the contrary, tensile bodies, along with courage and intelligence. In the environmental determinism adopted by Aristotle and, later, Hellenistic and Roman writers, the right to empire comes to be explicitly underwritten by the respective climates nurturing Greeks and Romans; other ethnic groups are thought to grow up in lands that breed slavishness.³⁸

Yet whether bodies conform to the categories of the master and the mastered because of the gods or nature, the principle of total conformity is rarely upheld outside of the shared fantasies that sustain ideological commitments to dominance. In practice, extrapolating character from bodily features requires a semiotics. If, in archaic Greek poetry, the weightiest sign of a community's health is the son who looks like his father,³⁹ the art of physiognomy is an open acknowledgment that no one, so to speak, much resembles his father. The author of the pseudo-Aristotelian *Physiognomica*, our earliest extant physiognomic treatise, written in the Peripatetic tradition in the late fourth or early third century B.C., adopts the premise that the body, and especially the face and the eyes, are in fundamental sympathy with the soul. Hence, they will betray its true nature and dispositions.⁴⁰ From that premise, he develops a classification of corporeal signs according to which any individual may be judged as healthy or sick, that is, more or less *fit to rule*.

Physiognomy breaks the iconic male body down into its component parts and assigns them values. Lapses in masculinity are weighted as feminine, for in both the Greek and Roman discourses of elite male self-fashioning, anxieties about class and ethnicity tend to find expression in the language of gender. The taxonomy of bodies is thus formalized as a continuum of decreasing masculinity, which Maud Gleason has described as "an achieved state, radically underdetermined by anatomical sex."⁴¹

At the same time, every body contains elements of its categorical opposite, so that "something masculine" and "something feminine" can be found in both male and female bodies. As a result, masculinity is rarely diagnosed by a single sign. Rather, the physiognomist uses all the signs available to construct an "overall impression" (*epiprepeia*). He then views this artful edifice as if it were a "seal of the whole," although he may concede that this seal "does not provide an account in its own right, but each sign in itself, both those in the eyes and the others, together comprise the whole appearance of the man; for when all these have been gathered together the reliable truth emerges."⁴² By determining the prevailing quality, a physiognomist can identify true masculinity.

Bodies are basically incoherent sites of identity, then, to the untrained eye. Another way of putting this would be to say that they become legible only in the presence of readers. And given that elite masculine identity is so tied to the exercise of power, it comes as no surprise that in practice diagnosing manliness was a high-stakes game in the agonistic public arenas of the ancient world.[43] In a fine fourth-century B.C. specimen of political sabotage, the rhetor Aeschines accuses a rival, Timarchus, of indulging in excessive pleasures and, more specifically, for-hire, passive homosexual sex, the most feminizing transgression in ancient Athens.[44] In his speech, Aeschines recalls a time when Timarchus threw off his cloak in the assembly. The body revealed was so misshapen by debauchery that those present covered their faces in shame, performing the emotion that Timarchus himself could no longer muster.[45] But whereas Aeschines assumes a body laid bare before a group of citizen judges, the physiognomist always suspects dissembling: "As to those who are not able to keep their eyes straight or their brows level, but have trembling in them along with a look that moves slightly—these are androgynous, but make an effort to be men."[46] The surface of the body is thus the site of both the truth and a cover-up perpetuated as a form of self-defense in a hostile and charged public arena.

Dissonance between the inside and the outside of a body is itself cued as feminine in the Greco-Roman, and later the Christian, traditions, from Pandora, the original "beautiful evil," to the made-up woman of Ovid's elegies, whose creams and paints screen the lover from the naked and unbearable truths of the female body, to the whores who, according to Tertullian, sin against God with every circle of rouge and every lengthened eyelash.[47] Anxieties about deceptive surfaces assume some kind of human interference in the production of corporeal signs. Instead of the soul automatically marking its nature onto the visible body, a wily, feminine mind gets between truth and its revelation. The practice of physiognomy means being able to sort out the false signs from the true ones. The famed Stoic Cleanthes, for example, identified a hairy man garbed in rustic gear as a passive homosexual, a *cinaedus*, by his sneeze.[48]

In the case of Cleanthes, the *cinaedus*'s attempts at masculinity may be dismissed as so many ruses. Yet other influential discourses vested in policing masculinity in Greco-Roman antiquity openly recognize elite men as active participants in shaping their bodies. Whereas physiognomy seeks to expose the scandal of two genders in a single deviant specimen, elite education (*paideia*) and its recursive enactment in ethical self-mastery presume the schism between male bodies and masculinity as a necessary gap within *all* elite men. These arts find their *raison d'être* in the belief that this innate uncertainty can be managed. Pleasures in classical Athens were to be carefully regulated, lest the male citizen become their slave and descend into the lethargy and cowardice of the barbarian. Emotions, too, required restraint. Gestures of mourning in Athenian vase painting are sharply distinguished for adult males—who exhibit coordinated, simple gestures—and women, who may be joined in their disarray by children and old men, that is, those outside the parameters of normative masculinity,[49] while on Attic gravestones, the bodies of slaves sometimes perform the emotion denied to citizen mourners.[50] Plato's *Protagoras*, in the eponymous dialogue, embeds the regulation of virtue in corporeal discipline: "People send their sons to a trainer, that having improved their bodies they may perform the orders of their minds, which are now in fit condition, and that they may not be forced by bodily faults to play the coward in wars and other duties."[51] Indeed, management of the body through diet and exercise was often seen as critical to keeping the soul in the proper condition—dry, warm, rarefied—for rational self-mastery and the mastery of others.[52]

The imperative to train oneself in masculinity appears with equal force in the medical, ethical, and rhetorical treatises of late Republican Rome and the first centuries A.D. Much attention has been paid in recent years to Greco-Roman theorists of elite education, who aim to inculcate political ideology in the register of the body through training in deportment, gesture, and voice. These writers target the physiognomic signs of masculinity not only as symptoms but also as sites of active production. The "good man" (*vir bonus*) has a steady gait; an erect, but comfortable demeanor; and a direct gaze; flamboyance is avoided at all costs.[53] Authority reverberates through a deep voice, while a high-pitched tone signals cowardice and sexual incontinence.[54] Yet this voice required training, as well as adequate diet, walks, and sexual abstinence. Failing this care, it might atrophy into the soft squeak of a woman, a eunuch, or an invalid.[55] The voice was an instrument of the soul, an index of self-control: "Shouting at high pitch . . . it has in it something unbefitting a free man, a quality that is more suited to female screaming than to speech of manly dignity."[56] Gesture, too, is subject to stringent regulations in Cicero and Quintilian, who helpfully offers a list of appropriate hand movements.[57]

At any moment, however, with a servile shrug of the shoulders, a stagy expression of emotion, or a glance that solicits or promises a favor, the public male body might be seen to lapse into the corporeal habits of its opposites.[58] Yet discourses dedicated to the self-formation of a public persona were also acutely aware that the lapse could also simply be an off-key note in the performance of masculinity. The student of rhetoric's uncanny double is the actor, whose melodrama is to be deplored but who remains a seductive model for learning how to mimic one's true nature.[59] Anxiety about deceit does not, then, disappear from the semiotics of masculinity. Rather, within the practice of self-mastery, that anxiety creates a kernel of uncertainty about which signs are natural and which are artificial.

We have seen, then, that not only is a roster of marks required to identify the normative, unmarked body, but these marks may be controlled and produced by the embodied subject. These discourses of *askesis*, that is, self-training, in both Greece and Rome are directed toward free men, who alone are seen as capable of achieving full selfhood.[60] Can we conclude from this that, for hegemonic subjects, the power to mark the body in the manner befitting a free man lies entirely within the self? In Foucault's last two published volumes of *The History of Sexuality*, which examine the practices of elite self-fashioning in Athens and Rome, he, in fact, treats the techniques of the self elaborated by antiquity's major regulatory discourses as facilitating the exercise of freedom. The very labor of this project—"intentional and voluntary"—challenges the notion of passive inscription suggested earlier by Gastron's tattoo.[61]

Classical scholars, however, have in turn contested the idea of creative self-marking, emphasizing how unstable elite masculinity functions as a political and psychic liability, as we saw was the case, for example, in Aeschines's *Against Timarchus*.[62] Moreover, it is not simply the subject who marks himself. Marking begins from infancy. Wet nurses were advised, according to Soranus's *Gynecology*, to "mold every part according to its natural shape" and to swaddle the baby in such a way as to give it "firmness and an undistorted figure" in accordance with its sex; Plato thought the expectant mother might, through exercise, form the fetus properly in the womb.[63] Soranus also includes instructions for tying the male infant's foreskin over the glans to ensure it will stay there in adult life.[64] (Both Greeks and Romans found the exposed glans unnatural and offensive—it is barbarian characters who are circumcised in Attic Comedy[65]—just as they reviled too large a member.) These external, social pressures are inextricable from acts of self-creation: Paul of Aegina, a physician active in Alexandria in the seventh century A.D., includes instructions for the surgical reduction of overly large breasts in a man, "which bring the reproach of femininity."[66] Another operation was de-circumcision, undertaken by Jews seeking assimilation in cultures that insisted on civic inspections of the young male body (Hellenistic Alexandria) and transacted business in the public baths (imperial Rome).[67] The daily life of men was presumably composed of such operations at the micro level.

The divergence of these modern interpretive strategies—Foucault's and that of classicists inspired by the work of Pierre Bourdieu and Judith Butler—faithfully reproduces the ambivalence of ancient authors regarding the scope and the power of self-invention, as we will see further. But both approaches, in any case, recognize marks of identity as unstable. In a myriad of ways, the ancient writers do, too, despite their strident appeals to nature. It is worth examining briefly the reasons they give for the instability of embodied identity, before taking a look at the implications of this instability for the materialization of specifically nonnormative identities.

LABILE MATTERS: METAMORPHOSIS AND MIMESIS

We have seen that physiognomy acknowledges the presence of both masculine and feminine elements in a given character. The stigmatized *androgunoi*, men-women, exaggerate this innate indeterminateness. Explanations of how such figures come to be reflect the fundamental sexual difference that is constitutive of human life according to the medical models that persist through the Middle Ages. Lactantius, writing early in the fourth century A.D., traces internally conflicted natures to accidents befalling the seed in the womb: "When it chances that a seed from a male parent falls into the left part of the uterus, the opinion is that a male is begotten, but since it is conceived in the female part, it suffers some female characteristics to hold sway in it more than its masculine splendor: either a beautiful figure, or exceeding whiteness or lightness of the body, or delicate limbs, or short stature, or a soft voice, or a weak mind." Female seed falling to the right side of the uterus may produce a woman marked by something masculine—strong limbs, height, ruddy complexion, a hairy face, an unlovely countenance, a heavy voice, or a daring spirit.[68] Not all ancient writers believe that women contribute seed,[69] although the left side of the womb may still cause trouble for the male fetus. Those writers who do believe in a female contribution of seed, who are in the majority in late antiquity, think that indeterminate sex may also be due to the unsuccessful mixtures of male and female seed, the latter naturally weaker and less defined.[70] Aristotle traces a number of congenital deformities to Pyrrhic victories in the womb, when (male) form does not completely master its (female) material, or when the heat necessary to imprint that form is in some way defective.[71] Female embryos are formed through similar failures.[72]

All embryos, then, take shape on the battlefield of the sexes. Moreover, the small body that emerges is resolutely composite in its mixture of different fluids and qualities (hot, cold, wet, dry), with the result that it harbors the potential for multiple identities. The very process of aging ensures that these will be played out. Most Greco-Roman writers imagine that the body grows increasingly colder with the onset of years, although they differ on whether it grows dryer or wetter.[73] Either way, aging is seen as a process of degeneration, as the body loses its tautness and form.[74] Women have something of a head start, since the ostensibly poorer quality of their flesh makes them take shape more slowly in the womb and age faster than men once outside it.[75]

The failure to maintain the body's proper constitution may also induce change. In a fourth-century B.C. medical treatise, for example, we find a case involving two widows. Many of the early medical writers saw sex as critical to keeping passages open and excess fluids moving in the female body. In the case of the widows, their celibacy is seen to result in the onset of masculine traits (deep voices, shaggy beards).[76] Under such circumstances, however, masculinity is pathological: unable to menstruate, the women die through a failure to achieve either gender.[77] The outcome is unsurprising insofar as incoherent identities, as we have seen, are viewed by regulatory disciplines as pathological, if not always fatal. Seneca, modifying the Greco-Roman *topos* that makes disease a symptom of cultural decay and extravagance, rants that by rivaling men in their drinking and partying, a generation of women has contracted male diseases (gout and baldness), their natures not so much changed as conquered by their debauchery.[78]

Both the transformation of the widows and the change to the Roman matrons' nature confirms the body as responsive to habit and practice, although with negative results in both cases. The malleability of the body is arguably even more damaging to men. When the Christian moralist Clement of Alexandria attacks men who depilate, he argues that removing hair, a key indication of a man's innate heat,[79] will, in fact, *cause* effeminacy: "If such people do not decontaminate themselves by getting rid of these embellishments, they cease to enjoy sound health and decline in the direction of greater softness until they play the woman's part."[80] Transformation may require even less work. Two centuries later, another Christian moralist, John Chrysostom, comes out against the cohabitation of male ascetics and female virgins on the grounds that feminine habits and speech might "rub off" on the men's souls.[81] Perception for these thinkers presents a particular danger, for the percipient is always at risk of being contaminated and seduced by images that are always peeling off of other bodies, according to materialist theories of perception developed in the fifth century B.C. So John worries, too, about what impact images of harlots might have on an otherwise virtuous man's soul; the concern about the mechanisms of vision is echoed by Tertullian in his arguments against women going without the veil.[82] All bodies turn out to be receptive, vulnerable, and promiscuous in their interactions with the external world. Such a world poses a threat to paternity. That the queen of Ethiopia could give birth to a lily-white daughter is explained in Heliodorus's *Ethiopian Story* by the fact that she was looking at a portrait of the fair-skinned Andromeda while the king made love to her.[83] Children, whom Galen likened to wax,[84] were particularly susceptible to imprints, not only the orthopedics of elite subject formation but also the evil eye and the poor habits of their minders, who were often foreign slaves.[85] The second-century A.D. sophist Favorinus warned of slavishness and barbarity being passed on to the child through the wet nurse's milk.[86]

The fluid relationship between inside and outside lies behind fears of mimesis in the ancient world. "Repeated imitation," Quintilian opined, arguing against the impersonation of lesser bodies (women, slaves, and so on), "passes into habit."[87] Plato famously bans citizens of his ideal city from imitating women—and the traits and behaviors associated with them, such as cowardice and grief—on stage (or off) and leaves acting to those already degraded (slaves and foreigners).[88] At Rome, actors, who were in most cases slaves or freedmen, were classed with others whose bodies were not their own (prostitutes, criminals, gladiators, slaves) and could be legally beaten. Mimes are mocked as womanish by pagans and church fathers alike.[89]

Mimesis, then, is a powerful instrument of transformation. As such, however, it might also enable desired becomings. The myth of Iphis, told in Ovid's *Metamorphoses*, is that of a baby girl spared exposure by her mother's decision to disguise and raise her as a boy. Growing up, she falls in love with a female playmate, to whom she is eventually betrothed by her unwitting father. As the marriage approaches, with its promise to unveil the sexed body, Iphis prays to Isis to make her body confirm her life's performance thus far and is granted her wish. The description of the transformation attends not to anatomy but to the public corporeal signs of masculinity—a longer stride, darker face, increased strength and vigor, sharper features, and shorter hair.[90] Iphis's metamorphosis thus locates divine authentication of her gender at the most contested corporeal sites.

The most provocative mimeses of masculinity are found in the stories of cross-dressing female saints, defiant female martyrs, and militant virgins in the first centuries of Christianity. Embodied identity was an electrified zone in the early churches. Certainly from one perspective, Christianity resolutely denied naturalized hierarchies, as well as the signs through which they materialized: Christians were made, not born. The rite of baptism, for example, stressed the dissolution of the differences (in ethnicity, status, gender) that were so central to social hierarchies in the ancient world. Like the plain and inexpensive *himation* adopted in classical Athenian democratic iconography,[91] the simple garment assumed after baptism marked a leveling of rank and thus stood in diametrical opposition to the Roman toga, a clear signifier of social and political power.[92] The sign of the cross affirmed baptism's enduring power and thus superseded the bodily mark of the covenant between God and the Jewish people, which was now seen as a false barrier between fellow Christians.[93] For Paul had declared that "neither circumcision counts for anything nor uncircumcision, but a new creation," insisting that circumcision was henceforth spiritual, rather than physical.[94] Bodies were to be transformed at the level of thoughts and desires, a domain available to all Christians. In one Coptic text, the young Mary is imagined as a model of autonomy, controlled not by men but by the "holy thought" inside of her.[95] A female slave might become a vessel of divine strength in order to stare down the wild animals of the Roman arena.[96]

In its challenge to conventional social ordering, early Christianity democratized the promise of self-mastery. Yet, even within the church, the nature of this challenge was complex. More fraught than the transformation of circumcision into a metaphor or the elimination of class differences was the question of sexual difference. Paul had declared the uniformity of baptized bodies. But when the women of Corinth saw fit to discard their veils in celebration of this androgyny, he reaffirmed the principle of sexual asymmetry—"a man ought not to cover his head, since he is the image and glory of God; but woman is the glory of man ... that is why a woman ought to have a veil on her head"[97]—and sought confirmation of the custom in corporeal signs, citing women's "naturally" long hair as prescriptive. Nevertheless, Paul seems to have allocated a significant amount of responsibility to women and worked alongside them on egalitarian terms in the early days of the Church.[98] Moreover, sexuality could be renounced, and female virgins, as well as continent widows, were powerful figures in the early churches. By holding aloof from sex, they set themselves on equal footing with men, a status they sometimes sought to confirm by rejecting the marks of their gender, as was the case with women who stood unveiled, or hoped to see confirmed by a higher power. For example, although cross-dressing was a pragmatic decision for women wishing to join ascetic communities, the transformations that it entailed were complex. For, often a woman's decision to adopt men's clothing or cut her hair, as the female saint Thecla does to follow Paul,[99] is both a prelude to, and sign of, her body's repudiation of femininity. When Hilaria, daughter of the Byzantine emperor Zeno, assumes the appearance of, first, a knight, then a monk to become an ascetic in Egypt, her initial metamorphosis is echoed by the shriveling of her breasts and her amenorrhea.[100] A martyr's embrace of masculine courage might also eliminate the telltale signs of femininity. As the day of her death approaches, Perpetua, a young citizen wife martyred in 203 A.D., dreams she becomes a man in order to wrestle dragons in the arena; her breasts stop yielding milk for her baby, about whom she ceases to feel anxiety.[101]

To become "like a man" in these cases is, from the perspective of marked and unmarked bodies, to lose gender. Yet the evidence of this achievement still requires the signs of gender.[102] If we imagine that the tattoo Bitinna wishes to place on Gastron's forehead in Herodas's mime is, in fact, the Socratic imperative "know yourself," at a basic level this is a demand, addressed to both hegemonic and stigmatized subjects, to know one's place. Knowing one's place, however, depends at least in part on the signs through which any identity is created and sustained. For even though marks themselves might circulate, appear, or disappear, individual bodies necessarily materialize only through legible marks and categories.

There remains the possibility that figures seeking to gain freedom from their gendered identities might invite the charge of teratology or criminality instead of indexing an unsexed soul.[103] Tertullian, for example, insisted that virgins at Carthage wear veils "unless a virgin is some monstrous third sex with her own head,"[104] while the *Codex Theodosianus* barred women with short hair from entering churches.[105] Tertullian's insistence that virgins wear veils, like Bitinna's desire to mark her slave-lover with a sign of her choosing, emphasizes the complex relationship between self-marking and the desire to mark others as a means of control. I would like to return to the question of the control of others, but it is first worth taking a closer look at the ways in which one can manipulate marks to affirm or to contest distinctions of gender, class, and ethnicity.

MOBILE SIGNS: FROM CLOTHING TO PLACARDS

The semiotic power of the veil and the *pallium* bespeak the importance of clothing and other forms of adornment as markers of status and identity. The manly (*virilis*) toga is the quintessential Roman garment, which could be worn only by (male) citizens. Its symbolic power was, as a result, considerable. Freedmen often chose to depict themselves and their freeborn children in togas on funerary monuments.[106] Its counterpart was the *stola*, a long, sleeveless white garment worn by citizen women over a tunic that, together with woolen bands (*vittae*) binding the hair and the *palla*, a rectangular cloth covering the head, signified the integrity of both the wearer's body and the household it represented.[107] By the end of the Republic, however, these garments were not always donned in daily life.[108] Augustus attempts to revive their use as part of his renewal of traditional Roman morality, by making it illegal, for example, for a Roman citizen to enter the Forum sans toga and establishing clothing as an integral feature of imperial iconography.[109] Freeborn children of both sexes wore the *toga praetexta*, whose purple border marked the wearer as sexually inviolate—slaves of any age were considered penetrable—and the *bulla*, an apotropaic amulet that was another popular symbol for freedmen to adopt on their children's funerary monuments.[110]

From the fourth to sixth centuries A.D., however, the basic dress template shifted together with the changing dynamics of power in the Empire; the toga and the traditional Roman tunic were gradually replaced with long-sleeved, snugly fitted tunics and leggings or trousers—precisely the garments that once marked the barbarian in the eyes of the Romans.[111] A law passed in 399 A.D. banning leggings and a certain kind of foreign boot (*tzangae*) from the city of Rome has the air of a belated and futile gesture:[112] fashions were changing in favor of the erstwhile barbarian. Of course, in the trade hubs and imperial capitals across the ancient world, cosmopolitanism among the elites had frequently favored the import over the homegrown, making a hard line between native and barbarian difficult to uphold.[113] Indeed, this line is difficult to uphold within barbarian groups.[114]

The difficulty of wearing the toga—the rhetorical textbooks provide extensive instruction in keeping them appropriately draped[115]—underscores the inextricability of clothing from deportment. Already in Homeric poetry, we can observe a relationship between class and ease of movement in female dress, with the most constraining garments reserved for the wealthiest women. Keeping a toga clean was, in itself, a considerable amount of work, and dark clothing is associated with the poor throughout antiquity, albeit for a host of ideological reasons. On the stage of New Comedy, poverty is cued by the small cloaks of slaves, which allow them to move freely in the service of others.[116] The contrast between leisure and labor was underscored further by the use of labor-intensive cloths, such as linen, and detailing for high-status garments.[117] Indeed, some of the highest markers of status in the ancient world cue the difficulty required to produce and obtain them. Here we can point to the elaborate hairstyles of Roman and late antique women, which required a fleet of hairdressing slaves (*ornatrices*);[118] Byzantine silk, a favorite luxury for ascetics to give up, was expensive and labor-intensive;[119] the purples that marked Roman senatorial dress and, from the time of Constantine, were associated exclusively by the imperial family, were notoriously difficult to produce.

At the same time, however, the interchangeability of clothing points to the contingency of social status, its ungroundedness in the body. The archaic Greek lyric poet Anacreon grumbles about a man who once went about with "buttons of wood hung in his ears for rings, and the hide of a threadbare ox scrubbed from a cast-off shield," and now parades down the street with "gold on his arms, gold on his neck."[120] (Characteristically the charge of luxury carries with it the charge of effeminacy—"like some dame in some society," Anacreon continues). Romans like to tell themselves stories of slaves who loyally put on their masters' clothing in order to be killed in their places during times of turmoil.[121] Aristophanes's *Frogs* plays on the way in which the right to corporeal integrity may be traded as a piece of clothing: the god Dionysus is beaten in place of his slave as soon as he assumes the slave's costume.[122] Clothing is key to persuasive, and indeed, too persuasive, mimesis: in late antiquity, it was illegal for actresses to dress as nuns or members of the royal family.[123]

Scent, too, is an ambiguous sign. A Roman aristocrat in hiding might be betrayed by his perfume, as in the case of Lucius Plotius, who had been proscribed by the Triumvirate.[124] Yet anyone with money could purchase the scent of power.[125] As a result, among Roman elites in particular, perfume was a topic that inspired ambivalence. Indeed, scent, which often traveled to Rome from the eastern regions of the Empire, could be denounced as barbarian and feminine, and Pliny, who reports the story of Lucius Plotius with distaste, goes so far as to pronounce such an unguent wearer worthy of death.[126] Seneca insists that in the old days, true Romans smelled like warfare, hard work, and manliness,[127] although the vast quantities of perfume bottles yielded by archaeologists suggest that few people were willing to go au naturel when it came to scent.[128] Slaves and the lower classes, on the other hand, are said to smell.

Like clothing and scent, posture and position are also fluid markers of identity. With the passage of the law of Roscius Otho in 67 B.C., for example, knights in the public theatre were distinguished by where they sat (in the fourteen rows behind the patricians); later, men were separated from women, the young from the old, and married from unmarried, and senators were given reserved seats.[129] At private Roman banquets, the arrangement of different types of bodies acts as a visible marker of status: free men recline, while slaves usually stand at attention, their taut readiness contrasted with their master's leisurely posture.[130] The flexibility of these marks of difference means that free elite men may themselves be compelled to sit or stand as a result of a loss of status. Caligula, for example, is reported to have made senators stand while he dined and dress in tunics girt at the waist—a style so indicative of low status that, according to Philo, if freeborn men served as part of a communal duty, they deliberately left their tunics ungirt.[131]

The risk here is that, pace theories of natural hierarchies, without clothing or scent or deportment, bodies run together. The "Old Oligarch," writing in fifth-century B.C. Athens, complains that he cannot exercise his citizen right to violence because he fears hitting blindly: so far as clothing and general appearance are concerned, citizens look just the same as slaves and foreigners.[132] Even Aristotle admits that the body's tendency to blur form makes free men end up looking slavish, and slavish men free.[133] The ease with which the slave body blends in is quietly attested by the degree of detail (tall, skinny, clean-shaven, with a [small] wound on the left side of the head, honey-complexioned, rather pale, with a wispy beard—in fact, with no hair at all to his beard—smooth-skinned, narrow in the jaws, long-nosed . . .) in papyri notices of runaway slaves from Roman Egypt.[134] Michele George has pointed out that stories of masters disguising themselves as slaves in Rome focus more attention on the concealment of the markers of elite identity than on the adoption of markers of slave identity.[135] While helots at Sparta were forced to wear low-class dress (animal hides, dogskin caps),[136] in societies without such sartorial distinctions, slave disguise may simply be social invisibility—the true unmarked body. And in one sense, this is how masters want it. Seneca reports a debate in the Roman Senate about requiring slaves to wear uniforms: the motion is defeated out of fear that uniforms would only make visible the slaves' numeric strength, thereby provoking rebellion.[137]

The seating arrangements at Caligula's dinner table make it clear that compelling others to perform an identity is an exercise that confirms power. In another story from the imperial biographer Suetonius, Caligula decides to stage a triumph after a highly theatrical military campaign to Germany. He rounds up the tallest Gauls, makes them dye their hair red, grow it long, learn German, and adopt "barbaric" names.[138] The performance indicates not only some key features of Roman ethnic stereotyping but also the element of performance: the Gauls can dress up as Germans. Poets and historians, too, stress costume and language in distinguishing ethnic groups.[139] Greek and Roman artists, while developing a distinctive iconography for blacks—broad nose, full lips, corkscrew hair, dark skin—rely on clothing and weaponry for other ethnic groups. The famous Hellenistic statue of the Dying Gaul, surrounded by a Celtic trumpet, belt, and sword, wears a Celtic torque around his neck and sports the moustache associated with this ethnic group. Stereotypical details in Roman representations of conquered peoples work in concert with other iconographic tricks—such as the miniaturization of the barbarian, who is thus easily trampled underfoot by the emperor, or the representations of captured bodies as bound or desecrated[140]—in order to map clearly the opposition victor-victim onto the opposition Roman-Other.

Walter Pohl has stressed, however, that in later antiquity, when relations between Romans and non-Romans are particularly unstable, "the relationship between outward signs and ethnicity . . . is less well attested than ethnographic theory assumes."[141] Pohl argues that Roman perceptions of the telltale signs of ethnic identity were rarely, it appears, shared by the barbarians themselves.[142] The difficulty of interpreting grave goods, our main evidence, in terms of ethnicity—status and age, for example, are always complicating factors—further frustrates our attempts to draw clear connections between markers and ethnic identities.[143]

If the more interaction there is between peoples, the harder it is to assign distinctive marks to each, it is another truism that distance can breed an exaggerated sense of difference. Pliny speaks of people lacking noses, lips, and tongues in southwestern Ethiopia, where, he believes, the fire of the sun is strong enough to deform bodies; the geographer Pomponius Mela offers reports of the distant Blemyes, a race with faces in their chests.[144] But, in cases of greater proximity, extant representations of ethnicity, a "culturally constructed way of categorizing people who might differ a lot among each other, and might not be so different at all from people who do not fall into that category,"[145] rely on symbolic codification as much as on corporeal and cultural difference, as we have just seen. The barbarian thus requires ethnography, iconography, and theater in order to crystallize in the cultural imagination. The fourth-century A.D. *Historia Augustae* reports a triumph of Aurelian in which the barbarian captives are paraded through the city with identifying placards around their necks.[146]

Slaves need their placards, too, as we have seen. New Comedy, a genre where Nature finally gets what it wants, uses a strict typology of masks to ensure that slaves are always recognized, for example, by their grotesque mouths, arched (read: roguish) brows, tawny hair, bulbous eyes, trumpetlike beards, and snub noses. Ancient biography, another genre that aimed to report things as they should be, described Aesop, a slave from Thrace or Syria by tradition, as potbellied, misshapen of head, snub-nosed, swarthy, dwarfish, bandy-legged, short-armed, squint-eyed, and liver-lipped.[147] Iconography supports ideology; ideology underwrites ethical categories. One can imagine that having a mascot like Socrates—ugly on the outside, godlike within—places philosophy in a complex position vis-à-vis the relationship between beauty and goodness.[148] Indeed, philosophy challenges the corporeal semiotics of normative (ethical) subjectivity in manifold ways, thereby rewriting the signs of masculinity and, in some cases, opening them up to women and slaves. To the extent philosophy does democratize signs of mastery, it anticipates the more pronounced egalitarian strategies of early Christianity, which also challenges the conventional physiognomic masculinity.[149] Yet, when Plato wants to represent the struggle between reason and passion, he draws a vivid portrait of a soul drawn by a beautiful white horse and a snub-nosed black one, crooked of frame, grey-eyed, shaggy-eared—a massive jumble of a creature with a host of anti-ideal features.[150] In such an image we come full circle to the reciprocal bind between the iconography of stigmatized identities and the appeal to nature as justifying the subjugation of stigmatized people. For this black horse is none other than the embodiment of the passion that dominates the woman, the child, the aged, the barbarian. To map "deformity" onto nonhegemonic bodies or to classify corporeal difference as degeneration is an attempt to make it obvious why such bodies need masters. In closing, I would like to examine how this mapping bleeds into violence, and to consider the multiple significations of the marks it leaves.

CONTROLLING MARKS: VIOLENCE, SPECTACLE, REAPPROPRIATION

The desire to control the representation of others is not benign, but rather participates in a complex dynamics of dominance that operates at both the ideological and the material levels. If it is difficult to pinpoint natural features of the slave, he or she may be easily recognized by the scarred back.[151] Nor is violence foreign to the production of elite male subjects. Schoolmasters at Rome were equipped with canes, whips, and sticks.[152] Indeed, it is education, *paideia*, Maud Gleason has suggested, that confers on the citizen immunity to corporal punishment, which is to say that by learning reason, rhetoric, and corporeal control, the elite male subject earns the right to stop answering with his body.[153] Indeed, upon the assumption of the manly toga, he becomes the law, which he henceforth inscribes into those without the wherewithal to control themselves.

The tattoo testifies as well to a need to mark a slippery body through force: one scholiast tells us that slaves were tattooed with the phrase "Stop me, I am running away."[154] Bitinna's plan to first beat, then tattoo Gastron is designed to compel his self-knowledge: either the very act of forcible inscription is sufficient to remind him of his low status, at which point the words "know yourself"—if indeed these are the words inscribed—become redundant. Or, if Bitinna inscribes the price she paid for Gastron (three minae) on his forehead, she undoes her act of "making him a man"—the way she describes her taking him qua love object—in the name of reasserting his identity in purely economic terms. In both cases, the mark is a strategy to keep Gastron from getting away again. Bitinna's fear that he might escape her grasp belies Cassandra's description of the captive slave, none other than herself, in Aeschylus's *Agamemnon*: "a small thing, lightly killed."[155]

Yet who reads the signs marked on the body? Kafka's famous short story "The Penal Colony" describes a punishment machine designed to inscribe a criminal's sentence into his skin over and over until he "deciphers it with his wounds." Ancient masters had their own way of marking sentences in the body (e.g., amputated hands for thieves).[156] But the message that lies within the imperative "know yourself" or the price tag of a slave seems addressed as much to Bitinna as to her lover, insofar as it affirms her own economic and social power. At the same time, we can see her wish to hold Gastron down as arising from a need to stabilize her own sense of self, destabilized by desire. To mark another's body would thus signify both the right to mastery and its unstable foundations, reassuring Bitinna of her power while also memorializing Gastron's capacity to slip away. The call for the whip smacks of bravura: "You will find, now, a Bitinna less foolish than you think."[157] Likewise, the perpetual failure of elite subjects to achieve masculine identity (or fully reject it) compels not only the repetitive rituals of self-mastery but also the rituals of dominance.

Force thus intertwines with desire as hegemonic subjects attempt to define themselves by corralling those who shore up those definitions. This intertwining of control and its loss flares up powerfully in the figure of the monster in late Republican and imperial Rome.[158] In the early Roman world, monstrous prodigies, such as hermaphrodites, were abhorred and exposed to die.[159] This practice appears to stop in the first century B.C., and Romans begin collecting human abnormalities for personal use. Julia, the granddaughter of Augustus, kept a dwarf as a pet; the emperor Elagabalus was said to have so many dwarves, eunuchs, and other abnormal specimens that his successor simply had no idea know what to do with them all upon assuming power and ended up distributing them to the public.[160] Plutarch reports a market of the deformed at Rome, and Quintilian makes it clear that the clientele was moneyed: "We see that some people set a higher value on human bodies which are crippled or somehow deformed than on those which have lost none of the blessings of normality."[161] A household's mastery of the exotic and the strange cued luxury. Black slaves were also valuable as status symbols. One text from the late Republic speaks of a middle-class youth pining for an Ethiopian to accompany him to the baths, while the preponderance of small bronzes and terra-cottas featuring blacks involved in household tasks suggests that households without the means for live slaves might have used such objects as surrogates.[162]

An interest in representing nonideal somatypes—the poor, the elderly, the disabled—is a defining, although still puzzling feature of art in the Hellenistic period. It is believed that for the Greeks and the Romans, blacks, dwarves, hunchbacks, and Pygmies, together with representations of them, functioned as apotropaic devices: fascinating bodies were thought to distract the evil eye, thus sparing those around them from the corrosive power of envy.[163] Mosaics of ithyphallic blacks have been found in Roman baths, which constituted particularly dangerous terrain, given the number of bodies on display and the greediness of the gazes.[164] Belief in the fascinating power of the phallus ensured its strong association with the grotesque, and "excessive" bodies often go hand in hand with untrammeled sexuality. Not only blacks but also dwarves, hunchbacks, and Pygmies are often represented with large, erect phalluses, which had also been an integral part of Attic Comedy's costume, fleshed out with a padded rump and belly and topped by a mask marked by squinting eyes, a snub nose, and a gaping grin.

Theater and spectacle are, in fact, integral to the role played by abnormal bodies in the public spaces of the ancient world. Representations of dwarves dancing on Greek vases suggests that they formed part of the *aklêtoi*, the "uninvited," the disadvantaged who provided entertainment at elite dinner parties in archaic Greece by "perform[ing] themselves as physically or morally imperfect," thereby reinforcing the positive ideal of *kalokagathia*.[165] Hellenistic artists frequently depicte Pygmies in comic scripts—fighting crocodiles, for example—or engaged in transgressive sex, and we hear of Hellenistic kings keeping dwarfs and mimes on hand at court. Entertainers went out of their way to exaggerate their nonhegemonic bodies, like the clown described by Lucian who shaves his head and dances himself into contortions.[166] The popularity of the grotesque mimic and the exhibition of human curiosities at Rome, such as in the images of marvels set up by Pompey the Great in his theater for public consumption,[167] attest to the Roman desire to be seduced by marginalized bodies.

And fascination not infrequently leads to mimesis on the part of the ostensibly hegemonic subjects: men dress up as women on the Attic stage;[168] masters become clever slaves in the comedies of Plautus; Athenians adopt the costumes and ways of lie of their "barbarian" enemies;[169] Romans, emperors and plebs alike, had a passion for dressing up as gladiators. Corporeal difference dissolves in the mime, "faithless to his face,"[170] as well as in the grotesque body, which mirrors every body as monstrous.[171] Yet the deviant body is not only imitated but also abused, often in the context of public spectacle: we can recall Odysseus rallying the troops in the *Iliad* by beating the bowlegged Thersites.[172] Romans found entertainment in the production of deformed and mutilated bodies, whether in the "snuff" plays staged in the arena,[173] or in the other spectacles found there, such as gladiator fights or battles pitting wild beasts against Christian martyrs. These performances demonstrate the right of some bodies to consolidate their power through the fragmentation of others.

At the same time, it is precisely within these asymmetrical relationships of power that we find corporeal inscription being appropriated into a new signifying system. A classic case would be the violently marked body of the martyr, a dead serious mimesis of Christ—who was played by the *stupidus* on the stage of Roman mime—that treats the violation of the body's integrity as an illustration of mastery through patience: the martyr's wounds are a way of writing Christ's name.[174] The valuation of bodily position is transformed, too, in a discourse that celebrates being "low, base, prone, and exposed."[175] And Christianity gives a jolt to the signifying potential of the diseased and disabled body, which becomes the site where the new religion's power is authenticated, rather than a sign of, or incitement to, divine displeasure. Over two dozen miracles involving the blind, the deaf, the lame, the dumb, and the leprous are ascribed to Jesus in the New Testament. These signs establish his messianic credibility, as well as that of the disciples who take up the work of healing in the following decades.[176] Visual representations of the disabled are integrated into Christian narratives of miracle and salvation.[177]

The changing signification of marked bodies within Christianity underscores not only diachronic shifts in perceptions of corporeal difference but also the importance of context to the interpretation of any sign: bodies are always overdetermined sites of meaning. Moreover, context is created out of multiple, overlapping, and yet discrete relationships. Given the role of gender in thematizing Otherness, it may seem strange that the model of the hegemonic subject to which we have returned throughout this essay is a slave-owning woman. On the one hand, Bitinna's position of power is true to life. Class threads each of the categories we have explored, transforming how marked bodies were lived and represented.[178] Power and status are highly relational in many of the ancient societies under discussion.[179] On the other hand, the Fifth Mime is not written by a woman, nor is it a particularly flattering picture of Bitinna. Indeed, the stereotype of the impassioned, erratic mistress who might take out her sexual frustration on her slaves was often used to illustrate the unsuitability of women, themselves lacking in self-control, as masters.[180] Gastron's tattoo, then, also indexes his mistress's enslavement to bodily passions. Such excesses are often seen as tyrantlike when they appear in a male subject. Thus, in a scene from a Greek novel that mirrors the Fifth Mime, a master's anger at being unable to seduce his (in fact, freeborn) slave explodes into violence—a slap across the face—and claims of dominance: "Since you will not receive me as a lover, you experience me as a master," to which the defiant victim responds by calling her tormentor a tyrant.[181]

Herodas's mime reminds us of how lopsided our view of corporeal difference and identity in the ancient world is as a result of the paucity of sources written by women, slaves, dwarves, or "barbarians." We are not, however, entirely at a loss when it comes to the competing significations attached to given marks. The Roman elegists and satirists develop a vile symptomatology of female old age—white and thinning hair, rotting teeth, sagging breasts, wrinkled face, crooked eyes, pendulous belly, scrawny thighs, fetid vagina.[182] Yet in funerary portraits of older woman, sagging flesh and crow's feet seem to signify a *matrona*'s lifelong commitment to "a Republican ideal of virtue."[183] Admittedly, we should be wary of understanding these representations as more authentic, given their own participation in societal expectations about a woman's later years.[184] Yet they can help us gain a more complex appreciation of how female old age was represented at Rome. Portraits of older men, too, often wear their wrinkles proudly as the etchings of heavy responsibility and a lifetime of service.[185] We might note, too, that the ritually tattooed bodies of Thracian women or Britannic warriors were incised to mean something quite different than the degrading interpretations given to them by Greeks and Romans, as Herodotus observed (among the Thracians, "to be tattooed is considered a mark of good birth, and not to be is a mark of bad").[186]

Shifting dynamics of power and context also affect the relationship between corporeal signs and status. Eunuchs are widely reviled by Greco-Roman writers for the challenges they pose to the two-gender system.[187] While invariably classified as male, they bear the telltale signs (stiffness, a shrill voice, sickly constitution, raised eyebrows, mincing steps, shifty eyes, upturned hands) of the *semivir*, the "half-man," and thus his vices.[188] Yet, when the poet Claudian, writing in the fourth-century A.D. court of the Latin West, has a personified Roma attack the powerful Byzantine eunuch Eutropius by declaring "the majesty of Rome cannot devolve upon a degenerate," he is speaking from a seat of dwindling influence.[189] Eunuchs occupied significant positions of power in the Byzantine court, where they were often assimilated to angels, and hence, signs of sacred sexlessness, rather than monstrosity. In self-representations, such as the donor miniature of the tenth-century A.D. Leo Bible, given by a court eunuch, eunuchs mark themselves by clothing that represents their status, as well as their beardlessness—a pair of signs that conjoins their corporeal identity to power, rather than its absence.[190] A similar logic governs official iconography. In a late ninth- or early tenth-century account of a miracle in Constantinople, the author describes the Archangel Michael himself as garbed in the clothes of a eunuch court official.[191]

Eunuchs' rise to power, however, did not mean that the significance of the mark was univocal. Indeed, so ambivalent was the figure of the eunuch within Christianity that the church fathers went to some trouble to dissuade men from castrating themselves, arguing for the importance of "spiritual" castration alongside spiritual circumcision.[192] The legality of castration was also troubled. Since castration was outlawed in the Empire, most eunuchs appear to have been castrated at birth in border states before being sold into Roman or Byzantine hands. The decision of Leo I (457–474 A.D.) to uphold anticastration laws, while permitting barbarian eunuchs to be traded, attests to both the widespread desire for eunuchs and official unease with the violence required to produce them.

In the end, the mystery of what Bitinna inscribes on her lover's forehead is solved not because we are told what is written. Rather, in ostensible recognition of the festival day, she finally decides to write nothing. Pens poised for the final word, we are in a similar position standing before the bodies of the past, worried that what we write on them will be only the sterile text of a master discourse, or perhaps the text that we ourselves wish to read there, a refracted answer to the ancient imperative "know yourself."

Our own uncertainty here does not mean, of course, that the bodies themselves were left unmarked. Two skeletons preserved in Herculaneum following the eruption of Mount Vesuvius in 79 A.D., both aged forty-six, seem to tell quite a simple story. The first, Erc86, was a healthy man with thick, solid bones, a man whose body appears to have been deliberately formed through regular exercise in accordance with an aesthetic norm. Ray Laurence describes the other as follows:

> Erc27 is short—163.5 cm—with spindly flattened bones; he had acute dental problems, having lost 7 teeth, and had 4 caries and 4 abscesses painful enough to cause him to chew only on one side of his mouth. Seven of his thoracic vertebrae were fused, and display osteoarthritis caused by Forestier's disease. His body had been exposed to years of hard labor and had been worked beyond its strength.[193]

It is tempting to take the flattening of these bones and the fusion of these vertebrae as true marks, indices of the "lived bodies" that some specialists in material culture seek beyond representation and the "superficiality" of constructivist approaches.[194] And yet, in the end, these signs are no easier or more reliable to read than Aristotle. We have firm evidence of the costs of the ancient world's commitment to a hegemonic subjectivity and the hierarchies required to uphold it. The psychic costs, together with traces of the complex negotiations at both the margins of this subjectivity and its center, register equally powerfully in our texts and images, as well as in their blind spots. But as for the fertile field between semiotics and silence? The bodies have simply slipped away.

CHAPTER EIGHT

Marked Bodies

Divine, Human, and Bestial

MARGUERITE JOHNSON

The means by which human bodies possess and perform divinity involve complex paradigms in discussion with the intellect, the spirit, and the body itself. From earliest times the ancients considered the concept of the divine/human nexus, sometimes poetically and thereby subconsciously, sometimes philosophically and thereby consciously. Today, armed with a bevy of theoretical tools with which to tease out and test these cultural tenets, viewpoints, beliefs, and doubts, we can reconsider the artifacts of antiquity, predominantly writing in the case of this analysis, to explore how humans related their bodies to the divinities that governed them and the degrees to which they viewed their own corporeal constituents as possessing godlike traits. If the Greeks and the Romans thought about themselves in relation to their gods (which they did), the inevitable association between humanity and the bestial arises. This all reflects an ancient tendency to view humankind as part of a universal scale, a hierarchy, with gods clearly at the top and animals clearly at the bottom. As intermediary beings, humans were not, however, all the same, and there was a sharp series of boundaries privileging the adult male citizen as distinct from lesser humans such as women, children, barbarians, and slaves. These ideas of binary opposites that defined the Greek and Roman male by corporeal differentiations between himself and other humans and that designated the degrees of connectivity between the gods and the bestial were expressed in thousands of ancient texts to varying levels dependent, of course, on textual preoccupations. Herein are a few selections that reflect and attest to this cultural thought process.

Beginning with a canonical text, Ovid's *Metamorphoses*, the theme of corporeal vulnerability in a harsh and decidedly uncertain world is established in order to situate the ancient body in a cultural framework characterized by a divine hegemony that dominates the bodies of underlings—human and beast. These transformation stories are indicative of the power of myth to reveal many things; in this instance, the underlying anxieties concerning the human body in relation to the divine. The ancient construction of the Self and Other based on a system of contrast furnished a principal discussion point for numerous philosophical treatises, and thus the works of Plato and Aristotle are necessary complements to the stories of Ovid. While the Roman poet presents the modern reader with binary systems—active/passive, male/female, divine/mortal, human/animal, subject/object—it is the philosophers who cogently systematize them in order to, like Ovid, make sense of the world. This philosophical follow-on from the brief encounter with the *Metamorphoses* then leads to a discussion of an individual body par excellence—that of Achilles—and the inherent problems it represents; expressly the difficulties entailed in reading the hybrid body. As a body marked by an entanglement with language, Achilles's body in the *Iliad* is representative of all major analyses of superlative bodies in ancient writings, namely the nigh impossible task of interpretation, curtly placing the modern commentator back in the position of the ancient philosophers endeavoring to unravel the triad of god/man/beast.

All bodies are marked, but some more so than others, and the primary focus of this analysis is to consider those bodies marked in particularly spectacular and often menacing ways. To this end, it is imperative to consider the carnivalesque and monstrous body to further interrogate the minds of ancient men. From the animal choruses of Old Comedy and the decimated bodies of the Arena, to the bogey(wo)men foreshadowed by Archaic age writers such as Semonides, these deformed examples of corporeal abjection serve to valorize the earthly adult male, rendering him not only superior to those around him but at times an almost divine presence to the multitude beneath him.

THE BODY IN A FEARSOME WORLD: AN OVIDIAN OVERVIEW

My mind proposes to speak of forms changed into new
bodies; you gods (since you also changed them)
inspire my undertakings and bring forth a perpetual song
from the first origins of the world to my own times.[1]

The divine and mortal states of which Ovid sang in the *Metamorphoses* dramatize the vulnerability of humanity in an uncertain universe. Complementarily, divine authority is emphasized by reference to the transformative powers of the gods as early as line two of the first book. In such ruminations on the dynamic between deities and humankind that underpin almost every major narrative of the *Metamorphoses*, the poet returns again and again to the related topos of the human body, the *corpus*, in conflation with the divine and the bestial. In this sense, the text is one of the most important treatises in the Greek and Roman literary canon on the dehumanizing processes that render the body other than what it was originally.

It is the specific act of metamorphosis that underscores the fragility of the human body and renders it Other in Ovid's work, yet the very concepts inherent in the theme of transformation are symptomatic of a broader cultural awareness, Greek as well as Roman, of the body's malleability, its intrinsic openness, and its perishable nature. Vernant labels this cultural construction of the body in antiquity the "sub-body" or "sublimated body,"[2] terms used to codify the human body in relation to the divine or "super-body" and, accordingly, as a means of self-designation. In such a system, the mortal body is found to be "diminished, derivative, faltering and precarious."[3]

Ovid's *Metamorphoses* is a systematic catalogue of predominantly Greek mythology that culminates in Roman foundation myths, early legends, and history, nearing its end with an exposition on Pythagoreanism. The Hellenic origins of many of the tales of transformation attest to the existence of Greek concerns about the arbitrariness of the human condition and the ambiguity of identity long before the poet gave them a Roman voice. A Greek myth such as the tale of Tiresias, which is a superlative example of the transformation narrative, is illustrative of such psychological impulses. In the tale of Tiresias's transformation from male to female and back again, the essence of humankind's bodily vulnerability and, in turn, volatility of identity, is categorically represented and done so in direct association with divine power; an example, then, of Vernant's "sub-body" and "super-body" dialectic. Ovid mentions the story[4] but remains silent about the deity responsible for the transformation; nevertheless, the miraculous nature of the incident—like that of the metamorphosis of Philomela, Procne, and Tereus[5]—points exclusively to divine intervention. That the gods can and do subject humans to transformation anonymously is in some ways more threatening than the more direct representations of their power in the myriad of metamorphosis stories in which they openly feature. In stories such as that of Tiresias, in which he impetuously strikes at two coupling snakes and is subsequently transformed into a woman, the message seems to be that the gods are ever-present, humanity is constantly subject to their unpredictable wrath or pleasure, that one's mortal body runs the risk of sudden change or violation by omnipotent and often unseen forces.

The world of the *Metamorphoses* is one fraught with almost continual incidents of divine inscriptions on the mortal body, thus reiterating the Greek and Roman cultural understanding that the bodies of the gods are by far superior to those of humankind. So spectacular in their perfection are these bodies that mere mortals cannot gaze directly at them without severe physiological penalties such as those incurred by Tiresias (yet again), who, according to some accounts, loses his sight as punishment for accidently viewing a naked Athena/Minerva (see *Pherecydes* 3F92).[6] Others are less fortunate; Actaeon, for example, also encounters a naked goddess, Artemis/Diana, who, like Athena, enjoys bathing in earthly streams. Actaeon's punishment is transformation into a stag that is subsequently torn apart by his own hunting dogs. Callimachus's *Hymn 5* records the story in conjunction with the principal tale of Tiresias and Athena; Ovid also includes it, describing the terrible punishment of metamorphosis in vivid detail:

> She [Diana] gives to his head the horns of a living stag, bedewed,
> she gives length to his neck and she makes the tips of his ears pointed,
> she changes his hands with his feet, his arms with long legs,
> and cloaks his body with speckled hide;
> terror is also added: the hero of the house of Autonoe flees
> and in his very course he marvels that he is so swift.
> But when he saw his features and his horns in the water,
> "Wretched me!" he was about to utter: but no voice followed!
> He moaned: that was his voice, and tears streamed over a face
> not his own; only his mind remained his own.[7]

The passage is illustrative of not only the impermanency and vulnerability of the human body in connection with divinity but also the distinction the ancients made between mind and body. Actaeon's physical being experiences terrible violence and alteration—one that cruelly toys with the possibility that humankind is closer to the bestial than the godly—but his *mens* (usually mind or intellect but also one's disposition or soul) remains untouched.

BETTER THAN A BEAST, BETTER THAN A WOMAN—THE PHILOSOPHERS (AND SEMONIDES) SPEAK

The Greeks were well aware of the natural order of things and humankind's position therein. Plato, for example, specifies that among all animals, humankind is superior to the rest in terms of logistical skills, a concept of justice and an understanding of the gods.[8] Similarly, Aristotle, particularly in texts that examine ethical and political concepts, specifies that humans are fundamentally different from animals in their advanced reasoning and capacity for belief. Commenting on the views expressed in the *Politics*, particularly 1256b, Steiner discusses Aristotle's notion of the human being's privileged position on a teleological grid that is at variance to his stance in the works on naturalism:

> There is a cosmic scheme of things, and human beings are superior to animals in that scheme because only humans possess the contemplative ability that likens us to the gods. In the zoological treatises, Aristotle treats human beings and animals as existing on a natural continuum and differing primarily in degree rather than in kind with respect to capacities such as intelligence.[9]

In either Aristotelian system, mortals are superior to animals—and therefore closer to the gods—but, as the Greeks (and Romans) were well aware, humans remain eternally inferior to their deities.[10] Therefore, humanity's understanding and identification of their body, their reasoning (or mind) and their soul in relation to beasts, is a complementary counterbalance to the system of self-realization based on comparison with the divine. Such a philosophical basis is related to later systems of inquiry such as those developed by the Stoics in which animals remain beneath humanity as servants to their needs. Animals are inferior to humans because they are not naturally designed to achieve a sense of rationality and logic—qualities that again raise the human to a position closer to the gods.

Humanity's closeness to the essence of the nonhuman animal is acknowledged by both Greek and Roman philosophers in their discussions of children. Plato writes of young children and animals possessing *thumos* but not *logismos*.[11] While *logismos* denotes reasoning power and is thereby clearly aligned with the concept of logistics, *thumos* has several meanings; in the context of the *Republic*, it is not so much the "soul" that is designated but strong feelings or passionate emotions, the idea of spiritedness. Aristotle also argues that the soul of the human child is virtually the same as that of the wild animal.[12] In Stoic thought (acknowledging variations and nuances), animals are associated with "naturalness" in the sense of demonstrating reactive sensory skills in an immediate way, much the same as an infant or young child does; however, the child has the potential to acquire rational thought and also virtue, which is indicative of the significant difference between human and beast. The wildness of the child's emotions is a strong factor in the equation the Stoics make between children and animals as revealed in a report of Posidonius's views as found in Galen:

> We see [children] being angry (*thumousthai*) and kicking and biting and wishing (*ethelein*) to win (*nikān*) and dominate (*kratein*) their own kind,

like some of the animals, when no prize is offered besides winning itself. Such things are clearly apparent in quail, cocks, partridges, the ichneumon, the Egyptian cobra, the crocodile, and ten thousand others.[13]

Yet it is not only children who are associated with animals in antiquity; women were likewise compared to beasts in a long cultural tradition that existed before the teachings of Greek philosophers like Socrates, Plato, and Aristotle. Semonides of Amorgos, for example, in the seventh century B.C., composed a satire that described specific female types in conjunction with certain animals. The poem begins with the stark maxim:

From the beginning the gods made women different.[14]

The following line immediately introduces the audience to the woman/animal comparison with an evocation of the "pig woman":

One type is from a pig—a bristly sow
whose house is like a rolling hill of filth;
and she herself, unbathed, in unwashed clothes,
rests on a dung heap, growing fat.[15]

The satire continues to list six more animal exempla (fox, dog, ass, weasel, horse, and ape), all of which are negative stereotypes for women. There is one insect exemplum, the bee, and she is the only commendable creature/woman on the list. The word *gune*, used in the opening line of the satire, denotes not only "woman" but also "wife"; additionally it designates a mortal woman as opposed to a goddess.[16]

In addition to a quite substantial amount of poetry, in both Greek and Roman culture, which makes the woman/animal comparison, the concept is prevalent in numerous other literary genres from tragedy to philosophy. As children are connected with nature and thus with animals, which are deemed to be closer to it, so too are women depicted. A common metaphor for the natural (or precultural or precivilized) state of womanhood is her equation with the earth; Semonides even has an "earth-woman" (22–27) who also represents a decidedly negative female category. Nietzsche, in a piece from 1871, succinctly presents this ancient Greek attitude on the *gune* and earth:

Woman is more closely related to nature than man and in all her essentials she remains ever herself. Culture is with her always something external and something which does not touch the kernel that is eternally faithful to Nature.[17]

As a being "eternally faithful to nature," woman remains beholden to the earth and the natural forces of the universe, removed from logic (to varying degrees, depending on the writer) and existing on a continuum that privileges the divine forces, followed by men, then—in varied order—women/children/barbarians/slaves, and finally animals.[18] Aristotle, in *On the Generation of Animals*, expounds this opinion on women in a manner that clearly inspired Nietzsche some two thousand years later:

> By a "male" animal we mean one which generates in another, by "female" one which generates in itself. This is why in cosmology too they speak of the nature of the earth as something female and call it "mother," while they give to the heaven and the sun and anything else of that kind the title of "generator," and "father."[19]

Animals, lacking in logic and a system of values or ethics, make for effective metaphors to depict women in antiquity, particularly in texts defaming the female of the species. While a feminist approach to the ancient sources may emphasize the sheer glut of material that equates the female with the beast, it would be a scholarly oversight to play down the equally proportioned amount of texts that make the same analogy between the male and the beast. Rather than an overemphasis on women as animals, it is more useful to consider how the beast/human nexus operates in reference to each sex before any conclusions are drawn in relation to roles and representations of gender in antiquity.

HOMER'S ACHILLES AND OTHER "ANIMALS"

Homer's *Iliad*, fecund with human/animal similes, is arguably one of the strongest pieces of literary evidence to illustrate the male/animal analogue. While these do not focus on the physical similarities between animals and men or make the comparison on the basis of *logos*, the associations are, nevertheless, problematic and, on numerous occasions, far from unambiguously positive or negative. The famous simile that compares Achilles's men, the Myrmidons, to a pack of wolves is a case in point:

> And they, as wolves
> who tear flesh raw, in whose hearts the battle fury is tireless,
> who have brought down a great horned stag in the mountains, and then feed
> on him, till the jowls of every wolf run blood, and then go
> all in a pack to drink from a spring of dark-running water,
> lapping with their lean tongues along the black edge of the surface

and belching up clotted blood; in the heart of each one
is a spirit untremulous, but their bellies are full and groaning;
as such the lords of the Myrmidons and their men of counsel
around the brave henchman of swift-footed Aiakides swarmed.[20]

As a simile indicative of the ferocity of the Myrmidons en masse it is highly successful; the passage neither extols the men nor judges them but presents them as they are: Homeric warriors, fierce and likened to, but not the same as, wolves in their savage bloodlust. Such a simile is thereby useful in introducing the distinction between being compared to an animal and originating from one, or possessing innate animal traits: a demarcation point that is significant in the Archaic (and the Mycenaean) system of thought concerning human and bestial states. Perhaps the distinction is of further use in understanding warrior machismo in the *Iliad* in an unprejudiced way, pointing out as it does that while the beast analogy is appropriate in relation to heroes such as Achilles and is at times used to condemn a certain fighter, there is an inevitable sense of impermanency about it, as if the idea of the male being rendered animal-like indefinitely is not only a destabilizing thought but an intolerable one. Therefore, while Apollo admonishes Achilles for his treatment of Hector's corpse (Figure 8.1), comparing him to a savage lion (*Il.* 24.41), the simile highlights the hero's lack of humanity but only in relation to this instance; Achilles is not irredeemable, as his return to the world of humanity reveals in the scenes with Priam at the end of the work (perhaps most powerfully presented in the symbolic sharing of a communal meal at *Il.* 24.601–42). Additionally, one of the most commonly cited animal images used of Achilles to define his character as partly bestial does not, under closer scrutiny, define him as such at all; as Hector is about to die, Achilles speaks to him:

No more entreating of me, you dog, by knees or parents.
I wish only that my spirit and fury would drive me
to hack your meat away and eat it raw for the things that
you have done to me. So there is no one who can hold the dogs off
from your head, not if they bring here and set before me ten times
and twenty times the ransom, and promise more in addition,
not if Priam son of Dardanos should offer to weigh out
your bulk in gold: not even so shall the lady your mother
who herself bore you lay you on the death-bed and mourn you:
no, but the dogs and the birds will have you for all their feasting.[21]

Scholars intent on arguing for the bestial Achilles overlook that he cannot or does not eat Hector's flesh because he *will not*; as he explicitly states,

FIGURE 8.1: Achilles dragging the body of Hector around Troy. Mezzotint after G. Hamilton, 1794. Wellcome Library, London.

his spirit (*thumos*) and his fury (*menos*) prevent him. Through his words to the dying Hector, Achilles therefore—somewhat ironically—embraces the very essence of humanity by adhering to an intense distinction between man and beast, namely the devouring of human flesh.[22] Also overlooked is a repetition of the same desire by Hecuba, mother of Hector, and matriarch par excellence, in relation to Achilles in Book 24:

> I wish I could set teeth
> in the middle of his liver and eat it. That would be vengeance
> for what he did to my son.[23]

Yet Hecuba is no more prepared to enact her desires than Achilles, and in this respect they both anticipate the Platonic tenet that humankind differs from animals in regard to their superior logic and reverence for the gods.

Achilles is a difficult figure to assess in terms of the human/bestial nexus alone, for the coalescence of the celestial body and the earthly one is also present in him. Born of a mortal father (Peleus) and an immortal mother (Thetis),

Achilles is a complex individual whose body, mind, and soul—a hybrid mix of divine and mortal—combine to render him closer to gods than animals. In his momentary stated desire to eat human flesh raw as a verbal marker of his outrage and fury, Achilles's superhuman nature is linguistically juxtaposed to Hera, whose hatred for the Trojans is expressed by an equivalent image of omophagia (*Il*.4.34–36). Indeed, there are numerous examples of the gods' fury as well as other excessive emotional states throughout the text, for to be divine in Homer is to express outrageously disproportionate sensitivities, and in this respect Achilles demonstrates his hybrid heritage.

SOCRATES'S BEAST WITHIN AND ARISTOPHANES'S TWO LITTLE PIGS

The common Iliadic simile that compares a warrior to an animal is frequent and powerful, highlighting as it does the acknowledgment of, but also the anxiety associated with, the man/animal connection by the very blurring of this natural, hierarchical boundary. The thin veil that is sophisticated culture is not lost on Homer, who through the use of animal similes to depict raging warriors, "shows that the investments of civilization fit loosely, usually falling away in the pitch and roll of battle."[24] The collation of these distinct but associative bodies, man and beast, is, as the Homeric text reveals, regularly associated with particular circumstances such as the battlefield. Later texts deal with the emergence of humanity's animal-like nature in matters concerning anger, which is also a dominant theme of the *Iliad* (the word *menis*, relating to the anger of Achilles, being the first word of the poem) and lust. This Archaic concept is developed by the Platonic construct of "the beast within," a creature that lurks within humankind and assumes the form of dangerous desires that come in the night, as Socrates explains:

> "Those kind," he replied, "awakened in dreams when part of the soul [*psuche*] rests—the logical, tame and governing part—but the beastly part, the savage part, glutted on food and wine, bounds forth and driving sleep away, desires to set out and satisfy its own instincts. You know on such occasions it is daring enough to do anything: it unyokes and sheds every trace of shame and prudence."[25]

Socrates continues with examples of what the beast may do, including random acts of unbridled lust and bloody deeds. These emotions or "beasts" can be tamed by education, especially in philosophy, which concurrently instills *sophrosune*, "the state of soul in which intellect rules securely over the other elements."[26] Socrates later contends that the beast can be subdued by the individual's active engagement with the good or the beautiful:

beautiful things being those that subject the bestial part of our nature to the human part or, rather, maybe, to the divine [*theios*] part.[27]

The Platonic Socrates saw the human soul as possessing both a bestial and divine component and that it was this divine spark, no matter how imperfect or intrinsically ungodly, that made the human *human*.

In relation to the *Iliad* one detects a different approach to the interpretation of humanity in contrast to the divine and bestial discussions in the *Republic* and one closer to, but clearly more philosophically explicit than, the ideas that may themselves be lurking in a post-Homeric but nevertheless Archaic work like Semonides's *Poem 7*. In the latter, groups of women are described as originating from specific animal species, reflecting an "us versus them" intellectual and cultural binary endemic in the construction of identity. Within this system of "polarized oppositions,"[28] women are consistently relegated to a position between male citizens and animals, which may explain why animal metaphors in connection with them are less open to favorable or even ambiguous meanings and regularly indicative of the female's need of subjugation (by a male). Perhaps the exception to this, particularly in Greek culture, is the use of metaphorical obscenity that utilizes animal imagery to denote sexual organs. The most famous or infamous example in Greek is surely *choiros* (pig, a young pig, a porker), which translates as "cunt," and is found in Old Comedy, as evidenced in Aristophanes's *Acharnians* 773. The latter play extends the woman/pig analogy to the extent of having it "materialize" on stage in the vignette of Dikaiopolis (Just City), a poor man from Acharnia, who opens a marketplace for the Megarians and takes his two little daughters there in the hope of selling them. As *human*s, the little girls are unsalable, so Dikaiopolis bundles them into a sack and advertises them as *animals* (pigs). The daughters are advertised as pigs to be sacrificed at a religious mystery rite and there ensues a series of jokes and double entendres based on the word *choiros*. The gendered nature of the situation is overt; not only are women (or, in this instance, girls) defined by their ripe, youthful pudenda (or cunts), but they are represented as commodities, goods literally belonging to a male, in this instance their father. The association between women and pigs was cited earlier in the passage from Semonides, and it remains an iconography of signifying the female as Other, particularly in classical Greek society.[29] Yet Aristophanes is also a rich source for euphemisms for the penis that make use of a wide range of animals to evoke a particular mental image; these include the horse, ram, bull, and dog.[30] Likewise, Roman authors imbued the penis with an animalistic life of its own and were particularly fond of bird imagery to denote it, although there are scarce examples of an equivalently fulsome use of consistent types of animals as euphemisms for female genitalia.[31]

THE CARNIVALESQUE BODY—ANIMAL CHORUSES, THE COLOSSEUM, AND DIVINE MEN

Humanity's connection with the animal materializes most overtly and dramatically in two public contexts, the animal (and insect) choruses of Old Comedy and the performances in the Roman Colosseum. In regard to the former, there are varied scholarly theories as to the origins of animal choruses, ranging from fertility ceremonies, agricultural rituals, rites-of-passage customs, and/or archaic symposia entertainment. In these pre-Classical settings, as Rothwell notes:

> The species of animals were often hybrid creatures. The bull-men or cock-satyrs, for example, combine the physical characteristics of two creatures, but even some of the "pure" animals that are ridden (dolphins and ostriches) were hybrid in that as "dualizers" they did not fit any single taxonomical category.[32]

In the animal choruses of Old Comedy and the aptly named Satyr Plays, the hybrid also reigns supreme. The Satyr is arguably one of the most overt examples of the divine/man/beast dichotomy. Coming from an obscure history of origin, it seems likely that these creatures were, like the hero Achilles, semi-divine, taking their heritage from Hermes and the princess Iphthime, herself with divine connections on her father's side. Hansen notes that the "god's role here as begetter of partly anthropomorphic and partly theriomorphic beings agrees in spirit with the tradition in which Hermes and a nymph are the parents of Pan, a mix of man and goat."[33] The Satyr's profound connection to divinity, namely the god Dionysus, expressed publicly, on stage, in a ritualized context epitomizes the divine/bestial construct that underpinned both the conscious and subconscious process of self-identification of many Greeks of both the Archaic and Classical eras. Situated between deity and beast, the Satyr symbolizes humanity's precarious positioning between these two states of being; this is arguably the case for the Greeks of the Classical age in particular, endowed as they were with the alarming possibility (as voiced by Socrates) that in addition to a divine spark one possesses a not-so-divine beast. In this sense, the Satyr embodies humankind's divine/bestial duality in materialized form. As an example of a hybrid being that is, to all intents and purposes, monstrous, the Satyr functions as a cultural allegory, revealing "what the ancients feared and also found fascinating, what worried them, and what (in contrast to monstrosity) was felt to be good and normal."[34] Humanity, expressly male humanity, was involved in this divine/bestial dynamic as well, transforming a binary paradigm into a triad via the audience's understanding that, despite the conventions of the surrealism of theater, these strange creatures are men in disguise.[35] The immediacy and impact of this acknowledgment accentuates individual and collective classification and demarcation: as undisciplined and hedonistic beings, Satyrs not only remind mankind of their own rusticity, the temptations and joys of the flesh, and the delights of unbridled imbibing, they reiterate by contrast the citizen's civic and military responsibilities and potential.[36]

As members of a Greek theatrical production, the Satyr chorus resembles its animal peers in Old Comedy. It is in the context of productions such as those by Aristophanes that humanity's bestial qualities are dramatized for all to see in the form of choruses comprised of birds, frogs, and wasps. As with the deceptive reality of the Satyrs on stage—as mythical beings who are in fact men playing mythical beings—the animal choruses elicit the same response from the audience, who cannot help making the distinction between actuality and fantasy—an example of Plato's *logismos* in practice. Aristophanes's *Birds* (414 B.C.), "the finest surviving example of an animal chorus,"[37] directly addresses one of the underlying motifs of the varying constructions embedded in the myths, art, and literature that feature the mortal/bestial theme, namely the civilized/savage dichotomy. The static rusticity of the Satyrs is readjusted in the play as the birds move from a state of precivilization to one of social organization; hence the *Birds* dramatizes the concept of the "theory of progress" that began to develop in fifth-century Athenian thought[38] as explored by Plato in particular.[39] The birds onstage, endowed with speech, logic, and other vital characteristics that define humanity, varied in species and dynamic and beautiful to observe, stand as a mirror to the human audience with their own speckled individualism and personal striving for progress. These are not the birds regularly mentioned by Homer as the scavengers of human flesh[40] or the objects of sacrifice to the gods, but the bird-as-man, perhaps a clever commentary on the philosophical construct of the soul (*psuche*) as winged and/or an oblique deference to the special connection the Greeks (and Romans) acknowledged as existing between the gods, humanity, and birds, believing as they did that deities could communicate to mortals through birds.[41] The figure of Tereus highlights humanity's proximity to not only the animal kingdom (here, expressly birds) but also to the gods, owing to the famous story of metamorphosis (clearly divinely orchestrated) that underpins his mythical history. A hoopoe, once a man, Tereus functions as intermediary between birds and men just as Nephelokokkygia (Cloud-Cuckoo-Land) does between the world of the polis and the realm of the gods. Built by birds under the guidance of Peisetairos, Nephelokokkygia is an avian kingdom, ruled by a man, Peisetairos, and designed in part to challenge the gods in order to reestablish world domination by birds.[42]

Theriomorphic choruses were preserved ichnographically on Attic pottery from the sixth century onward and provide a glimpse of the powerful visual impact of not only the costuming but the movement and drama of these masquerades. An arresting Athenian black-figure *oinochoe* (wine jug), attributed to the Gela Painter, from c. 480 B.C., now held in the British Museum (B 509), shows men dressed as birds that may well resemble the later Aristophanic chorus. Heavily costumed in attire that suggests feathers, winged, masked, and barefooted, the bird-men cavort around the *oinochoe* to the accompaniment of a flute player. Such images with their evocations of the marked (human) body as bestial and decidedly alien are repeated some two thousand years later in an etching by Henry Gillard Glindoni depicting a production of the *Birds* at Cambridge University[43] in 1883 (Figure 8.2). Here the chorus dominates each side of the stage, with each individual member depicting a distinct species of bird; unmasked and wearing sandals, the chorus retains a human element that emphasizes a sense of joyful yet chaotic disfigurement as men embrace their marked bodies that symbolize their inner-beast or, more specifically, their inner-bird. The jubilant nature of these animal choruses of Greek antiquity celebrates humanity in a way far removed from the anxious tales of transformation that imbue Greek mythology and the powerful retellings of Ovid. Even Tereus, that frightening figure of Thracian alterity, unbridled violence, and ultimate object of transformation, is depicted by the comic poet as nonthreatening and humorous, as befits the genre.

The performances of the Romans, however, expressly in the Colosseum, offer a far bleaker series of bestial/human representations for the delectation of their audiences. In this situation, Vernant's "sublimated body" in all its devalued glory is on display for the Roman world to see as the "super-body" domi-

FIGURE 8.2: A performance of the play *Birds* by Aristophanes: a man is performing on a stage attended by a man with wings and a young boy; other people dressed in bird costumes are gathered around the front of the stage. Etching by Henry Gillard Glindoni. Wellcome Library, London.

nates in dual forms: the fake gods onstage and the divinely graced emperor whose imprimatur governs life and death. The stories of Ovid's *Metamorphoses* and perhaps also Homer's *Iliad* come to life in a public environment in which the Romans could engage in culturally ingrained notions of the classification of the human body based on the elite/marginalized praxis through enactments of their myths, legends, and history. The marginalized body was ripe for the imposed and very public mark of deviancy in the Colosseum, where the *corpus* of the perceived Other became the object of abjection through contact with the bestial (and sometimes the enacted divine).

The literature and art of antiquity demonstrate a fixation with mating practices far from normative; for example, Jupiter/Zeus mating with various mortal women utilizing a range of animal disguises from swan to bull. Until the opening of the Flavian Amphitheater in A.D. 80 these stories were painted, composed, and at times acted. With the Colosseum they came to life; thus, in Martial's *Liber Spectaculorum*, a book of epigrams about the events in the Arena, we read of the real-life reenactment of Pasiphae and the bull:

> Believe it that Pasiphae was joined to the bull of Dicte:
> we've seen it, the ancient tale has gained authenticity.
> Nor should long-lived Antiquity marvel about herself, o Caesar:
> whatever Fame sings, the arena presents to you.[44]

Coleman discusses the possibility of actual bestiality underpinning this epigram, suggesting that the Roman aspiration to "realize the impossible, combined with scant regard for those human lives that were deemed dispensable (chiefly slaves and prisoners of war),"[45] points to its likelihood. As the woman to whom Martial refers was clearly a condemned prisoner whose sentence evidently involved execution by being thrown to wild beasts (or *a* wild beast), the penetration by the bull was clearly anticipated to be a lethal form of punishment.[46] As a condemned body, a nameless body in Martial's epigram (save for the pseudonym), the woman is punished by being rendered no more than a *corpus*, no more than an open hole to be penetrated in the most ghastly and inhuman of ways.

Presiding over the proceedings was, generically speaking, Caesar, the closest embodiment of a god on earth, whose power was displayed in the reiteration of the natural order of things—that sublime cosmological classification that arranged living beings, from immortals down to beasts, in a stringent chain of command. In the Colosseum he was a god (of sorts):

> Every detail of the displays in the amphitheatre is given an encomiastic twist: nature—both animal and human—bows to Caesar's command, so that he can conjure up such novelties as *bestiaria* . . . and display the

unpredictable behaviour of that most exotic of beasts, the rhinoceros . . . Caesar's aura casts a spell over wild animals, who instinctively recognize both his might and his capacity to protect the weak . . . And he is a miracle-worker who can bring the stories of mythology to life . . . or even trump the canonical version . . . He gives criminals their fabled deserts . . . His protagonists surpass the feats of their mythological prototypes . . . His displays combine the paradox of birth from death . . . His technology defies what we know today as the laws of gravity.[47]

This image of an earthly ruler with divine associations, when examined in the environment that gave rise to the *Liber Spectaculorum*, has a cultural context that is inherently Roman in terms of imperial specificity. Nevertheless, its heritage has its origins in earlier Greek and Roman mythology, legend and history, which, as Boak observed (in relation to the Hellenic world), may well have begun with the "demigods and heroes" who "formed a sort of easy transition from the human to the divine."[48] Indeed, Greek mythology and, to a lesser extent, history, is characterized by such "easy transition[s]" as evidenced in the case of Philip II and, more overtly, his son, Alexander III. A famous image of Alexander the Great, for example, depicts him with the ram's horns of the god Ammon.[49] Coins such as the one in the Metropolitan Museum of Art (52.127.4) show a profile of Alexander wearing a horned diadem, with the goddesses Athena and Nike on the reverse; the two images combine to produce a powerful attestation of his postmortem deification, explicitly as the son of Ammon.[50]

While Roman legends bespeak of deification of heroes and kings such as Hercules and Romulus, respectively, historically speaking, deification became prominent only after the establishment of a divine cult to Julius Caesar at the end of his political career. Before Caesar's death, Mark Antony was his designated *flamen,* and after his assassination, the Senate consecrated him as *Divus Iulius* in 42 B.C.[51] Caesar's deification and the associated religious fanfare established a precedent for the future line of emperors beginning with Augustus, and it is in this context that Martial's book is situated and therefore must be interpreted. Martial's evocation of "Caesar" (be it Titus or Domitian) bespeaks of an established cultural, religious and political tradition instituted (perhaps inadvertently) with the creation of *Divus Iulius* during the death throes of the essentially dead republic. The title "Caesar," therefore, when mentioned in the *Liber Spectaculorum*, is imbued with divine connotations.

Unfortunately for the members of the imperial family, to be truly *divine* necessitated being truly *dead.*[52] Nevertheless, the living emperor could still flex his godlike muscles in a variety of ways in preparation for future apotheosis, and participation in the activities of the Colosseum was an excellent avenue for rehearsal of divine privilege and status. Accordingly, "Caesar" is explicitly and implicitly evoked throughout the *Liber Spectaculorum* as a constant reminder (albeit an unnecessary one) of his patronage and omnipotent presence (Coleman's "encomiastic twist"); his preeminent involvement truly rendering him as close to being divine as humanly possible. Not only does the emperor materialize mythology in the case of Pasiphae and the bull, initiating a miraculous human-bestial amalgam; he also presides over the marvel of human-divine metamorphosis, as exhibited in the following vignette:

> That bellicose Mars serves you in unconquered arms
> is not enough, o Caesar: Venus herself serves you too.[53]

As with the spectacle that was the Greek theater with its subversion of reality as evidenced in the chorus of birds from Aristophanes's comedy, the Colosseum also creates a suspension of belief by blurring the same lines of demarcation between immortal/human/bestial. While Weinreich, Carratello, and Moretti interpret the couplet as a reference to combatants dressed as gods,[54] Coleman is unconvinced: "It is hard to imagine how a gladiator's garb could be so distinctive."[55] Nonetheless, in view of the extravagance of the Arena per se, it is unnecessary to dismiss Martial's direct reference to mortals actually being dressed as gods; in fact, if taken as a testimonial, it would augment the emperor's divine status (if not his pure divinity at this stage).[56]

The living body—visceral, expressive of the inner spirit, demonstrative of social and civilized indicators of power (or lack thereof) and inscribed through culture—becomes a formidable embodiment of ultimate control when the emperor aligns corporeally with divinity. While a material "thing" and thereby a signifier of reality, the emperor's body as marked for impending apotheosis stands as a symbol of transforming reality, whereby a collective sense of "social fact" is obliterated and replaced with a new, albeit surreal, fact.[57] In this sense, a reading of the divine (or near-to-divine) body on earth as represented by an Augustus or a Titus or a Domitian recalls the phenomenon of Alexander, who presented himself as a god (at least some of the time).[58] Alexander, at times, fused his understanding of his mortal body with that of an immortal, conveying his divinity through lineage, prophecy, and, perhaps most significantly, action. To be superhuman, which is to be close to being a god, is to achieve and thereby to astonish. Alexander's conquest of virtually the known world[59] is an action that relates to the body, expressly, to the "super-body," which in turn enables a reformulation of a collective sense of reality: Alexander *is* a god.

Inherent in such systems of bodily analysis is the importance of ritual; in the spring of 327 B.C., for example, Alexander introduced the Persian practice of *proskynesis* (homage to a king by prostration). Here is the physical submission of the individual's body, the passive body, to the active, elite body. The rituals that were a part of the carnivalesque nature of the Colosseum likewise institute ritual whereby the masses witness the emperor's miracles and rites and respond with due homage. The scapegoat ritual is an obvious example of the sacred and profane rites of the Arena, involving the purging of Roman society of its corrupt bodies. In addition to Martial's *Liber Spectaculorum*, Roman historians describe the persecution of minority groups such as the Christians prior to the completion of the Arena. In the *Annals*, Tacitus discusses the public torture of Christians during the reign of Nero; he writes of the fire at Rome in A.D. 64 and explains how Nero accused the sect members of lighting it in order to deflect blame:

> To remove the rumor, Nero ascribed the guilt and inflicted the most exacting tortures against those the populace called Christians, shunned on account of their shameful acts. . . . Therefore, those found guilty were arrested, and then, upon their evidence, a vast mass were charged accordingly, not so much with the crime of arson as of hatred of the human race. Additionally, humiliations were added to their destruction, such as dying by being covered with the skins of wild animals, to be torn to pieces by dogs, or being nailed to crosses or set on fire and burned so as to serve as nocturnal lighting when the day had expired. Nero had offered his own gardens for the spectacle and was putting on a show in the circus dressed in the costume of a charioteer, mingling among the common people, or taking a position in a chariot. Accordingly, even though the criminals deserved the most novel of punishments, pity was stirred up towards them in as much as it was not in the public interest that they were consumed but in the savagery of one man.[60]

Tacitus's passage is, arguably, among the strongest evocations of the persecution of the body-perceived-as-contaminated in the canon of ancient literature. The passage powerfully evokes a world turned upside down, as befitting Bakhtin's definition of the carnivalesque landscape, in which the marked or socially sanctioned (or imperially sanctioned) bodies of the outcast are further inscribed with grotesque stigmata in the form of the debased animals they are forced to become. As Nero assumes the garb and the energy of a deranged charioteer, signifying himself as the circus master par excellence, he embodies Jupiter inasmuch as he metes out the god's celestially ordained justice. Ironically for the Romans, these debased Christian bodies are bestowed with postmortem sanctity, becoming martyrs for the cause. Even the Colosseum, once the site of public destruction of polluted bodies, became a shrine of sorts for Christians by the sixteenth and seventeenth centuries.[61] Ovid's wild world of vulnerable bodies, changing and re-forming, is compellingly brought to life by the experiences of such victimized people; people condemned by one (earthly) divinity, only to be resurrected by another.

(RE)ENTER THE (MONSTROUS) BODY OF WOMAN

The persecuted, the Other of society, has his or her "outside" marked often because the "inside" is judged to be in need of purification. This outside/inside binary is endemic in ancient thought and may be regarded as a foregrounding influence on Aristotle's outline of the Pythagorean system of opposing principles in *Metaphysics* 986a.[62] To the Greeks in particular, what is on the outside may be beautiful and desirable, but it does not necessarily mean that the inside (home of mind and soul) is good. Pandora, the archetypal first woman of Greek creation mythology, is the ultimate symbol of this dangerous dualism:

> Zeus commanded famed Hephaestus to mix earth with water
> immediately, and to add human speech
> and strength, and to make a face similar to immortal goddesses
> on a beautiful, lovely body of a girl. And Zeus commanded Athena
> to teach her handicraft, to weave the cunning web.
> And Zeus commanded golden Aphrodite to pour charm on her head
> and cruel longing and cares that gnaw the limbs;
> Zeus commanded Hermes, the messenger, the slayer of Argus,
> to put in her a dog's mind and a thievish disposition.[63]

Thus the body of Pandora, indeed Pandora herself, is a semiotic danger zone, primarily because she represents the most precarious of hybrids: the creation that has links with both the divine and the bestial. While Pandora is created as the first woman and is thereby mortal, Hesiod's emphasis on the role of the gods in her manufacture accentuates a divine presence that translates to something akin to Freud's concept of the uncanny. This combined with the decidedly bestial aspect of her mind and character, expressly a canine quality, ensures a product of ultimate despair, as the poet goes on to describe (*WD* 61–89). Pandora herself is not purified via a harmful scapegoat ritual or some form of persecution, but her descendants, women per se, are regularly subjected to derision or suspicion based on the inside/outside dichotomy and the cognitive tension it provokes among men.

While women are not tied to the human/bestial convergence any more than men are in the multitude of animal similes found throughout ancient literature, they are often associated with animals as a rhetorical means of establishing a prevalent tenet that advocates their limited mental and spiritual capacities (in comparison to men). Thus, Aristotle is confident to utilize not merely an animal metaphor to categorize a woman's physiognomy but moves on to describe her as a monstrosity (*GA* 767b), a mistake of nature that is nevertheless necessary for the continuation of the human race: the requirement of a womb for the housing of the male's super-seed, from which all life emanates. Braidotti, on the woman as monster, writes: "The woman's body can change shape in pregnancy and childbearing; it is therefore capable of defeating the notion of fixed *bodily form*, of visible, recognizable, clear, and distinct shapes as that which marks the contour of the body. She is morphologically dubious."[64] As a "morphologically dubious" being, woman regularly "lends" her gender to monstrous creatures, be they human or otherwise. From Harpies, Sirens, and Gorgons to Amazons, Maenads, and witches, women embody numerous forms of grotesque alterity.

The most fascinating manifestations of female corporeal inversion involve instances where a woman's body merges with that of a beast. Overt cases include Harpies, Sirens, and Gorgons. Harpies are mentioned by Homer in both the *Iliad* and the *Odyssey*, where they are the personification of storm winds; this concept is then developed in Hesiod's *Theogony* with the description of them as women with wings and lovely hair (a mark of goddesses and aristocrats).[65] Later depictions, however, emphasize a decidedly negative and threatening creature: birds with the faces of girls bearing long, nasty talons. In Apollonius's *Voyage of the Argo* they assume the form of divine retribution and accordingly punish the blind prophet Phineus, guilty of offending the gods: every time Phineus attempts to eat, the creatures swoop down, devour the food, and leave a stench in their wake (2.262–72). The power and fear embodied by the Harpies are evoked by a focus on their physical exploits rather than on their physical appearance[66]—an example of the monstrous body-in-action. These divine beings[67] become grosser as time goes on; in the *Aeneid*, for example, Virgil's hero narrates his encounter with them and therein ensures their most ghastly depiction:

> No more offensive monstrosity than they, no more savage
> pestilence did god drag up from the Stygian swamp itself.
> Maidenly and winged in form: the filthiest excrements
> from their paunches, and hooked hands, and pallid
> faces from constant hunger.[68]

After introducing the Harpies thus, Aeneas goes on to describe their intervention in a sacred feast shared by himself and his crew:

> Next thing, swooping down from the mountains, horribly,
> the Harpies appear, and they shake their wings with a mighty racket,
> and they rip apart the solemn feast, and they befoul everything by filthy
> contact; then, an ill-omened cry among the repulsive smell.[69]

The men attempt to fight the creatures, and, although armed, they discover that the Harpies cannot be wounded; these super-bodies are immortal and therefore indestructible. Nevertheless, they finally take their leave, except for Celaeno, that *infelix vates* (seer of ill-omen; 246), who remains to deliver a prophecy—or curse—to Aeneas. Celaeno describes herself as *maxima Furiarum* (the supreme Fury; 252), which signifies the connection between the Harpies and the Furies in both Roman and Greek mythology. Like the Furies of Greek myth, the Harpies are chthonic deities associated with an archaic justice system (as illustrated in their involvement in the punishment of Phineus).

The voice of woman is a site of contention in antiquity, particularly in Greek society of the Classical era whereby a female's silence, a mark of her *sophrosune*, is ranked among her greatest qualities. Goddesses may be outspoken, although they are kept under the control of Zeus as much as possible, and hybrids such as the Harpies are also clearly able to vocalize without male intrusion. In the case of monsters such as Celaeno and her sisters, as well as the Sirens, the fear of the feminine voice, notably the "open" or unchallenged feminine voice, symbolizes an enormous sense of male anxiety. As hybrids with animal parts, these creatures' vocalization is, to a certain extent, understood by their innate connection with the unbridled essence of nature, a topos explicated by Aristotle in his connection between women and earth. Like the portentous Celaeno, the Sirens are also associated with dangerous words. As with the Harpies, the Sirens are first mentioned in Archaic literature, appearing in Homer's *Odyssey*. According to the goddess-witch, Circe, in words quoted by Odysseus himself, the Sirens are defined most notably by their voices:

> You will come first of all to the Sirens, who are enchanters
> of all mankind and whoever comes their way; and that man
> who unsuspecting approaches them, and listens to the Sirens
> singing, has no prospect of coming home and delighting
> his wife and little children as they stand about him in greeting,
> but the Sirens by the melody of their singing enchant him.
> They sit in a meadow, but the beach before it is piled with boneheaps
> of men now rotted away, and the skins shrivel upon them.[70]

As advised by Circe, Odysseus stops the ears of his crew with wax—but not his own—and has his men affix him to the ship's mast that he may hear the enticing words of these creatures but escape death (Figure 8.3). When the poet comes to describe Odysseus's encounter with them, utilizing the direct words of the hero again, the creatures come across as "attractive and repulsive":[71]

> Come this way, honoured Odysseus, great glory of the Achaians,
> and stay your ship, so that you can listen here to our singing;
> for no one else has ever sailed past this place in his black ship
> until he has listened to the honey-sweet voice that issues
> from our lips; then goes on, well pleased, knowing more than ever

he did; for we know everything that the Argives and Trojans
did and suffered in wide Troy through the gods' despite.
Over all the generous earth we know everything that happens.[72]

Like the Harpies, the Sirens represent unguarded feminine words and like Aeneas, Odysseus demonstrates linguistic mastery over them by personally narrating the encounter. Yet it is not merely by words that the heroes overcome them; it is also imperative to demonstrate the masculine trait of self-mastery by dominating these examples of nonconforming female entities either physically or with other means. Aeneas and his men drive off the Harpies with weaponry, and even though the beasts cannot be wounded, this does not stop the poet from resorting to a diminutive battle scene in which the hero can demonstrate his corporeal prowess. Odysseus is less macho on this occasion, having been instructed by Circe on how to deceive the Sirens; his body is rendered immobile and thereby symbolically castrated. Nevertheless Homer saves the day—or the hero—by placing narratorial emphasis on Odysseus's crew, namely Perimedes and Eurylochus, who perform heroic companionship and loyalty by keeping their leader under control, despite his pleas to be set free.

The reestablishment and reinforcement of male hegemony are common motifs in myths involving the defeat of the monstrous feminine. When Perseus overcomes Medusa, the only mortal of the sisterly triad of Gorgons, he decapitates her—an act that unleashed considerable rage among her feminist successors in twentieth-century academia—an act imbued with "phallologocentric sublation."[73] Medusa's body is already grotesquely marked before Per-

FIGURE 8.3: Ulysses [Odysseus] and the Sirens. Etching by P. Aquila after G. Rossi after Annibale Carracci. Wellcome Library, London.

seus finishes it off: inscribed with scales, hair entwined with snakes, tusked, bronze-handed, and winged, Medusa is formidable in her monstrosity. Cixous reads this body as not only "beautiful"[74] but as a text of female oppression, a beheaded corpse of feminine silencing: "We've been turned away from our bodies, shamefully taught to ignore them, to strike them with that stupid sexual modesty."[75] Yet she also reads it as a signifier of the potentiality of the female's corporeal power: "body without end," or, perhaps, "a whole composed of parts that are wholes, not simple partial objects but a moving, limitlessly changing ensemble."[76]

Ovid read Medusa's body differently. He inscribed it with his own design and thereby transformed her into a "Galatea-of-sorts" to his own Pygmalion. In so doing, by reinscribing her story and removing the original monstrosity, replacing it with a latent, externally inflicted one, Ovid marks Medusa's body as a site of compassion:

> The most glorious shape
> that girl possessed, and of numerous suitors the longed for hope:
> and alone of all her parts, none was more worthy of attention
> than her hair. I have encountered a man who remembers having seen her.
> This girl, the master of the sea violated in the temple of Minerva,
> it is said. The daughter of Jove turned away and covered her
> pure face with her aegis; and lest this had remained unpunished,
> she transformed the hair of the Gorgon into snakes.
> And now also, in order to frighten her enemies, terrifying them with fear,
> as a breastplate, which she fashioned, she wears the snakes.[77]

Perseus, as narrator of Medusa's tale, may embody Cixous's "phallologocentric sublation," yet the story per se nevertheless remains a testament to the pain entailed in the manufacture of the monstrous body. Ovid is also light on the details of Perseus's decapitation of Medusa and, even more significantly, ends Book 4 with this final passage (quoted above), so as to emphasize the tragedy of Medusa's metamorphosis rather than reveling in the hero's victory over her.

Amazons and Maenads are liminal figures, their bodies existing somewhere betwixt and between reality and fantasy, sometimes real and sometimes imagined. Whereas Medusa and her monstrous counterparts are clearly designated to the world of mythology, Amazons and Maenads as conceivably factual beings represent a far greater threat to men. The Athenians in particular were obsessed with Amazons, requiring their existence in order to conquer them over and over again. Later Greek writers may have questioned the truth behind the numerous stories about them,[78] but this could not quell the fascination. It has been suggested that the essentially mythological nature of the Amazon may have originated in some modicum of reality in the form of Greek contact with Scythians, a nomadic people from the Black Sea region: Plato references Scythian women in the *Laws* (805a, 806b) to illustrate woman's capacity for fighting, and Herodotus before him describes an encounter between Scythians and Amazons.[79]

To evoke the Amazon body, the Greeks inscribed it regularly with the iconographic stereotyping of barbarians, Scythian as well as Persian. On the connection between Amazons and Persians in popular Greek thought of the fifth century B.C., Hall comments:

> The close relationship between the Amazon and the Persian is demonstrated by the way in which, after the wars, Persian details creep into the traditional type of the Amazonomachy scene, thus turning the mythological conflict into the archetype, with profound patriotic significance, of the Greeks' subordination of the Persian barbarian.[80]

Not only did the Greeks mark the Amazon body with the corporeal inscriptions of the Persians, revealed most strikingly in the artworks of the Classical era, they also marked it with a more powerful stigma in the form of the cauterized right breast. As a sign of the rejection of femininity, albeit for the very practical purpose of fighting—itself a rejection of womanhood in antiquity—the removal of the breast was interpreted as part of the Amazons' status as antithetical to men. On the reappearance of the fascination for the Amazon in Early Modern Europe, Paster makes an observation that highlights the continuity between ancient Greek and Medieval (male) anxieties concerning the corporeal oddity that is the Amazon: "The Amazons' significance as ambivalently powerful figures of aggressive, self-determining desire is epitomized by their self-mutilation."[81] In this sense, Paster's words evoke the concept of the "super-body" in relation to the Amazons, an interpretation augmented by her use of Montrose's scholarship that defines these bodies as symptomatic of a cultural apprehension imbedded in the realization that "men are in fact dependent on women."[82]

The rejection of the rigid demarcation points of gender definition also linked the Amazons with the bestial; as with their more fantastical sisters, be they Harpies, Sirens, or Gorgons, the Amazons' wild refusal to be conquered by men saw them not only as partly androgynous but as untamed animals—a concept reiterated by the emphasis in the written sources on Amazons and their horses. Added to this was one particularly awe-inspiring myth of origin: the Amazons as the daughters of Ares.[83] When combined, this series of images and histories represents a frightening Other that merges divine, bestial, and woman.

As icons of apprehension, the Amazons proved to be formidable, although they were safely situated on the edges of the world, in the region of nowhere, problematic only when they came into civilization (and were conquered). In terms of geography, a more immediate symbolic threat to the Greek male was the Maenad, for she came from within the polis. Additionally, while the Amazons were quintessentially mythical in nature to varying degrees depending on source context and historical era, the Maenads were real, as Reeder explains: "In contrast with Nymphs and Silens, who are semi-divine members of Dionysos' retinue, a Maenad was not envisioned as a permanent state of being but rather a mortal woman in a temporary state of madness."[84] As with the Amazons, there exists in the construct of the Maenad the tripartite division of being: divine, human, and bestial, for although they do not come directly from a divine source as do the Amazons, the Maenads possess a godly energy in the form of Dionysian-inspired madness. As women in the service of Dionysus, the Maenads experienced a mode of the ecstatic sublime in yearly festivals in honor of the god; to render it all very stereotypically, they would leave the private domain of the home; embrace their sacred group; and, under the influence of wine (one of the Dionysus's most precious gifts to humanity), would enter the wilds that beckoned beyond the city limit. Then, through the agency of wine and, esoterically speaking, divine possession, the women embraced the god and nurtured the beast within—the latter demonstrated by adornment in animal skins, unrestrained and wild movements and dancing, the handling of wild creatures including snakes, and the practice of omophagy. As a frightening example of a marginalized entity as evidenced in the preceding scenario, the Maenad is the living signifier of the woman-as-nature dynamic, presenting an image of a divinely insane devotee whose body challenges attempts to inscribe it by unambiguous and straightforward means. The process of decoding this type of body becomes, therefore, one characterized by hyperbole and fantasy that sometimes verges on sexual fascination, best illustrated by Euripides's *Bacchae* of 406 B.C. While the Romans attempted to legislate such larger-than-life bodies out of existence in rulings such as the *senatus consultum ultimum* of 186 B.C., which prohibited Bacchanalian rites, the Greeks' preference was to duly respect the female rituals in honor of the god while simultaneously remaining in a state of hyperanxiety about them—at least in their writings and artwork. In reality, the Maenads of Greece and the Bacchantes of Rome may have been far tamer and, as scholars have noted, an individual woman's involvement in the cult, expressly in the Greek world, provided prestige for her family. In this context, being a Maenad offered the Greek woman a means of escape, albeit temporarily, from a rigidly controlled life, just as the fantastical stories (composed, painted, and sculpted) about what she was doing functioned as an outlet for worried men.

Men did worry, largely because they invented different types of women to worry about. If they were not worrying about Maenads or Amazons, they could always worry about witches. While historical artifacts reveal that a higher percentage of men practiced magic than women, men remained undeterred in their insistence on the prevalence of the wicked witch in both Greek and Roman societies. Witches were not marked by the Devil in these societies because they had not invented him, and it was not until the predominance of Christianity in Europe in the early Middle Ages that he was able to make his presence felt *on* and *in* the bodies of such depraved women. Nevertheless, the witch's body was stained in antiquity by other means. While Circe was marked by her beauty in Homer's *Odyssey*, the witches who succeeded her were not so fortunate, inscribed as they were with powerful stigmata that aligned their bodies with animals and immersed them in contagion.

Among the most infamous depictions of witches in Greek and Latin literature is the portrait of Medea in Euripides's play of 431 B.C., and the grotesque portraits by Horace and Lucan several hundred years later. In all three contexts, the witches that emerge are characterized by varying associations with the divine, human, and bestial triad. In her long mythological history, Medea is regularly defined as the granddaughter of the Greek sun god, Helios, who mates with the Oceanid, Perse, to produce her father, Aeetes, and aunts Circe and Pasiphae. As the child of a mortal mother, Medea, like Achilles, is a heroic hybrid—a status that, again like his, is the source of much of her troubles. As Achilles's hybridity marks him and thereby defines him as outside the normative, Medea's does the same, restricting her to an existence between two worlds, that of gods and humankind. Added to her mixed genealogy, Medea is trapped between the barbaric and the Greek; originating from Colchis and travelling to Hellas, Medea's body is consistently reduced to an object of suspicion if not loathing; in short, she has the worst of it: semidivine, female, witch, and barbarian.

Euripides presents all four of these competing constituents in his portrait of Medea and adds conventional animal and nature similes, such as the comparison to a lioness (187, 1342) and the numerous references to the sea (such as 768–70), to accentuate her *thumos* (animal imagery) and her predicament (sea imagery in the preceding lines). Euripides also makes productive use of monsters to develop his heroine's character: she is a Fury (1260) and also Scylla (1342–43; here there is reference to both lioness and monster).[85] She is both natural and unnatural—thus embodying the conflicting states of womanhood in Classical Greece. As woman-as-unnatural, Medea is a killer, destroying numerous bodies, including her sons in pursuit of survival, success, and revenge.

Part of her unnaturalness is her semidivine disposition; her body is marked by the actions or "performances" of the gods, best illustrated in her exodus at the conclusion of the play: on the roof of the *skene*, sitting in Helios's chariot pulled by dragons, with the corpses of her boys, Medea enacts the terrifying "super-body" of the divine, powerfully evoked by contrast to Jason's "sub-body." Her body has, in the past, been inscribed by the language of others, the Nurse, Creon, Jason, and the Chorus; in the exodus it is reinscribed by Medea herself as she eliminates her objectified body and becomes the self-imposed subject. This powerful reassertion of the Self-as-Subject is perfectly encapsulated by her maxim in Seneca's tragedy: "Medea nunc sum!"—"Now I am Medea!"[86]

The ancients' preoccupation with witches is partly based on what they perceive to be the innate differences between their bodies and the normative bodies of other women. As the alleged witches of Early Modern Europe were supposedly marked by the Devil, the witches of antiquity were also marked, albeit by different corporeal signs. Medea's body is marked by cultural signifiers, all of which point to her outsider status, but her successive sisters-in-evil are designated rather more crudely by overt deformities of the body that stress their intrinsic mortality as well as their social alienation. Horace's witches are a case in point. In *Epode 5*, Horace describes some of the most infamous witches of Western literature:

> Canidia, hair entangled with short vipers
> and disheveled head,
>
> demands wild fig trees ripped from sepulchres,
> demands funereal cypresses,
>
> as well as eggs and feathers of nocturnal screech-owls
> smeared with the blood of a hideous toad
>
> along with herbs, which Iolchus and Hiberia
> fertile in poisons despatch,
>
> and the bones ripped from the jaw of a starving dog,
> to be burned in Colchian flames.
>
> Meanwhile, unencumbered Sagana, sprinkles
> water from Avernus throughout the house,
>
> with hair that bristles like a spiky sea urchin
> or a wild boar.
>
> Veia, held back by no conscience,
> with a strong mattock dug at the
>
> earth, groaning over her labours.[87]

Both Canidia and Sagana are marked with bestial attributes, with hair that recalls not only the uncivilized nature of animals but also the snake-bedraggled Medusa. They are *in* and *of* nature, separated from society and intent only on the use of the organic world to implement transformation in the form of magic. In their grotesque way, especially Canidia, the witches are a deformed version of the divine "super-body," able as they are to affect change. The sheer power Horace attributes to these deviant bodies is indicative of the cultural abhorrence toward them; as the Romans outlawed the Bacchanalia some 150 years earlier, so too did they attempt to discipline the bodies of magic practitioners via a series of new laws under the Augustan regime. To reinforce the public dislike of witches, Horace conjures up a frightening brew of images and metaphors to depict this most revolting icon of the subaltern, to which he adds a series of rituals that further damn the already damned.

The body's enactment of rituals is a decisive marker of deviancy; hence the emphasis the poet places on the actions of the witches in *Epode 5* (they are, to bowdlerize Butler, performing witchhood) (Figure 8.4).[88] Herein, Horace tells of Canidia and her cronies' capture of a well-bred Roman youth, his burial to the neck, starvation, and resultant demise. The object of the rite hinges on the belief that once deceased, the victim's innards, suitably scarred and thereby magically enhanced by the tortuous death, can be utilized in an erotic potion. The description of this act of deviancy underscores the elite/nonelite political message of the piece; Horace is keen to champion the power of the youth despite the fact that he cannot fight back or answer with his body. As with the Greek preoccupation with the "silence" of animals—and, to a lesser degree, women and children—Horace advocates and indeed celebrates the victory of the speaking male subject by giving the youth the "last say" in the form of his own curse, which verbally marks the bodies and the blackened souls of the witches with an indelible stain. The boy's curse concludes with a powerful threat:

> The throng will crush you—you filthy hags—pelting you with stones
> this way and that, throughout the streets;
>
> afterwards the wolves and birds of the Esquiline will scatter
> your unburied limbs,
>
> and my parents, who—alas—will survive me,
> will not miss that sight.[89]

The fear of witches was a real one, and while they could be dealt with by legislation, ostracism, and, in the case of *Epode 5*, counter-cursing, it could also be laughed away. This is exemplified by Horace's lighthearted treatment of Canidia and Sagana in *Satire 1.8*, in which a flatulent statue of Priapus causes the witches to take flight during another nocturnal ritual. As the curses of the youth represent the power of language to overcome the Other, the

FIGURE 8.4: Three old hags surround a basket of newborn babies, with bats in the distance. Etching by F. Goya, 1796/98. Wellcome Library, London.

sounds of the flatulent garden god function as a form of apotropaic "sound-magic," reducing the witches to something akin to prehistoric creatures, early humans, startled by the sound of a thunderclap and running scared. Less humorous is Lucan's Erictho, the witch par excellence of Latin literature. The dominant force of Book 6 of the *Civil War*, Lucan's Neronian age epic, Erictho is essentially defined by her body and the spaces it occupies (the land of witches, Thessaly itself; graveyards; the war-ravaged, blood-soaked Pharsalian fields).

> Erictho, primarily a necromancer, has a body inscribed with death:
> The appearance of this profanity possessed
>
> a loathsome, deathly atrophy and, ignorant of the serene sky,
> her horrid face was oppressed by a Stygian pallor,

> burdened by uncombed locks. If storms and gloomy
> clouds subdue the stars, then the Thessalian
> emerges from empty graves and tries to seize nocturnal lightning.
>
> She has scorched the fertile seeds of the field with trampling and
> destroyed the breezes, not fatal, by breathing.[90]

In the works codified as Silver Latin, no better example of abjection exists than the lengthy evocation of Erictho that comprises the bulk of Book 6 (of which the preceding is but one small illustration). Erictho embodies and emits death. A walking corpse, similar in this way to those she reanimates, she is the materialization of *profanum* (6.515), that which is entirely unholy and unclean. The female's association with the profane is a regular topos in both Greek and Latin literature, reflecting, as it does, yet another cultural anxiety about the bodily site that is woman. Carson writes of the impurities that mark the boundaries of woman in the areas of "hygiene, physical and moral."[91] As illustrated by their mysterious bleeding, their Aristotelian wetness and their emission of disgusting matter during childbirth, women were viewed as "pollutable, polluted, and polluting in several ways at once."[92] Such outer signs of bodily impurity, of unadulterated abjection, transfer themselves to the moral arena, as Carson signposts, and there they become emblematic of woman's inner state of pollution. Erictho, even by ancient standards of contamination, is beyond the limits of decency.

Like her monstrous sisters, be they witches or hybrids, Erictho has her fair share of animal metaphors as Lucan depicts her ripping, tearing, scratching, biting, and gnawing human flesh (5.560–68). She, like them, comes close to a bastardized form of the divine "super-body" in her seeming defiance of death and her remarkable sorcery. Like the gods who transform themselves and hapless others, this super-witch also possesses the power to effect physical change on her environment and on random humans. Her polluting power is evidenced in the infamous necromancy scene, where Erictho reanimates a soldier recently slain during the civil war (6.588–830). Her violation of others is also seen in earlier passages in the book, where Lucan describes her misuse of human bodies to secure parts for various spells:

> Nor do her hands refrain from slaughter, if living blood
> is needed, as when a throat first slit gushes,
> and if her ghoulish feasts demands palpitating entrails.
> With a wound thus to the belly, the birth, which nature has not called forth,
> is dragged out to be placed on the fiery altar;
> and whenever the deed demands a savage and bold shade,
> she herself makes the ghost.[93]

> Every death of a human is utilised.
> That woman has torn at the bloom of cheeks belonging to a youthful body,
> that woman has cut away the hair from the dying youth with her left hand.
> Often, even, the cruel Thessalian, during a funeral,
> has brooded over the loved one and, kissing the face, pierces it
> and has mutilated the head and, pressing with her teeth
> she has relaxed the closed mouth and,
> biting the tongue clinging to the dry throat,
> she has poured arcane murmurs into icy lips
> and has delivered impiety to the Stygian shades.[94]

In such scenes where the body is described as being violated, both before birth (6.554–559) and after death (6.560–68), Erictho "performs" the evil of her own body, of her own being, unleashing and keeping outside the beast of the inside. As with the Harpies, the Sirens, Medusa, the Amazons, and the Maenads, Erictho's vicious alterity is most powerfully called forth in her ability to transform others. All of these monstrous females possess and enact the power to render men polluted, near death, dead, petrified, and decapitated. They inscribe men's bodies, and men are the worse for it.

With the rise of Stoic philosophy, which pushed the boundaries between man and beast further and further apart, a figure such as Erictho seems to be a particularly striking aberration, crossing a threshold that even Achilles refrains from: the consumption of human flesh. As discussed at the beginning of this chapter, deformed bodies serve to valorize the adult male citizen, which can be witnessed in an imaginative context with constructions of fantastical figures like Erictho, and also in real-life situations, via the bodies on display in carnivalesque settings such as the Colosseum. While one may experience empowerment at the distance between Subject and (abject) Object in either circumstance, it is also imperative to consider the symbolic significance such bodies possess in reminding the citizen of his own bodily vulnerability, his precarious existence and his inevitable death. The citizen will die one day, and in this he is far closer to beasts than he is to the gods, whose bodies go on forever.

CHAPTER NINE

The Body of a Hero

Images of Herakles and Their Political Use in Antiquity

AMALIA AVRAMIDOU

In ancient Greece and Rome the representation of the human body was not a mere reflection of naturalistic form but an image laden with cultural traits.[1] Whether rendered in three-dimensional sculptures of bronze or stone, or in two-dimensional compositions on reliefs, frescoes, coins, or vase paintings, the body of the protagonists, legendary or historical, became an integral part and often the focal point of each synthesis, carrying further connotations once viewed within the larger context. In anthropocentric cultures such as the Graeco-Roman civilization, the human body became the basic form upon which the appearance and physique of gods and heroes were shaped. In a strictly hierarchical society, the body was the one common factor that equalized all, and any deviation from its normal representation resulted in the creation of monsters, beasts, and mixed creatures.

Studying the hero's body and how its form and conception develops through the centuries offers a unique opportunity to examine not only the changes in the depiction of the human body itself, but also the use and appropriation of the hero's physical and symbolic qualities. The case of Herakles lends itself to a continuous exploration of the iconography and reception of a well-known hero whose imagery and myth had a lasting impact in the society and arts of antiquity. The popularity of his image will help us understand further the significance of the human body and its multifaceted use in art.

In addition, the study of Herakles's exploitation for political purposes will shed some light on the use of the hero's body and attributes by generations of rulers. By appropriating his image, they aimed to acquire some of the hero's physical properties, such as his athletic, strong body, his virtues, and, ultimately, his immortality.

Herakles enjoys a special place in ancient art and literature mainly because of his complex status as a mortal/hero/god. He certainly has the body of a mortal man but is gifted with divine qualities, such as supernatural strength and endurance, since he was the son of Zeus, breast-fed by Hera herself.[2] The legends surrounding Herakles crafted a particular visual representation of the hero, which rendered him easily identifiable and familiar to all social strata, shaping a uniquely overarching figure between the domain of the living, the dead, and the immortal.

Herakles was perhaps the most famous hero of ancient Greece. Son of Olympian Zeus and the mortal Alkmene, queen of Thebes, he was the protagonist of fascinating myths that inspired poets and artists, and a model of athletic excellence and physical vigor to men and heroes alike. Hated by Hera since his birth, Herakles had to face the rage of the goddess, along with numerous challenging hardships that came his way. Struck by Hera and blinded by the Furies, Herakles killed his first wife and children, and to expiate his crime, he offered himself to the service of Eurystheus, ruler of Argos. This episode signals the beginning of the Twelve Labors of Herakles, a series of deeds impossible for ordinary mortals that ranged from fighting monsters and taming wild animals to visiting the Underworld and the extremes of the world.[3] Apart from his club and arrow, Herakles's most characteristic attribute became the lion skin, which he took as a prize after killing the Nemean lion.[4] Along with the divine guidance of Athena,[5] Herakles accomplished his labors and numerous smaller deeds that granted him immortality and eventually established him as a protector of mankind.[6]

Such a unique, powerful character became the prototype of masculinity, heroic behavior, and athletic excellence.[7] At the same time, his apotheosis and final ascent to Olympus associated him with funerary beliefs and rendered him a comforting figure for those about to begin the journey to the Underworld.[8]

Even though he was originally a Peloponnesian hero, Herakles's popularity exceeded the boundaries of the Greek world, and his statues, paintings, and reliefs decorated public and private spaces in large cities and smaller towns throughout antiquity. The cult of Herakles flourished all around the Mediterranean, while temples, shrines, and altars were built in his honor.[9] This chapter traces the representation of Herakles's body in the visual arts from the Archaic to the Byzantine era and focuses on the reception of his image through case-specific works, indicative of his influence, his extraordinary qualities, and their periodic appropriation by kings and rulers.[10]

HERAKLES'S APPEARANCE AND PHYSIQUE

There are numerous descriptions of Herakles's strength and courage, and both narrative and visual accounts of his labors confirm his endurance and invincibility:[11] even though he is a mortal son of Zeus, there is hardly any mention of him being wounded by an opponent.[12] Surprisingly, the sources offer rather limited and sometimes contradictory descriptions of his body and age. Herodotus (4.82) discusses a huge footprint, two cubits long, that Herakles allegedly left at Scythia near the river Tyras. A late source (Clem. Al. *protr.* 2.30) claims that Herakles lived to the age of fifty-two and elaborates on a fourth-century B.C. inscription by Dikaiarchos (frag. 54 Wehrli) that portrays Herakles as a sinewy, rather unattractive type, with a hook nose, blue eyes, and straight hair. Herakles's height is also a matter a debate in the sources: some say he was taller than most men (Paus. 5.8.8; Plutarch frag. 7 Sandbach); others report that he was rather short and stocky (Pindar 1.4. 71).[13] However, we do get a consistent description of his eyes: they were fiery ever since he was a child.[14] By the first half of the fifth century, Herakles's hairstyle becomes standardized and functions as another of his attributes. The short, curly locks of hair decorating his forehead and the matching style of his well-trimmed beard render him easily recognizable from other heroes.[15] Modern scholars have questioned Herakles's masculinity based on certain episodes where he puts on female clothes, for example, during his service to the Oriental queen Omphale, and they have also pondered his allegedly melancholic nature.[16]

Herakles is often described in not so flattering colors as an individualistic hero whose excessive drinking, gluttony, and lack of sexual control render him almost antisocial;[17] and yet, despite all these demeaning qualities, he manages to become a symbol of power and triumph versus barbarism and even death itself.[18] This image of Herakles as a model of virtue and passionate defender of justice was first crafted by Pindar, whose hymns present a well-behaved, self-controlled hero.[19]

Herakles is mentioned by Homer and Hesiod,[20] but regrettably, poems dedicated completely to his life, such as the *Herakleia* composed by Peisandros of Rhodes in the late seventh to early sixth century B.C. are no longer preserved. It is noteworthy that until the Classical period there was a distinct emphasis on his violent, rough aspect that highlighted his performance as a warrior, monster slayer, and successful fighter. Around the beginning of the fifth century B.C., writers and artists became more interested in portraying his psychology, the influence of his deeds, and his quest for immortality. Thus the body of Herakles is treated as a canvas upon which qualities of his character and reflections of his exemplary life are painted. Since Herakles ended his life with an ascent to Olympus, he had no earthly tomb, in contrast to other Greek heroes such as Theseus, Oedipus, and Amphiaraos, thus enhancing his divine status.

One of the most interesting subjects depicted on Attic vases from 490 to 450 B.C. is the encounter of Herakles with Gêras, the personification of Old Age (Figure 9.1). A mature, bearded Herakles is shown either addressing or trying to capture Gêras, depicted as an ugly, skinny, deformed figure, either with white hair or entirely bald, in contrast to the tall, handsome, well-shaped hero. This is the most eloquent display of all the qualities that Herakles stands for and all the things he fights against. It is the pursuit of eternal youth, strength, and beauty undertaken by the single most extraordinary mortal capable of subduing Old Age itself, thus paving his way for overcoming Death as well.[21]

Herakles was a favorite character of tragedies, satyr dramas, and comedies, but his persona is thought to have inspired Ionian philosophers and later on Sophists and Cynics to take a more critical view of myth and history. The heavy-muscled body of Herakles becomes a metaphor for the strength of character.[22] In Hellenistic literature, Herakles's portrait as an invincible hero gives way to the image of an ordinary man, with passions, excesses, and soft spots,

FIGURE 9.1: Herakles and Gêras. Attic red-figure pelike, ca. 480–470 B.C. Rome, Villa Giulia 48238, *ARV*² 284.1 Archivio Fotografico della Soprintendenza Beni Archeologici Etruria Meridionale.

while in Vergil Hercules receives only limited attention, as his image was not always necessary in the promotion of Republican values.[23]

Herakles becomes a noble paradigm for Hellenistic and Roman rulers, who commission portraits and artworks, based largely upon the hero's illustrious example. Despite his popularity and the plethora of references in Latin literature, there is no detailed surviving account of Herakles's life and labors from the Roman era.[24]

HERAKLES: EDUCATION AND MORALITY

Herakles played a significant role in the education of young men, and paradoxically he was also perceived as a model of morality. From the Late Classical period and on, Herakles's image gradually changed from fighter of monsters to savior of mankind, and artists dared to represent him weary of his labors and remorseful of killing his family.[25] Herakles's contribution to a young man's education was valued more than the hero's own artistic achievements. Attic vase paintings attest to his flair for music recitals, but they are equally telling about his unorthodox relation with his teachers.[26] Such representations softened the hero's character, allowing details from his personal life to become sources of artistic inspiration, and cultivated his profile as a proponent of education. In addition, through his association with waters and fountains and because of his legendary power and stamina, Herakles served as a model for young men, and his statues decorated palaestras and gymnasia throughout antiquity.[27]

One story that had a lasting effect and carried a moralizing message is Herakles's Choice in front of the path of Vice and Virtue. In this legend, which is surprisingly absent in earlier artistic production, the monster fighter transforms into a soul searcher, contemplating the consequences following his decision on which road to take in life.[28] Herakles's wise decision to walk the path of Virtue indicates the importance of self-control and moderation and the value in education and disciplined living. The paradox, of course, is that there are dozens of other myths that show Herakles deviating from his chosen path, as well as artistic representations inspired by the hero's self-indulgence.[29] Despite his less than virtuous lifestyle, the story of Herakles choosing Virtue over Vice was very popular in antiquity, as it is evident for example from Cicero (*De Officiis* 1.118), who uses the same didactic technique to inspire his audience to follow the right role models. It appears even in fourth-century church fathers, such as St. Basil, who tries to educate his nephews using the moral of a pagan story.[30]

HERAKLES, EROS, AND FERTILITY

Surprisingly for Herakles's temper, his iconography includes relatively limited abduction or rape scenes, and even fewer representations of the hero with a sex partner. Equally rare are depictions of Herakles among family members.³¹ The list of his official consorts includes Megara, Iole, Deianera, Hebe, and Omphale, while numerous others are left anonymous. One of the most revealing representations of the hero with a sex partner is found on a marble relief in Boston (Figure 9.2). It dates to the turn of the second to the first century B.C. and shows the hero naked, lying on his lion skin, mounted by a woman who wears nothing but her breast band. A Priapus column enhances the eroticism of the scene, while a curtain hanging from a tree and a pillar completes the setting for the couple's union. Herakles is depicted beardless, in control, and fully concentrated on the act. He is almost twice the size of the girl who straddles him and utterly engulfed in the tension of the scene, as his body reaction reveals.³²

Strikingly absent are sexual unions of Herakles with another youth. This lack of representations of sexual engagements may be indicative of a systematic de-

FIGURE 9.2: Marble relief showing Herakles mounted by a young woman. Boston, MFA 08.34d, second/first century B.C. © 2010 Museum of Fine Arts, Boston.

piction of the virtuous side of Herakles. Even though equipped with a male body, as if he were an ordinary man, Herakles is clearly not an average family man. Despite his attraction to the female body, as one induces from his encounters, Herakles does not seek out the pleasures of sexual activity or violent unions, unless there is a reason that would legitimize such a quest. His service to Omphale was dictated after the killing of Eurystheus's son; the rape of Auge occurred under the influence of wine, and his marriage to Hebe, Eternal Youth, was not a matter of choice but the acceptance of a prize. Herakles's sexuality was never an issue, and his occasional sexual encounters are secondary motifs that tell us more about the hero's character; his inescapable, tragic fate; his self-indulgence; and his final gratification rather than his masculinity and pleasures of the flesh.[33]

HERAKLES AS A MODEL FOR RULERS

After this brief overview of Herakles's significance in the shaping of a young man's character and education, let us proceed to the examination of his role as a model for the ruling class. In the Archaic and Early Classical eras the use of Herakles as a propaganda tool is hard to prove, but from the late Classical through the Byzantine periods, we can trace several tyrants, kings, and rulers who eagerly assumed Herakles's characteristics to promote their own plans. Additionally, they advanced Herakles's cult and introduced coins stamped with Herakles's image.[34]

Even though in the Early Archaic period representations of Herakles are equally common in Boeotia, Sparta, Corinth, and Athens, as the century wanes the hero's popularity begins to fade in most places except for Athens, where he becomes a dominating figure in the visual arts. In the Archaic period, the club and lion skin are established as Herakles's attributes, and episodes from his life and labors become popular subjects of decoration, even though there is not yet a canonical cycle of labors.[35]

On the significance of Herakles in Sparta, we know from Herodotus (7.204.1) that the Spartan kings traced their lineage to Herakles and that king Leonidas is considered a descendant of the Heraklid clan (7.208.1; 8.131.2). However, there is no clear indication in the Archaic and Classical art of Sparta that the figure of Herakles was appropriated and exploited solely by the royal families. On the contrary, the proliferation of Herakles's scenes on Laconian vases as a hoplite indicates a promotion of Herakles's image as an ideal soldier, regardless of divine, royal, or common descent.[36] In Archaic Corinth the dynastic family of the Kypselids supported its connection with Zeus more than other gods and heroes, a practice evident in the dedications at Zeus's sanctuary in Olympia. However, Herakles was the protagonist of a plethora of vase paintings produced during this period,[37] while the hero's legendary travels can be compared to the colonizing campaigns of the city in the West. Corinthian vases decorated with the typical animal friezes, the rosette, and Herakles's adventures, reached Greek and non-Greek cities of the Mediterranean alike and may have functioned as a trademark of the motherland.

A more direct case of Herakles's political exploitation occurred in Archaic Athens. The significance of Herakles in Athenian artistic production is such that more than 40 percent of the black-figure vases produced between 560 and 510 B.C. involve scenes with Herakles. What is more intriguing is that Herakles is represented on several Archaic pediments from the Acropolis, and yet it seems that he was not even worshipped there. John Boardman explained Herakles's predominance in Archaic Athens as a well-constructed state-symbol, on account of the hero's association with the city-goddess Athena. Moreover, Boardman suggested an identification of the Athenian tyrant Peisistratos with Herakles and argued for a purposeful exploitation of the hero's representations and a promotion of his own image through the abundant and versified iconography of Herakles.[38]

It is interesting to point out that in Athens artists seemed to focus more on the aftermath of Herakles's labors and his apotheosis, while in the Peloponnese Herakles was perceived more as a hero and less as a newly introduced god of Olympus. The Peloponnesian view emphasizes the hero's qualities as a hoplite in contrast to the Athenian perception of a rustic demigod. His imagery on Attic vase paintings of the late Archaic period is overwhelming, and despite the lack of consensus on the subject, Boardman's interpretation of Herakles as an alter ego of the Athenian tyrant Peisistratos is an attractive one. As the Panhellenic hero enjoyed the support and company of Athena, so did Peisistratos claim to be a protégé of the goddess.[39] The tyrant fashioned himself a caring leader of the common people, who provided them with fountain houses, temples, and festivals, just as Herakles is considered to be an archetypical engineer and reliant role model.[40] Herakles is also depicted in Attic vase paintings as a music performer, and judging from his importance in Peisistratid Athens, this may imply certain changes in the musical events of Athens under the tyrants' reign.[41] The rustic yet popular character of Herakles as a fighter and a bon vivant, along with his dual status, mortal and divine, qualified him as the ideal model Peisistratos aspired to imitate.

The labors of Herakles were a popular subject in the arts and architecture of the Archaic period, not only in Athens, but also in other areas of the Greek world.[42] Panhellenic religious centers such as the sanctuary of Apollo at Delphi or of Zeus in Olympia attracted a tremendous amount of dedications and thriving building enterprises, which more than usually included Herakles and his deeds in their iconographic program.[43]

The most characteristic example of this tendency is the Siphnian Treasury, built by the prosperous island of Siphnos in the sanctuary of Apollo at Delphi around 525 B.C., shortly before the gold and silver mines of the island were inundated and Siphnos was overpowered by Samos.⁴⁴ This jewel of architectural sculpture encompasses a variety of mythological scenes in its repertory, including a rather unexpected episode with Herakles and Apollo, placed on the most prominent part of the building, the eastern pediment: Zeus as an arbitrator imposes himself between Herakles on the right and Apollo on the left as they fight for the Delphic tripod (Figure 9.3). Zeus is represented as the tallest, most authoritative figure, distinguished by his height, larger scale, and luxurious dress. Comparing the figure type of Herakles to Apollo, one notices the sheer volume and plasticity of the hero's body, along with his large scale, second only to Zeus. Despite the loss of the original paint, Herakles's tight muscles signal the tension of the scene. His thick, heavy body type contrasts to the narrow, slim figure of Apollo, who tries to pull the tripod from Herakles's hands. Such an effort appears in vain once we observe the solid grasp of Herakles, who has already put the tripod on his shoulders and leaves the scene with a wide pace. The torsion of Herakles's body reveals in full frontality his sculpted abdomen, broad chest, and powerful arms, while the tilting and turning of his head to the left directs our gaze to the object of dispute. The heads of Apollo and Zeus are missing but that of Herakles is still preserved; the hero has longish hair, arranged in elegant braids covering his neck and curly locks on his forehead. A well-trimmed beard completes the image of the usurper. All figures are curved almost in the round, and it is interesting to note how Herakles occupies more space than Zeus and almost equals his height with the tripod on his shoulders. With his wide pace and impressive body mass, Herakles becomes the central figure of this group.

When Siphnos refused to pay the tenth of its profits from the mines to Apollo, the island was heavily punished and never recovered afterward. The myth that decorates the pediment of the Siphnian treasury recounts a moment of doubt and uproar at the Oracle. The Peloponnesian hero decides to create his own mouth of prophesies and goes as far as to steal the very seat of mantic Apollo: his tripod. One may interpret this composition as an allegory of the Siphnian superiority and their disregard of Apollo and Delphi.⁴⁵ Herakles's solid body may be compared to the wealth the Siphnians had amassed before the Samian attack, while his firm grasp on the tripod reflects their own determination to maintain their profits. The mortal hero Herakles casts doubts on the behavior and demands of the Olympian Apollo; the heavily-muscled body of Herakles represents the firm stand of the Siphnians, who remind the Delphic priesthood of Zeus' supervision and intervention when the god of justice himself, Apollo, makes an unfair (to their standards) claim.

A similar approach can be pursued for the architectural decoration of the Athenian Treasury at Delphi, built in the early fifth century B.C.⁴⁶ The rather modest building was set on a prominent elevated area of the sanctuary, and this allowed an unobstructed view of three of its four sides, which carried painted

FIGURE 9.3: The struggle for the Delphic tripod. Pediment of the Siphnian Treasury, ca. 525 B.C. © EfA/G. de Miré

relief decoration. The metopes of the Athenian Treasury depict the adventures of the Panhellenic hero Herakles and the Athenian copycat, Theseus. Herakles's labors are carved on the north and west sides of the treasury and stand out from contemporary works for their innovative composition and detailed execution.

As leaders of the Greeks during the Persian Wars, the Athenians found inspiration and support in the heroic examples of both Herakles and Theseus. Having physical strength, courage, and qualities equal to heroes', the Greeks overpowered their enemies. They exceeded human abilities and achieved the impossible against an army much larger than theirs, and this explains why the labors of Herakles became a fitting metaphor for such an extraordinary victory. Their pride is manifested on the metopes of the Athenian Treasury, where the figure of Herakles (and Theseus) concentrates all these qualities, and his dramatic pose against his opponents reflects the strenuous contest of the Greeks against the Persians.

From these select examples, it becomes evident that because of Herakles's universal character every Greek city-state, every local ruler or exceptional athlete could shape the message of Herakles's achievements so that the hero's characteristics reflected on themselves, and use the hero's body as a platform advertising their own accomplishments.[47]

Herakles's political significance was not diminished with the arrival of democracy in Athens.[48] On the contrary, he continued to be a vehicle for political messages, used either as a metaphor of the phenomenal achievement of the Greeks in defeating the Persians (see, e.g., his central place in the Gigantomachies of this period) or as an Athenian hero by appropriation. He is thought to have assisted the Greeks at Thermopylae and particularly at Marathon, where the Herakleidai, the offspring of Herakles, receive special worship later on.[49] The Battle of Marathon was the subject of a painting at the Stoa Poikile at Athens, which, apart from battle scenes, represented the goddess Athena accompanied by Herakles and two other heroes. It has been suggested that the red-figure calyx-krater in the Louvre, attributed to the Niobid Painter, reflects the style and subject of the painting by Polygnotos.[50] Herakles's statuesque figure is placed in the center of the composition, his body being the only one represented frontally. This is no longer the image of a hero who comes to the rescue of the Athenians but an icon of a protector and ally. His immobility, frontal stance, and static posture indicate a new treatment of the hero's body: it now functions less like an idealized version of the mortal warriors who follow his example, and more as a malleable frame of the hero's characteristics, a figure so distant and beautiful, and yet so close and within reach as the numerous statues set in Athens.

Even though the Twelve Labors of Herakles become standardized in Classical art, largely thanks to the metopes of the temple of Zeus in Olympia, the hero's image itself ceases to be identified exclusively with the powerful albeit rustic male, and instead it reflects qualities descriptive of his character and personality.[51] Herakles assumes a new persona and even becomes a tragic hero, weary, and sorry for his madness, rewarded for a full life's work with immortality. This shift in Herakles's representations is most evident in the new dimension of the hero, particularly in Athens, where Herakles's association with the patron goddess Athena is still maintained but no longer dominates the artistic repertory.[52] The numerous depictions of Herakles as an earthly hero performing his labors give way to more frequent representations of him as a rejuvenated god accompanied by Hebe.[53] For this reason, the last labor of Herakles, the apples of the Hesperides (Figure 9.4),[54] becomes popular during the end of the fifth century and the fourth century B.C., along with scenes of Herakles's apotheosis.[55] Now Herakles acquires a younger, more graceful body and is often represented beardless, a convention that implies his rejuvenation and immortality. It is not the body of an athlete or a warrior depicted in the Late Classical and Hellenistic art, but rather an eloquent illustration of the transition of a mortal hero into an Olympian god. It is in these periods that some artists chose to emphasize the human nature of Herakles with all its vices and limitations, while others concentrated on his soul and the immortal body that enclosed it.[56] This development in Herakles's imagery coincides with the vigorous preoccupation with the afterlife and the flourishing of mystery cults. After the end of the Peloponnesian War (404 B.C.) and the rise of

Macedon, battles and conflicts became frequent phenomena and the call to the arms sounded louder than before. Adult males were required to sacrifice their bodies and youths to train for the same military purposes. Herakles's figure became an ideal, not only on account of his muscled body and unprecedented strength and vigor, but mainly because he served as a living example of success and the rewards one enjoys after leading a principled life.

The Macedonian rulers traced their family lineage to the Heraklids of Argos, a concept most evident in the coinage of Alexander I through the Successors.[57] In the fourth century, Alexander the Great not only acknowledged himself as a legendary descendant of Herakles, but he also showed particular interest in cultivating this connection by promoting the cult of Herakles. Alexander's special attention to Herakles became conspicuous after the conquest of Tyre and his sacrifice to Herakles-Melkart, a syncretized divinity of the Greek hero and the Phoenician god. With Alexander the Great begins a conscious attempt to use a divinity recognizable by many people to promote one's political agenda: Herakles shared common traits with Bes (Egypt), Melkart (Syria), and Gilgamesh (Babylon) and thus could be used as a commonly accepted figure.[58] Herakles's achievements are made commensurate with the dangers Alexander encountered, while his quest for immortality and final gratification became the goal of the Macedonian king. As a son of Zeus himself, Alexander could relate to the special status of Herakles, and being invincible like the Greek hero proved his own divine origins. It was a fitting appropriation of traits and equation of imagery, bound to have a lasting impact not only on his Successors but on Roman emperors as well. From Alexander the Great on, every ruler who claimed relation to Herakles or assimilated himself with him added to the persona of the Panhellenic hero, thus guaranteeing the continuation of Herakles's fame and the glorification of all previous "impersonators."[59]

FIGURE 9.4: Herakles at the Garden of the Hesperides. Attic red-figure hydria by the Meidias Painter, ca. 420 B.C. London, BM E 224, *ARV*² 1313.5 © The Trustees of the British Museum.

The Successors embraced the assimilation of Alexander/Herakles and continued to use similar means to promote their own agendas. Minting coins with their own image strikingly similar to Herakles was one such mechanism, while another involved exploiting local myths to justify their royal ambitions and enhance their relation to Alexander.[60] A new type of Herakles's representations, the herm, is found widely dispersed in gymnasia, palaestras, and baths of the Hellenistic period.[61] Herakles's imagery is reduced to the most basic features, and surprisingly his body is not valued as important an attribute as his lion skin or club. It is Herakles's face and hairstyle that appeal to the taste of this period and at the same time allow mass-produced images of the hero to be set in direct and continuous viewing and admiration by the people.

Despite the diversity of Herakles's iconography on coins, vase paintings, and architectural decoration, the most significant addition to Herakles's imagery in the Classical and Hellenistic periods was the proliferation of new statuary types. Most of these statues are known to us today through the literary sources and Roman copies of the Greek originals.[62] Set in sanctuaries and public and private spaces all over the Mediterranean, the variety of Herakles's statues resulted in the hero's extreme popularity among the Romans, while some of his images survived even in the Byzantine period.

Among the Classical and Hellenistic statues of Herakles stands out a type known as Herakles Alexikakos, dedicated as a votive to the hero responsible for saving Athens after the great plague of ca. 430–427 B.C. The Hope Herakles, as the statuary type is known today, is considered a copy of the late-fifth-century original by Hageladas.[63] Herakles was thought to have averted disease, pain, and death from Athens, a feat no less impressive than the rest of his labors and thus worthy of commemorating in the heart of the city. He offers his protection to the people, and his statue serves as an active reminder of what a mortal hero is capable of. It is perhaps one of the instances where Herakles's aspect as protector of mankind is so eloquently revealed.

The so-called Lenbach Herakles follows a similar type and shows the hero with the club turned downward used as a support, and the lion skin hanging from his arm. Herakles has short, curly hair with sideburns, and cauliflower ears. This type was extremely popular in antiquity and was copied in various materials and scale, often with exaggerated musculature and extreme anatomical details. A gilt version of this Herakles was set at the Forum Boarium in Rome, and its prototype is thought to be the bronze Herakles of Lysippos, created in the fourth century B.C. for Sikyon.[64]

An Apulian volute krater dated around 360–350 B.C. depicts a similar statue of Herakles, rendered in white and set on a high pedestal, while the artist puts the final strokes of color to the lion skin (Figure 9.5).[65] Even though the hero is portrayed as a bearded adult, his body is as toned and muscled as if he were an adolescent. He rests one hand on his club and holds his arrow with the other.

The S-curve of Herakles's body accentuates his physical qualities, while the torsion of his upper body and the tilting and turning of his head create an illusion of perspective. What is remarkable in this scene is that the whole episode is witnessed by Zeus; a young satyr burning incense; a winged female figure; and, most surprisingly, by young Herakles himself. The beardless youth looks with astonishment at the statue as he brings his right hand on his mouth in a gesture of bewilderment. The lion skin and club confirm the identification of the young hero with Herakles, but the whole scene remains a riddle. One could explain the young Herakles as a symbol of his new divine status: rejuvenated and now a member of the immortal Olympian family, the Greek hero admires his former self and wonders at the sight of this old, fatigued, mortal body. The presence of the young satyr implies a strong Dionysiac and even theatrical element, and one may attempt to explain the scene as the preparation of the setting of a dramatic performance related to Herakles.[66]

Quite interesting is the so-called Cherchel Herakles, a colossal statue taking its name from its find spot, the bath complex at Cherchel, Algeria. The hero appears aged without his lion skin, resting on his club and holding an arrow. The original was probably made in bronze and is thought to date to the second

FIGURE 9.5: Apulian red-figure column krater with artist painting a marble statue of Herakles, ca. 350–320 B.C. Attributed to the Group of Boston 00.348. New York, MMA Rogers Fund 1950 (50.11.4) © The Metropolitan Museum of Art/Art Resource, NY.

quarter of the fifth century B.C., perhaps a work of Onatas. In Hellenistic times this type became a particular favorite among the representations of monarchs and rulers.[67]

The one sculptor whose work gave new features to Herakles and had a tremendous impact in later periods was Lysippos of Sikyon, the court portraitist of Alexander the Great.[68] The larger-than-life bronze statue crafted by Lysippos around 330 B.C. represented the hero resting on his club after obtaining the apples from the Hesperides; the most famous copy of this type is known today as the Farnese Herakles. Tired from his last labor, this heavy-muscled male figure was termed Weary Herakles and made such a lasting impression that it was copied throughout antiquity in various media and scales. The hero leans again on his club placed under his armpit, but this time he is clearly more aged and his musculature far more accentuated. He looks down and to the right, while his right hand is lowered in the back holding the apples of the Hesperides. Portraying the hero as a weary mature athlete is a fitting convention to indicate the last of Herakles's labors, while at the same time giving a more human, even psychological description of the hero's physical and emotional condition. This Lysippean statuary type was particularly popular in gymnasia and baths and a favorite statue of Macedonian rulers and Roman emperors. The same type was used as a mint symbol in Alexandrine coins and enjoyed immense popularity even during the Byzantine years. There were numerous replicas in all the large cities of the Mediterranean, while smaller reproductions are found on other artifacts, such as coins, lamps, mirrors, statuettes, sarcophagi, and reliefs.[69] Colossal marble reproductions, miniature bronze copies, and numerous variations of this Lysippean prototype decorated private houses and public spaces from the Late Classical until the Byzantine period. Athens, Pergamon, and Constantinople were only some of the capitals that embraced the Farnese Herakles along with a variety of other statues of the hero.

The most often cited statue of Herakles is undoubtedly the colossal bronze that Lysippos created for the city of Taras. Elaborating on preexisting statuary types of a seated Herakles, Lysippos portrayed the hero weary and fatigued after completing one of his hardest labors, cleansing the Augean stables. The exhaustion and weariness of the hero is obvious in the looseness and withdrawal of his pose. Seated on the basket he used to clean the stables, Herakles can hardly sit up straight, barely supporting his head with his right hand, while the left collapses on his leg. The Lysippean Weary Herakles showed the hero as a suffering aged mortal, thus bringing him closer to his worshippers. In contrast to the previous depictions of an invincible hero, Lysippos chose to represent him in a human moment of contemplation and assessment of a hard day's work. Herakles's endurance in hardships and final gratification set the best example to follow both in pagan antiquity and even in early Christian times. The statue was taken to Rome after the fall of Taras and then later on to Constantinople, and its impact was such that this Lysippean type even decorates Byzantine ivory caskets of the tenth century A.D.[70]

A variant of the seated weary Herakles was produced again by Lysippos and is known as Herakles Epitrapezios. The epithet has a twofold meaning: it can either define the statue of Herakles as part of table decoration, or it means that the hero was depicted banqueting. The Lysippean original is thought to have been commissioned by Alexander the Great himself, and later on the type not only passed down to his Successors, who saw it as an additional link with the Macedonian king, but also was adopted by Hannibal and Sulla. The hero is shown seated on a rock covered with the lion skin, holding a cup in his right hand and resting the left on his club.[71] For this type Lysippos used a stereotypical concept of Herakles the glutton and transformed it into an image embraced by royalty.[72] Herakles, the half-witted character of Greek comedy, is now rendered in a positive light as an elegant symposiast. This carefree representation of Herakles allows for his body to relax and enjoy the banquet offered to him originally by Alexander the Great. The scheme echoes the earlier practice of *Theoxenia*, where mortals prepared banquets for heroes and gods to partake together. With Alexander the Great the equation between the immortal Herakles and the deified Macedonian king has already been achieved and is advertised to the world with yet another statue of Alexander's guardian.

Unusual in concept and style is the so-called Kavala Herakles, a type that represents the hero as an obese, stocky figure, covering his round belly with his lion skin as he presses his right hand against his chest. He has a square head crowned by a fillet, which keeps together the locks of hair on his forehead, and a long beard. The once heavy-muscled, athletic Herakles is now transformed into a chubby character with a rather sad expression. All the examples from this type are under life size and date to the Roman Imperial period, perhaps reflecting a third-century B.C. original.[73]

Another unflattering view of Herakles comes with the representations of a naked, aged, intoxicated hero, so utterly surrendered to his excesses that he struggles to stand or shamelessly urinates in plain view (Figure 9.6). The image of a drunken Herakles was already introduced in the fifth century B.C. with Euripides's *Alcestis* (782–89), but in the visual arts this motif became popular only after the middle Hellenistic period.[74] This was the period of exploration in art, allowing for the representation of all vices and the monumentalizing of the everyday, simple, and playful acts. It comes as no surprise that even the great hero is depicted as a common drunkard, possessing an unexceptional body that has long ceased to serve as an athletic model. The most mundane bodily functions are now appropriate for the divine Herakles: he may be a god but he still carries a human body with all the pleasures and limitations that this entails.

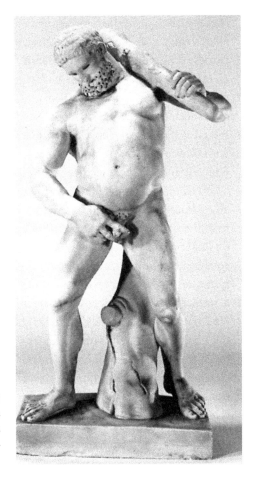

FIGURE 9.6: Marble statuette of the type of Herakles Mingens. Wörlitz Schloss II$_{1,2}$. After *LIMC* 4 Herakles 771, no. 898 © Kulturstiftung Dessau-Wörlitz, Bildarchiv, Heinz Fräßdorf

In the Hellenistic era the circulation of small-scale statuettes; variations of his statues; and representations of his labors on coins, gems, and plastic vases facilitated the transmission of Herakles's figure types. Herakles's image as an invincible athletic male continues to be popular throughout antiquity, but in the Hellenistic and Roman eras the artists begin to highlight more his mortal character, occasionally portraying him in an unflattering, even comic light.

The main difference between Herakles and Hercules is that in the Greek mentality Herakles is a hero, while for the Romans he is an Olympian.[75] Because of his association with Alexander and the Hellenistic ruler cult, Herakles becomes in the Roman period an integral part of the state rituals. In addition to his civic aspect, Herakles's presence in the funerary art increases not only because of his immortality, but also on account of his syncretized image with Dionysiac/Bacchic cults.[76] A significant addition to the multifaceted character of Hercules comes with the eclectic taste of Late Roman periods (second to fourth centuries A.D.), when Hercules becomes a god of abundance, of healing, and patron of the army.[77]

Even though in the first century B.C. Roman aristocrats appropriated Herakles's traits and depicted themselves as new Herakles/Alexander, the early emperors of Rome did not really identify themselves with the hero.[78] In contrast, Herakles's imagery had a particular appeal to Roman emperors of the second century A.D.[79] His uncontrolled gluttony, wine consumption, and bestiality are reflected on the life of Nero. Seneca's *Hercules Furens* is thought to have served a didactic purpose: to instruct and guide the young Roman emperor with power and vices similar to those of Herakles against such a frenetic government and outrageous lifestyle.[80]

Nero may have failed to portray himself as a virtuous, successful new Herakles, but this was not the case with Trajan, who enhanced his own association with the hero by encouraging the cult of Hercules. He placed the hero's image on imperial coins, and even created a statuary type of Trajan as Herakles, thus equating the body of the emperor with that of a divine hero with all the implications that such a process pertains.[81] Even though Roman authors pursued a similar comparison between Hercules and the emperor Domitian,[82] it was Hadrian who most closely followed the footsteps of Trajan. Hadrian issued his own coins decorated with Hercules's imagery, emphasizing, however, not so much the hero's fighting qualities, but rather the aspect of Hercules as a traveler in quest for immortality. Hadrian compared his own efforts to unify all nations under his rule with the Herculean deeds: the emperor as a new Hercules travels to the extremes of the world to bring the message of Roman greatness. A quite different view of Hercules was promoted by Antoninus Pius, who was very much centered in Rome and preferred to highlight the traditional type of Hercules Invictus as the focal point of his government.[83]

The next emperor to exploit Hercules was Commodus, who fashioned himself as a new Hercules, offspring of divine origin and therefore legitimate owner of the Roman throne and a god himself. Commodus assimilates Hercules to the Genius Augusti and sets the path for his own acclamation as Hercules Romanus.[84] Of all the second-century emperors, Commodus was the one who activated all possible promotion mechanisms, ranging from minting coins and sculpting statues to hosting large public ceremonies with the emperor dressed as Hercules. Among the most known statuary types that show the divine emperor Commodus-Hercules are the bust of Commodus in Rome (M. Capitolini) and the colossal Weary Hercules with the head of Commodus in Florence (Palazzo Pitti).[85] The equation between ruler and hero that was started by Alexander the Great was now completed and taken to another level with Commodus's total identification with Hercules. The emperor no longer seeks to imitate the virtues or the warlike, masculine qualities of the hero, but he actually appropriates Hercules's bodily characteristics, his attributes, and behavior as his own. The identity of the hero has been shaped in such a way as to accommodate the emperor's own views of Hercules, and his ultimate desire to become the hero himself.

Similar use of Hercules, although in smaller scale, occurs in the third and fourth centuries A.D. For example, Postumus introduced coinage with Hercules's labors and himself wearing a lion skin, holding a club on his shoulder.[86] The message that these images convey was radically changed from the divine aspirations of Commodus and the religious connotations of Trajan: Hercules now stands as a symbol of the security at borders. Other emperors focus more on the use of the signum, a name type, such as Constantius Herculius, or Maximian Herculius, that under the Tetrarchs assumes a more symbolic value, since it displays the qualities shared in common between the emperor and the divinity, in the same way an epithet assigns special traits to a deity.[87] The frequent appearance of Hercules (and Zeus) on coins issued by the Tetrarchs is due to their role as protectors of the Augusti, Diocletian and Maximianus Herculius.[88]

As the New Rome, the capital of Constantine was adorned with a large number of statues representing gods, heroes, philosophers, and animals. Constantinople was in essence a large metropolis like Rome, Alexandria, and Pergamon. The pagan roots of the city had not vanished even after the arrival of Christianity, and although these works are missing today, we still have literary accounts praising the emperor and his capital, which can be used as a guide of the ornate Christian city.

In Byzantium, pagan statues did not cease to carry a symbolic meaning, even after the advent of Christianity. What did change, however, was the interpretation of these pagan works. This new approach toward antiquity affected the understanding of Herakles's imagery, both his large-scale statues set in public spaces and smaller works for private use.[89] Depending on the colors Byzantine authors used to describe these ancient artifacts and the superstition of the crowd, the images of Herakles served a new purpose and reflected a different message.

Two of the greatest venues in Constantinople are particularly known for their exquisite decoration and pagan ornamentation: the famous Baths of Zeuxippos and the Hippodrome. The Baths of Zeuxippos stand as a bright example of the long-lasting influence of Greek education. They were the premiere bathing facilities of the capital, frequented by the empire's elite and dedicated by Constantine himself in 330 A.D.[90] As an epicenter of physical fitness and a masterpiece of architecture and engineering, the baths could not lack lavish decoration. Among the works of sculpture displayed in the baths the sources include a group statue of Herakles and Auge. It was an appropriate choice to set this group in the Baths, since Auge was actually bathing herself in a sacred spring when Herakles surprised her. One could argue that the Byzantines (at least the learned ones) were familiar with the myth and able to make the connection between the springs of Auge and the Baths of Zeuxippos, and even secretly smile at the sexual connotations of such a group statue.

The Hippodrome was the main venue for both athletic and political events in Constantinople. Large numbers of people gathered there for sporting competitions, religious and civic celebrations, and of course imperial ceremonies. Such a setting required extraordinary decoration, and it was precisely the vast quantities of sculptures that demonstrated the city's superiority and link with the former Roman capital. The Hippodrome collection of antiquities reflected the contradictions of the new city: it was Rome-like, but not Roman. The spolia proudly exhibited in prominent sites, including the Hippodrome, played a didactic role and promised a new vision for the future.[91]

The Hippodrome was built by Septimius Severus and its decoration began with Constantine and was continued by later emperors. The art gathered there was extremely impressive and preserved until 1204. Literary and graphic sources as well as archaeological evidence confirm the diversity of the collection. The Hippodrome decoration falls into four categories based on the message the monuments convey: apotropaia, victory, public, and glorification of Rome. The group of victory monuments includes images of demigods, mythical creatures, and heroes who served as exempla for competitors in the circus. This is why statues of Herakles (along with those of the Dioskouroi, Skylla, and an eagle) were prominently set in the Hippodrome and frequently quoted in the literary sources.[92]

At least three statues of Herakles stood at the Hippodrome, among them a statue of the hero fighting the Nemean lion, and a seated statue of Herakles resting after cleaning the Augean stables. It is possible that the seated Herakles was the colossal bronze of the fourth century B.C. that once stood in Taras. When the city fell to the Romans in 209 B.C., it was taken to Rome as spoils, only to be moved farther east to Constantinople with the rise of Constantine. Nicetas Choniates, a twelfth-century Byzantine historian, comments on the statues in the Hippodrome and provides a detailed description of this statue of Herakles. Thanks to Choniates's account we learn of a type missing from the repertory of surviving freestanding Greek statues.[93]

Herakles as model of male strength and physical stamina was an appropriate presence in the Hippodrome, particularly as the legendary patron of athletic contests and on account of his special relation with the Circus through his temple at the Circus Maximus.[94] In addition, Herakles stands as an *exemplum virtutis*. He embodies the moral strength and wits that empowered him to defy the odds and succeed in his labors. Apart from the well-known traits of the hero, however, the Lysippean Herakles carried a further connotation: being part of the spoils of Taras, it showed the greatness of the Roman and Byzantine Empire. It was a trademark of the city of Taras and Rome, and as such it degraded them both when removed from its setting. The Lysippean Herakles was also a symbol of the old city's traditions absorbed into the new one. The same way Herakles's imagery was appropriated by rulers and kings throughout the centuries, so artworks, symbols, and traditions were adopted by Constantinople in an attempt to assimilate power, authority, and prestige, and to explicitly evoke its succession from Rome. Thus, Herakles and the rest of the statues appear in a Rome-like setting in the heart of the New Rome.[95]

The myths and images of Herakles survived in the Byzantine and Christian tradition and often served as decorative motifs on ivory reliefs, caskets, and Coptic textiles.[96] The influence of Herakles's imagery was such that even St. Peter's throne in the Vatican was adorned with panels treating his life.[97] Granted that Herakles had a much lesser impact on the formation of Christian iconography than Apollo and Dionysos, still he is credited with a significant contribution: the scheme of Adam and Eve flanking the Tree of Knowledge is thought to originate from Herakles's visit at the Gardens of Hesperides.[98]

Herakles's myth and imagery were embraced by people from all social levels and worshipped in many different guises. This resulted in a lasting, consistent presence of Herakles throughout the centuries, which explains why he is one of the few pagan figures that did not need to be rediscovered after the fall of the Roman Empire.

As a case study of the cultural history of the human body, Herakles is a fascinating subject. He enjoys a special status in Greek myth and literature and this is reflected in the artistic representations. Originally, his appearance and physique barely differ from those of other Greek heroes, and it is thanks to his attributes, typical hairstyle, and context that he can be singled out. Later on, though, from the Classical and Hellenistic eras and beyond, the well-trained body of the hero will give way to the rejuvenated, graceful god and then to the worn body of a mortal. Herakles's image is canonical and yet easily shaped so that it can be used with equally plausible results as an example for athletes, warriors, kings, emperors, and common men. The values that he reflects and the messages his icon conveys render Herakles a unique figure that links the domain of the ordinary people with powerful rulers and the realm of the dead. In the Graeco-Roman world and the multiple gods and heroes of its pantheon, Herakles's image reflects effectively and diachronically the perception of the human body and its potential uses.

CHAPTER TEN

The Self from Homer to Charlemagne

MARC MASTRANGELO

With his tongue only half in his cheek, the philosopher Galen Strawson begins a recent essay by remarking that "human life is founded on three fundamental (and connected) illusions: the illusion of romantic love, the illusion of free will, and the illusion of the self."[1] With the establishment of Western democratic and capitalist society and its attendant prioritization of the individual person, the idea of the self has been a lively topic for scholars, philosophers, and intellectuals. The last 50 years have been marked by a multidisciplinary discussion focused on the nature and history of the self, a nontechnical term that can stand for the uniqueness of an individual, a psychological or psychophysical structure, or that which constitutes the essence of a human being.[2] Putting aside the difficult problem of whether the self exists, it is not controversial to assume that a *sense* of self exists in human beings. The articulation of a sense of self that a culture puts forth is the basis by which it sets norms and establishes ideals for how its members live as individuals and as a community. From Greco-Roman antiquity to the early medieval world, the various intellectual manifestations of a sense of self are grounded on norms that promote human flourishing.

When we reappraise past notions of what it is to be a human being, we subject our own contemporary ideas on the topic to a vigorous critique. One facet of this critique, especially with regard to Ancient Greece, has been that the post-Cartesian and post-Kantian view of the self as an autonomous, self-legislating, rational, and unified mental entity has been a philosophical and historical mistake; rather, as many have recently argued, we should understand our sense of self according to the ancient notion that the self is relational, that is, it exists in virtue of relationships to others, with multiple parts that dialogue with each other and with the world.[3] According to this critique, Descartes's *cogito*-self, or single source of consciousness, a unitary "I" from which all "mental processes and actions" originate,[4] should not be understood as the climax of a progressivist story in which early Greek views of the self represent a less developed stage of a self-consciously aware individual.

Recent scholarly work on the history of the self has rightly focused on debunking this progressivist story. According to the progressivist story, the ancient antique evidence indicates a self that lacks the autonomy to exercise moral agency or exhibit a unique personality, building blocks of the autonomous, individuated modern self.[5] However, this essay argues that Greek, Roman, and early medieval thought display variously articulated versions of a relational self that not only accommodates the idea of individual moral agency but also assumes a robust form of communal life.[6] A representative sample of ancient, late antique, and early medieval Western thinking on the self shows a diversity of views that support this claim.

There is no word in ancient Greek and Latin that can stand for the modern term "self."[7] The idea of the self has grown out of the "high value on personal individuality and uniqueness" present in intellectual thought after Descartes.[8] On the other hand, the ancients, according to Christopher Gill, began from the question of what it is to be a human being. For them, the self is best understood as a normative notion of what it is to be a rational human being, whose ethical life is defined according to interpersonal relations within a community that discusses the nature of those very relationships and roles.[9] Gill's formulation can be extended, with qualifications, to late antique and early medieval Western culture. From the time of Homer to the age of Charlemagne, the self is understood to exist in virtue of relationships to the earthly and divine worlds. Throughout the intellectual history of this period, these external bonds are normative in their nature and purpose; that is, they have an ethical character that becomes internalized. Consequently, the ancient and early medieval self has interiority, but an interiority and separateness predicated on its relationships.

TWO CONTEMPORARY VIEWS: GILL AND SORABJI

Gill has argued that the Greeks and Romans understood the self in two related ways: first, psychologically, as a structure of parts that dialogue with each other. A good example is Plato's tripartite soul with its appetitive, spirited, and rational parts, which constantly struggle for ascendancy; and second, the ancients view a self as having an ethical life, which is fundamentally "shared rather than private" or "individuated" and becomes salient by engaging in "dialectical debate rather than solitary introspection."[10]

Gill has posited two notions of self that he calls "objective-participant" and "subjective-individualist."[11] The objective-participant concept of self understands the person as a set of internal parts or organs, which dialogue with each other—hence, "objective." In addition, the person gains authenticity through participation in a community where bonds to family members, friends, and the city actually constitute the self—hence, "participant" or relational. The subjective-individualist concept of self is a nonrelational self, in other words, subject centered. It gains authenticity through abstracting itself from pertaining circumstance. It legislates its own moral principles and generally establishes an individuated stance and first-personal point of view. It is a unitary "I." These two categories, objective-participant and subjective-individualist (the singular terms can be separated or recombined), have helped us to recognize the value of the ancient relational self as well as certain weaknesses of the modern Cartesian-Kantian conception of the autonomous, individuated self. Moreover, we now have intellectual tools flexible enough to describe ideas of the self from the past to the present: An objective-participant sense of self with interiority persists in philosophy and literature through the early medieval period.

Richard Sorabji has written that there is an array of representations of the self in Greco-Roman antiquity—in fact, he lists sixteen.[12] Sorabji argues for an interest in what he calls "me-ness" as well as self-awareness in several versions of the ancient self. He argues that whatever model of self we might think up, whether it is the self as a bundle of qualities, a stream of mental activities united in various degrees for various amounts of time, there must be an owner of these qualities or mental activities.[13] Hence, Sorabji is open to seeing more individualist and subjective versions of the self in the Greco-Roman tradition. Unlike Gill, he argues that there was in fact a "shift in the thought of later antiquity towards a more subjective or individualist interest."[14]

The main differences between these two scholars lie in their assumptions regarding methodology. While both make Herculean efforts not to project inappropriate contemporary philosophical views onto the Greeks and Romans, Sorabji is willing to allow the first-personal point of view of "me-ness" and "ownership" to inform his arguments. Gill, at times, sees this as crossing a line in methodological propriety that, instead of "using insights from modern philosophy merely to clarify," they are in fact shaping "the philosophical agenda." Secondly, on the one hand, Sorabji's project has a fundamentally different purpose: he has set out to prove the existence of the self by means of philosophical argument and the use of both Eastern and Western intellectual history. Sorabji's work on the self centers on what it is to be a particular person, whereas Gill focuses exclusively on the Greek and Roman concept of self, which conceives the self as the essence or the core of human being.[15] Gill's project, although historical, is more descriptive—though even Gill harbors assumptions of what a self is or could be. One result of this scholarly back-and-forth has been to refine the distinction between objective-participant and subjective-individualist to the extent that individuality, understood as separateness and interiority, "can figure in both types of conception."[16] The following historical survey will show that the articulations of self, from Homer to Charlemagne, are objective-participant and possess a rich interiority.

THE SELF FROM 800 B.C.E. TO 900 C.E.

In the second half of the twentieth century, classical scholars told a progressivist story of the self. They began with Homer's *Iliad* and the *Odyssey*, which, they argued, marked the primitive beginning of the modern, autonomous, self-legislating self. According to these scholars, the Homeric person is a fragmented psyche lacking unity and individuality.[17] In the act of choosing a course of action or making a decision with moral implications, the Homeric character, so the argument went, makes a decision, but it is a decision that is devoid of self-awareness and ownership. In other words, on this view, the decider must be aware that he is deciding and that the act is his own in order to be a self.[18]

What seems to be lacking in these moments of decision, according to these arguments, is the consciousness of being a unitary self with a strong first-personal point of view. Examples in Homer portray characters, when faced with a choice, addressing their inner organs, which, according to earlier scholars, stand for atomized psychological functions and therefore indicate a self with little if any unity or first-personal point of view—at least from the standpoint of deliberating choices, that is, moral agency.[19] There are also scenes in which major characters appear to ignore important consequences of their decisions. Achilles's monologue in *Iliad* book 9 depicts Achilles's decision to stay out of the battle because of the slight he has endured at the hands of Agamemnon. In doing so, he sentences many of his Greek comrades to death and changes the trajectory of his own heroic status. In book 18 of the *Iliad*, Hector makes the disastrous decision in his monologue to engage Achilles in battle. His choice as well causes pain and suffering to many and cements his legacy in heroic myth. Some scholars have seen Achilles and Hector as reflecting a sense of self that is neither unified nor self-conscious—that is, in fact, childlike—because of either a lack of awareness of the consequences of their choices or because their world is completely determined.

However, more recently, scholars have argued that in fact there is unity in the Homeric self. Contemporary action theory and the postmodern idea of unity in multiplicity have been useful models of explanation to recent scholars.[20] With its roots in Aristotelian notions of psychological capacities, action theory holds that a person's actions can be understood according to the reasons they are performed. And these reasons are dependent on the person's beliefs and desires rather than on whether the person is conscious of them. This opens the door for a more positive appraisal of the Homeric self, because under this model, Homeric characters do deliberate for reasons, based on their beliefs and desires, and make decisions for which they can take responsibility. Suffice to say that Homeric people exhibit a self that is "objective," made up of parts, that dialogue with each other in deliberating choices and that achieve unity or self-consciousness by internalizing ethical attitudes of those in their community.[21] Homeric characters appear to be self-conscious during the moments when they are having an internal dialogue, a dialogue that they could just as easily be having with someone in their community.[22] The interiority explicit in the idea of internal dialogue is central to the sense of self of the cultures of this period.

THE SELF IN ARISTOTLE AND PLATO

From Homer to Plato, it was the conventional wisdom that the Greek concept of the self shifts its meaning. Since they predate the dissemination of Plato's thought, Homeric epic and Greek tragedy are believed to describe a thought world different from Plato's and Aristotle's writings. This pre-Platonic and pre-Aristotelian thought world was conceived as containing individuals with disconnected, multifaceted selves: selves unable to form and use notions of moral responsibility and unable to locate within an individual a persisting self that deliberates certain choices, acts as a result of these deliberations for reasons, and finally accepts responsibility for the act. We have seen that it is more plausible to argue that Greek epic (and tragedy) portray the self as unitary, relational, and even self-aware.

Christopher Gill, Bernard Williams, and Alisdair Macintyre have made powerful arguments that, for the most part, both Plato and Aristotle view the self as objective-participant with separateness and interiority. In the *Republic*, Plato at times speaks of a tripartite soul, a soul with a rational part (*logistikon*) for practical and theoretical reasoning as well as learning, an appetitive part (*epithumetikon*) for desires and fears regarding bodily and material pleasures, and the spirited part (*thumoeides*) for emotional responses, such as admiration and shame, to ethical ideals.[23] For Plato, the immaterial soul, made up of parts in dialogue, independent of the physical body, represents a person's essence or self.

When a person is engaged in deliberation, especially moral deliberation, Plato has been understood to grant to the self the capacity to abstract itself from its communal attachments, thus achieving an impartial rationality. This is reminiscent of a Kantian morally deliberating self,[24] which achieves an individual stance and universalized moral prescriptions through the domination of the non-rational by the rational.[25] However, Plato's soul structure reflects an objective-participant notion of the self. Plato is promoting the harmony of the parts of the self through its connection to "reasonable norms" that set standards of action and evaluation.[26] This version of the self coheres and is unified by being "reason-ruled" And defines the Platonic self without recourse to the Kantian and Cartesian concepts.

In fact Plato's self has more in common with Freud's psychological model, which can be summarized as possessing a three-part self, the contentious harmonization of these three parts, and an unconscious with powerful beliefs that ground human action. But even closer associations can be made between Plato and Daniel Dennett, whose functionalist model of the self focuses on the conditions that need to be met in order for the self to be reasonable.[27] Plato models a self that is objective with respect to having parts, and is participant with respect to having the capacity to engage in dialogue. That is, the parts of the self interact with each other to achieve a reason-ruled harmony. By extending by analogy the structure of the soul to the city, the self takes part in a communal dialogue with other selves to establish and follow ethical and political norms. Finally, because the Platonic self is capable of reflective debate when deliberating a course of action, it is able to recognize the virtues as divine, thus implying separateness and interiority.[28]

Aristotle, similarly to Plato, posits a soul with parts, in his case, a binary self, rational and non-rational. Aristotle speaks of these parts as unified, but not as a Cartesian or Kantian entity that is a locus of subjective and introspective experience above and beyond beliefs and desires. Instead, Aristotle argues that a human being is a composite of body and soul and that the soul cannot be separated from the body. This hylomorphic view eliminates the separation between form and matter, a separation from which Plato's whole system is derived. Instead, hylomorphism emphasizes "structures and functions in explaining the characteristics and activities of living things."[29] For Aristotle the self is a set of capacities, which are actualized through a certain kind of living. Examples include the capacity to engage in practical or reflective reasoning. More importantly, Aristotle's "functionalist" model posits reasons as causes for action. Decisions are based on reasons; that is, beliefs and desires exist for reasons. Aristotelian notions of action are not the conscious acts of a will, originating from an autonomous self.[30] As Aristotle himself says: "Choice is either desiderative reason or ratiocinative desire, and such a source of action is a person."[31] For Aristotle, reason is the best part of us, our divine core.[32]

THE HELLENISTIC SELF

Both Plato and Aristotle cast a long shadow over the following centuries of thinking about the self. The Hellenistic era of philosophy, which scholars traditionally see as beginning after the death of Aristotle (322 B.C.E.), extended to the reign of Marcus Aurelius in the late second century. Two of the major philosophical schools, Stoicism and Epicureanism, developed the Platonic-Aristotelian self in a different direction. Stoics and Epicureans departed from Platonic-Aristotelian notions that defined the self "as a combination of a psychic core and a body or a complex of distinct psychic parts."[33] Rather, Stoic-Epicurean thought understood humans (and animals) as structured and integrated psychophysical wholes. Our intellectual and emotional functions are conceived of as corporeal, or at least inseparable from the corporeal; hence, not dualistically as in Plato's *Phaedo* and *Republic*. The life force (*pneuma*) is fully integrated with matter (*hule*). Psychological functions are understood to cohere through principles of addition, transformation, and systematic coordination.[34] Motivations, virtues, and happiness are psychophysical states. Gill has argued that Stoics and Epicureans also adopted Socratic ethical ideals, such as "the attainability of virtue, the invulnerability of the wise person to passion, and the unity of virtue in knowledge of the good."[35] In combination, psychophysical holism and Socratic ethical ideals make for an ordered, stable self.

By the late first century, writers such as the philosopher emperor Marcus Aurelius and the slave turned philosopher Epictetus, write of internal deliberations concerning ethical norms. However, in addition, authors like Aelius Aristides, a Greek Sophist, and Ignatius, a Christian, represent the self as primarily directed toward union with a god. These developments have led scholars to posit for this era a move toward a subjective-individualist self. For instance, Michel Foucault locates a sense of individualism, or a subject-centered self in Stoic-Epicurean thought. The self of the early Roman empire, according to Foucault, is founded upon "the insistence on the attention that should be brought to bear on oneself," and more specifically, "an intensification of the relationship to oneself by which one constituted oneself as a subject of one's acts."[36] On this view, a trend emerges in which the self becomes the locus of transformation, correction, and purification.[37] Accordingly, self-critique and self-examination exhibited in the work of Epictetus and the Stoic emperor Marcus Aurelius have been taken to usher in a more self-reflexive and self-conscious concept of self.

Scholars have applied a Foucauldian approach beginning with Seneca in the middle of the first century. Two recent studies conclude that during this imperial Roman period there is a trend toward the representation of an autonomous and subject-centered self.[38] According to Judith Perkins, Ignatius, Marcus, and Aristides write as though the self does not exist until it has suffered. In his letters to Christians, Ignatius portrays himself as the ultimate tough guy when it comes to enduring privation and bodily pain. He and the Christian community he addresses achieve their most authentic selfhood when they are sick, poor, and afflicted. It is in this condition as sufferer that the self can meet the divine.[39] The pagan Aristides, in his *Sacrae Orationes*, similarly portrays himself as a sufferer whose self is subject to the "survey and management of the divine" and "is no longer controlled by either itself or society but directly by [Asclepius],"[40] the god of healing with whom Aristides desperately wanted to connect. Marcus Aurelius wrote his *Meditations* as an act of self-examination and quest for ethical truth. Unlike Ignatius and Aristides, Marcus does not turn his very self over to the divine. Moreover, rather than the body as the exclusive site of suffering, Marcus sees his suffering in terms of the pressure brought to bear on his mental and emotional life because of the presence of death and pain in the world. Marcus transforms his experience of the world into an intrapersonal dialogue about how to live. These texts, especially the *Meditations*, have been seen by scholars as a clear point on the trajectory toward the subject-centered self because they use introspective discourse to express self-scrutiny without interpersonal dialogue. However, it is preferable to view such discourse as a form or internalized dialogue about ethical life that derives from philosophical discourse between student and teacher, the typical method by which Marcus, Aristides, and Epictetus were trained.[41]

Foucault's version of ancient subjectivity has been challenged as an account of the self that is merely an aesthetic picture of self-fashioning in the thought of the early Roman Empire.[42] There is a lack of normative reference points in Foucault's argument; in other words, his whole approach rests on the projection of personal experience and preference rather than on the ancient preoccupation with norms for a good human life.[43] Consequently, Foucault and those who follow him have constructed for this period a subject-centered self that is the site for their critique of Western social and political systems and institutions.[44] Other scholars have doubted that this version of subjectivity is relevant to ancient thought.[45] As far as Epictetus and Marcus Aurelius are concerned, they are more interested in practical or normative ethics, thereby pushing the pendulum toward a more participant notion of the self. That is, although particularity and contingency are central to ethical deliberation, these authors are directed toward the capabilities of virtue needed for a good human life. Intrapersonal dialogue results in knowledge of how to live in the world and how to connect with the divine.

THE LATE ANTIQUE SELF

Similar views that a subject-centered, individualist self can be seen emerging have been made for the Roman Christian authors from the fourth to the sixth century: Prudentius, Paulinus of Nola, Augustine, Boethius. Yet, if examined more closely, one discovers that these authors as well exhibit in their writing a self that is objective and participatory, a self that takes on the ethical point of view of early Christian thought. Parallel to the Stoics and the Epicureans, the Christian conception of the self is articulated in new ways. Pagan moral agency develops into a notion of Christian agency in which a Christian adopts specific doctrines and undertakes a search for knowledge of God. The aforementioned authors articulate versions of the Christian self that seek a union with the divine. The quest for this union takes place "inside," through an internalization of doctrine, biblical stories, and universal salvation history.

The two most important innovations of the period are (a) the establishment of a typologically constructed soul in which the characters, events, and stories from the Bible are the foundation of ethical deliberation, and (b), which is related to (a), the idea of the memory as an inner space and a meeting point with the divine that is accessible only through an inward turn that triggers intra- and interpersonal dialogue. As the European medieval world takes shape, the inward turn accompanied by dialogue to achieve knowledge of God reflects a self that is individualist in its quest, but profoundly objective and participant in its form. Another consequence of this way of thinking is that an understanding of the universe can occur only through an understanding of the history of humankind and its possibility for salvation.[46]

The typologically constructed soul is on display in the poetry of Prudentius, who lived into the fifth century C.E. In his work the *Apotheosis*, Prudentius establishes the "first" typological relationship: "Christ is the form of the Father, we are the form and image of Christ."[47] In arguing for the necessity of Christ as integral to the nature of God the Father, Prudentius yokes together an archetypal and typological triad: God, Christ, and humans. This typology originates in the fundamental proposition from *Genesis* that God made humans in God's own image.[48] Prudentius applies the language of typology, *forma* and *imago*, to link God and human through the mediation of the Word, Christ. The self in its archetypal conception in history and ontology is relational. Both the reader and poet are assumed within the poetry to be seekers after difficult knowledge, knowledge of their own souls, God, and their relationship. The point of this search is for the reader to rediscover—or in Platonist terms, "to recollect"—his relationship to God through the typology God/Christ/human.[49]

Moreover, for Prudentius, recognitions of various narrative typologies play a central role in ethical deliberation and the portrayal of free will. The poet uses the typological example of Lot's escape from Sodom and Gomorrah.[50] Lot's wife chose to look back at the cities, thus turning into a pillar of salt. By confronting the reader with this story, Prudentius associates the reader with a choice between Lot, who did not look back at the cities being destroyed, and his wife, who did. The implied typological choice for the reader is either Lot or Lot's wife. The right choice, which proceeds from a deliberation about a narrative of two characters, Lot and his wife, affirms the typology of God/Christ/human because each character in the story reflects a person's relationship to the divine. Moreover, this typological choice between Lot and his wife indicates free will. In the story of Lot the choice is determined by whether the human being has the right sense of faith. Lot listens to God and his wife ignores God. These character-types gleaned from a narrative form the essence of an ethical deliberation. In the realm of poetry, Prudentius expands the explanatory power of typology, a literary trope, from its use in structuring stories to a use that provides the content of the memory in ethical deliberation. Memory forms the basis of Prudentius's use of typologies. The reader must remember Old Testament stories, understand the connections to the Gospel narratives, and enact a moral choice these connections represent.

For Augustine, a younger contemporary of Prudentius, the remembering self is the self: "the memory . . . is awe-inspiring in its profound and incalculable complexity . . . Yet this thing I am."[51] Augustine concludes that to confess one's thoughts and desires is to remember them, and remembering is a never-ending process where thoughts and desires are collected and united. It is an inner space accessible by turning inward and it is a place where God is: "Don't go outward; return within yourself. In the inward man dwells the truth."[52] And as with Prudentius, inwardness links a human being with the divine so that the final relation is human—Christ—God in ascending order.[53]

In the *Confessions*, Augustine defines memory as a concrete metaphor, an architectural object with an ethical dimension.[54] Furthermore, at *Confessions* 10.25.36, memory becomes a secret hiding place for God: "Where in my consciousness, Lord, do you dwell? . . . What kind of sanctuary have you built for yourself? You conferred this honour on my memory that you should dwell in it . . . But you were not there." Augustine's confessional writing implies and intensifies social relations; in exposing one's personal history of thoughts and desires, one invites the judgment and advice of others, thus bringing one's bodily desires and appetites under control and encouraging confession among others.[55] Internal reflection conceived as a form of intrapersonal dialogue among mental parts indicates an objective self once again. It is a sense of self that internalizes participation in a community that is undergoing the same debate about the nature of the God. Augustine appears to have furnished the explanatory power to describe the Greco-Roman sense of self without positing a self as a thing or a space sitting *behind* all thoughts, emotions, and decisions.

The letters of Paulinus of Nola, a contemporary of Augustine and a Roman senator who renounced his wealth and position to live as a Christian, highlights the importance of interpersonal dialogue and internal reflection as seen in Augustine's confessional writing. His letters assert and negotiate bonds between friends and associates in terms of their relationship to Christ. Often in his letters we encounter "Yours in Christ." The love between friends anticipates the conversion experience as the true love of Christ.[56] By exchanging letters, Paulinus establishes a connection with the addressee and the letter carrier. Moreover, Paulinus shares his conversion narrative, which, like Augustine's, implies an inward turn that facilitates access to a shared thing, the Trinity, rather than to something more personal.⁵⁷ For Paulinus, Augustine, and Prudentius, the self exists in virtue of relationships to others because Christ is "in" us. To participate in a spiritual community of human beings, who are focused on discovering God within, implies a lack of boundaries between people. Paradoxically, to discover one's true relational self, one must begin alone, introspectively.⁵⁸ Introspection can furnish knowledge of Christ; and that knowledge is the basis for friendship with others. In the same period, Neoplatonist thought parallels this Christian understanding of self, defined according to a community that separately reflects through an inward turn to seek the divine. For instance, Proclus (410 or 412–485 C.E.) carries on the Neoplatonic tradition of searching for the true self by beginning with Plato's emphasis on the Delphi command to know oneself. In his commentary on Plato's *First Alcibiades*, Proclus argues that the self endeavors to ascend to the divine intellect, *nous*. To know oneself is to be self-conscious, which makes possible a union with the divine.⁵⁹

Like Augustine, Prudentius, and Paulinus, the sixth-century writer Boethius turns inward and constructs an imagined dialogue that the reader is invited to observe in order to spur her own dialogue and deliberation. Boethius was a senior Roman official who was arrested and executed. While in prison, he wrote the influential *Consolation of Philosophy*, in which he carries on a dialogue with a personified Philosophy to free himself from the limitations of this world and achieve a relationship with God. A work of prose, punctuated with frequent poems, the narrative of Boethius's life can be understood as a description of every human life. This forms the basis for an ethical deliberation on wealth, self-sufficiency, freedom, evil, fate, and providence. While Philosophy is Boethius's interlocutor, she also represents the rationality necessary to shake Boethius out of his emotional and limited earthly perspective. Like the "Platonist books" of the *Confessions* that lead Augustine toward his final conversion experience, Philosophy personified, who is Platonic in her emphasis on reason, leads Boethius to shed his earthly constraints and accomplish a union with the divine⁶⁰—goals that both Augustine and Prudentius establish in their works. This internal dialogue allows Boethius to see the truth of his own life through rational reflection. However, as with Plato and Augustine, the limits of philosophy and rationality are exposed. Platonist concepts of unity, goodness, and beauty cannot achieve a transcendent vision of God. At the end of the *Consolation*, Boethius fails to consolidate the partial visions of God he has been able to achieve through philosophy. But he is certainly on his way.

Poetry's function in the *Consolation* confirms this reading. At times poetry is a hindrance to union with God because of its Platonic reputation for superficiality, earthly pleasure, and disregard for the truth. On the other hand, as with Plato, Prudentius, and Augustine, the narrative function of poetry can be useful, especially if it portrays the narrative of the ascent to God as heroic.⁶¹ And the presence of poetry throughout the *Consolation* does not reflect a "new"

(Platonic) poetry that would be acceptable to philosophy and to God.[62] Rather, the all too human perspective that Plato and philosophy habitually assign to poetry can never be eliminated. The self is represented as a perpetual dialogue between emotion, suffering, and mortality—the purview of poetry—and the unity and eternity of philosophical ideas. For Boethius, this is what we are.

THE EARLY MEDIEVAL SELF

From the late fifth to the middle of the eighth centuries Europe split into East and West. While political unity was somewhat the norm in the East, the West experienced a period of change and decentralized politics that was unabated until the regime of Charlemagne. Taking advantage of its administrative reach (which in part took the form of monasteries) and concerned about the lingering vestiges of paganism, the Carolingian dynasty emphasized the Christianization of villages and aristocratic villas of the countryside.[63] Monks and clerics were supported by powerful monasteries, which furnished the Carolingian potentates with an effective network to spread and reinforce Christian ideology.

The extreme situation of Boethius (i.e., in jail, awaiting execution) stimulated him to think intensely about himself and his relationship to the divine. This kind of literary engagement, minus the harrowing life circumstances, was not limited to an aristocrat such as Boethius. In the late antique period, Western aristocracies defined themselves according to a literary lifestyle that required a firm grounding in classical texts and originated in their civilian (not military) identity.[64] Boethius and other Roman aristocrats bequeathed this literary culture to the circle of scholars, intellectuals, and artists around Charlemagne. The revival of classical Latin among the intellectual elite (for example, Alcuin, and earlier, Bede) triggered, through their own works, the popularization of doctrines of the church fathers such as Ambrose, Augustine, and Jerome. Moreover, an important part of this literary culture was the heroic stories of the Passions of the saints, which also originate in fourth-century literature.[65] Thus, Carolingian literature looked to the Roman Christian thought world of the fourth, fifth, and sixth centuries for its sense of self.

However, the role of a broad monastic network, which disseminated Christian doctrines and Passion stories—often rewritten by Carolingian intellectuals—became central to the self's quest for divine union. Right up to the tenth century, monasteries capitalized on a new notion of "spiritual sacrifice/offering." In the fourth century, the age of Prudentius, Paulinus, and Augustine, "Christ's death was expounded as a surrender of self for mankind."[66] Early Christians understood a sacrifice to be personal and spiritual; for instance, one's own soul and body/life. In the Carolingian period, monasteries modified this concept: They accepted significant money and property from wealthy individuals and in return gave spiritual benefits including prayers for salvation. These spiritual rewards came as a result of earthly, material gifts that enriched monasteries. The Augustinian and Prudentian spiritual surrender of self in order to gain knowledge of and union with God had, by the time of Charlemagne, transformed into a more overt, reciprocal transaction in which the individual offered an earthly gift and received the "countergift" of a cleric's help in achieving divine knowledge and communion.[67]

These developments during the Carolingian dynasty do not radically alter the fundamental sense of self that winds its way through the history of the Greco-Roman world; that is, the sense of self that rests on an internal psychological dynamic capable of inter- and intrapersonal dialogue concerning norms, and, as a result, is relational by virtue of its attachments to other people and the divine. However, in the early medieval period, the Church explicitly inserted itself as the gatekeeper between God and the self. The church's infrastructure of monasteries, supported by political power, became the means for experiences that made one a true Christian. The Carolingians seemed to have moved on from the exemplary story of Augustine as portrayed in his *Confessions*, alone in the Milan garden, undergoing a conversion that is triggered through an inward turn. Rather, one could enlist the help of the church directly for one's quest for true Christian knowledge and salvation—for a price. The age of Charlemagne succeeded in reconnecting Western Europe's "townless society" of the seventh and early eighth centuries to a unifying political-religious presence that determined how individuals understood themselves. A sense of self becomes inextricably bound to this new institutional reality. One had to enlist the aid of the Church, which, as an advocate for the self, was necessary for individual salvation. While still internal and relational in its essence, the self's relation to God was negotiated through this material engagement with the Carolingian Church.

Bede (c. 673–735), a monk in Northumbria, wrote several Passions of Saints, including the *Life of Felix*, in which he looks back to Paulinus of Nola's series of poems on Felix as well as to Prudentius, whom he also mentions in this simple prose adaptation. Bede pays close attention to patterns of moral behavior and qualities, which are exhibited by Felix and which the audience of monks (and the layman) is encouraged to apply to themselves.[68] By reading these highly entertaining and inspirational works, the monk or layman would begin the process of turning inward to find God. And this in turn would promote relationships with fellow monks and their charges who were engaged in the same process, just as the addressees of Paulinus's letters, nearly three centuries earlier, experienced with each other. Moreover, Bede and other authors of such works seem to implicitly assume that God has bestowed superior gifts on certain people like the saints.[69] This assumption dovetails with the assumption of exceptionalism reflected in a layman offering a monastery money or property in return for spiritual benefits. A Christian layman internalized the narrative and the meaning of the unique spiritual sacrifice of a saint.

Bede's influence is felt directly in the Carolingian age through Alcuin (c. 730–804), another Northumbrian, who was a major figure in the circle of Charlemagne and brought with him Bede's scholarship to the court. In his work, Alcuin reinforces the idea of the self as an effect of an inward turn, a radical reflection involving one's mental parts or capacities (e.g., emotion, reason, the will). Alcuin indicates this sense of self in two areas of his work. First, in *On the Virtues and Vices*, Alcuin follows in the footsteps of Prudentius in writing about the battle of the virtues and vices within the soul.[70] And second, in his many letters to Charlemagne's political lieutenants and confidantes, including the emperor himself, Alcuin capitalizes on his ecclesiastical knowledge to give advice and carry on a dialogue of self-improvement. Both sets of texts carry on the importance of intra- and interpersonal dialogue that set the conditions for self-knowledge and interactions with one's peers, superiors, and even subjects. They also emphasize the inward turn, which the works of Augustine, Prudentius, and Boethius established for the early medieval Christian tradition. Alcuin's epitaph reinforces this way of thinking through its references to the work of Bede, Prudentius, Ovid, Vergil, and Christian and Roman inscriptions.[71] His conclusion is that pagan wisdom allows one to turn inward and discover the Christian soul, the only immortal part of us. Notwithstanding the Carolingian view of spiritual sacrifice, the Christian sense of self in the fourth and fifth centuries remained strong among the aristocratic and intellectual elite of the seventh and eighth centuries. Moreover, the addressees of Alcuin's letters are encouraged to undertake an inner dialogue that should result in behavior in line with ethical (Christian) norms. We have come full circle back to the epics of Homer in which deliberation of communal norms for correct individual behavior forms the basis of what a self is.

CONCLUSION

The late antique and early medieval self, like the Homeric, Platonic, and Hellenistic self, are capable of inner dialogue that mimics an intra- and interpersonal dialogue about norms and ideals, including the ideal of connecting with God. The self as an inner space to which one turns to find the divine is represented as a dialogue between author, an (imagined) interlocutor, and reader. If there is development in the history of the self in the West between 800 B.C.E. and 900 C.E., it is not a progression from a fragmented and nondeliberative self to a unitary and autonomous self. Instead, the complex and complicated story of the Greco-Roman self is stitched together by means of the conceptual thread that a self consists of a psychological structure that dialogues with itself as if with other persons in the world or with a god. This interpersonal dialogue indicates a self that participates in the ethical debates of the community to establish norms and ideals. Ancient, late antique, and early medieval thought on the self is anchored on a broad notion of moral agency, that is, the choice one makes concerning either an action that will affect others, or the doctrine one lives by, or even the god that one worships. Moral agency implies an inner, reflective life *and* vital relationships with others, including the divine. Interiority is also an integral part of the story of the self. Whether figured as a set of parts, a type of structure, an inner space, or a series of stories, before the Early Modern period, the West conceived of the self as separate from the world but simultaneously relational.

NOTES

Chapter 1

1. *Interpretation of Dreams* 1.13.
2. V. French, "Midwives and Maternity Care in the Roman World," in "Rescuing Creusa: New Methodological Approaches to Women in Antiquity," special issue, *Helios*, New Series 13, no. 2 (1986): 69–84.
3. Pliny the Elder *Letters* 4.21.
4. Pliny the Elder *Natural History* 7.16.72; Juvenal *Satire* 15.138–40; Plutarch *Consolation to his Wife* 11; K. Bradley, *Discovering the Roman Family* (Oxford: Oxford University Press, 1991), 28; V.M. Hope, *Death in Ancient Rome: A Sourcebook* (London: Routledge, 2007), 180–83.
5. W.V. Harris, "Child-Exposure in the Roman Empire," *Journal of Roman Studies* 54 (1994): 1–22; M. Corbier, "Child Exposure and Abandonment," in *Childhood, Class and Kin in the Roman World*, ed. S. Dixon, 52–73 (London: Routledge, 2001); E. Scott, "Unpicking a Myth: The Infanticide of Female and Disabled Infants in Antiquity," in *Proceedings of the Theoretical Roman Archaeology Conference 2000*, ed. G. Davies, A. Garner, and K. Lockyear, 143–51 (Oxford: Oxbow, 2001).
6. M. Golden, "Did the Ancients Care When Their Children Died?" *Greece and Rome* 35, no. 2 (1988): 152–63; and Golden, *Children and Childhood in Classical Athens* (Baltimore: Johns Hopkins University Press, 1990), 89; for a selection of evidence see Hope, *Death in Ancient Rome*, 183–88.
7. N. Demand, *Birth, Death and Motherhood in Classical Greece* (Baltimore: John Hopkins University Press, 1994), 63; Galen *On the Natural Faculties* 3.3.151.
8. Pliny *Natural History* 7.6.41.
9. Pliny *Natural History* 28.77.250; 30.143.124.
10. Pliny *Natural History* 30.44.129–30.
11. Augustine *City of God* 4.11; M. Harlow and R. Laurence, *Growing Up and Growing Old in Ancient Rome: A Life Course Approach* (London: Routledge, 2002), 39; M-L. Hänninen, "From Womb to Family: Rituals and Social Conventions Connected to Roman Birth," in *Hoping for Continuity: Childhood, Education and*

Death in Antiquity and the Middle Ages, ed. K. Mustakallio, J. Hanska, H-L Sainio, and V. Vuolanto, *Acta Instituti Romani Finlandiae* 33 (special issue, 2005): 39.

12. Varro *Ling* 5.69; Hänninen, "From Womb to Family," 50.
13. Demand, *Birth, Death and Motherhood,* 88–89; Hänninen, "From Womb to Family," 51–52.
14. Soranus *Gynaecology* 1.1.13.
15. French, "Midwives and Maternity Care."
16. Soranus *Gynaecology* 2.2.2.
17. Soranus *Gynaecology* 2.2.3.
18. For further images see Demand, *Birth, Death and Motherhood,* 123–24.
19. Plutarch *Lycurgus* 27.2–3.
20. Demand, *Birth, Death and Motherhood,* 129.
21. Hänninen, "From Womb to Family," 58.
22. Soranus *Gynaecology* 2.6.10.
23. B. Shaw, "Raising and Killing Children: Two Roman Myths," *Mnemosyne* 54, no. 1 (2001): 31–77.
24. Golden, *Children and Childhood,* 23–24; Demand, *Birth, Death and Motherhood,* 8–9.
25. Harlow and Laurence, *Growing Up,* 39–40.
26. Demand, *Birth, Death and Motherhood,* 19, 68.
27. R. Garland, *The Greek Way of Death* (London: Duckworth, 1985), 16; D. Noy, "Goodbye Livia: Dying in the Roman Home," in *Memory and Mourning: Studies on Roman Death,* ed. V.M. Hope and J. Huskinson (Oxford: Oxbow, forthcoming).
28. A.J.L. Van Hooff, "Ancient Euthanasia: 'Good Death' and the Doctor in the Graeco-Roman World," *Social Science and Medicine* 58, no. 5 (2004): 975–85; C. Edwards, *Death in Ancient Rome* (New Haven, CT: Yale University Press, 2007); V.M. Hope, *Roman Death: The Dying and the Dead in Ancient Rome* (London: Hambledon Continuum, 2009), 50–57.
29. For a summary of deathbed and display rites and a listing of relevant ancient sources see J.M.C. Toynbee, *Death and Burial in the Roman World* (London: Thames & Hudson, 1971), 43–45; D. Kurtz and J. Boardman, *Greek Burial Customs* (London: Thames & Hudson, 1971), 144; Garland, *Greek Way,* 23–31; K. Stears, "Death Becomes Her: Gender and Athenian Death Ritual," in *The Sacred and the Feminine in Ancient Greece,* ed. S. Blundell and M. Williamson (London: Routledge, 1998), 114–15; Hope, *Death in Ancient Rome,* 93–99; Hope, *Roman Death,* 71–74.
30. Garland, *Greek Way,* 38–47; H. Lindsay, "Death-Pollution and Funerals in the City of Rome," in *Death and Disease in the Ancient City,* ed. V.M. Hope and E. Marshall, 152–73 (London: Routledge, 2000).
31. Servius *On Virgil's Aeneid* 6, 8, 11.2.
32. Kurtz and Boardman, *Greek Burial Customs,* 144.
33. Garland, *Greek Way,* 45.
34. Plutarch *Sulla* 35.
35. Seneca the Younger *Consolation to Marcia* 15.3.
36. J. Bodel, "Graveyards and Groves: A Study of the *Lex Lucerina,*" *American Journal of Ancient History* 11 (1994); J. Bodel, "Dealing with the Dead: Undertakers, Executioners, and Potter's Fields in Ancient Rome," in *Death and Disease in the Ancient City,* ed. V. Hope and E. Marshall, 128–51 (London: Routledge, 2000); J. Bodel, "The Organisation of the Funerary Trade at Puteoli and Cumae," in *Libi-*

tina e Dintorni: Atti dell' XI Rencontre franco-italienne sur l'épigraphie (Libitina, 3), ed. S. Panciera, 149–70 (Rome: Quasar, 2004).
37. Garland, *Greek Way,* 32–33.
38. H. I. Flower, *Ancestor Masks and Aristocratic Power in Roman Culture* (Oxford: Clarendon Press, 1996); J. Bodel, "Death on Display: Looking at Roman Funerals," in *The Art of Ancient Spectacle,* ed. B. Bergmann and C. Kondoleon, 259–81 (New Haven, CT: Yale University Press, 1999); J. Pollini, "Ritualizing Death in Republican Rome: Memory, Religion, Class Struggle and the Wax Ancestral Mask Tradition's Origins and Influence on Veristic Portraiture," in *Performing Death: Social Analyses of Funerary Traditions in the Ancient Near East and Mediterranean,* ed. N. Laneri, 237–85 (Chicago: University of Chicago Press, 2007).
39. Polybius *Histories* 6.53–4; for funeral processions, biers, and speeches see Toynbee, *Death and Burial,* 46–47; Bodel, "Death on Display"; D. Noy, "Building a Roman Funeral Pyre," *Antichthon* 34 (2000): 30–45; Hope, *Death in Ancient Rome,* 100–107; Hope, *Roman Death,* 74–79.
40. Garland, *Greek Way,* 36; Hope, *Roman Death,* 77.
41. A. D. Nock, "Cremation and Burial in the Roman Empire," *Harvard Theological Review* 25 (1932): 321–59, reprinted in *Essays on Religion and the Ancient World,* ed. Z. Stewart, 2 vols. (Oxford: Clarendon Press, 1972); I. Morris, *Death-Ritual and Social Structure in Classical Antiquity* (Cambridge: Cambridge University Press, 1992).
42. Cicero *Tusculan Disputations* 1.45.108; Lucian *On Funerals* 21.
43. Cicero *On the Laws* 2.22.55; Petronius *Satyricon* 65.
44. Garland, *Greek Way,* 40.
45. See Kurtz and Boardman, *Greek Burial Customs;* Toynbee, *Death and Burial;* H. Von Hesberg, *Römische Grabbauten* (Darmstadt: Wissenschaftliche Buchgesellschaft, 1992).
46. For overview and relevant ancient sources see Toynbee, *Death and Burial,* 61–64; Garland, *Greek Way,* 105–20; Hope, *Roman Death,* 98–102.
47. B. Malinowski, "Magic, Science and Religion," in *Science, Religion and Reality,* ed. J. Needham (London: MacMillan, 1925), 19–84; J. Frazer, *The Fear of the Dead in Primitive Religion,* 3 vols. (London: Macmillan, 1934–1936).
48. Ulpian, *Digest* 11.7.14.7.
49. D. Felton, *Haunted Greece and Rome: Ghost Stories from Classical Antiquity* (Austin: University of Texas Press,1999), 9–12; S. I. Johnston, *Restless Dead: Encounters between the Living and the Dead in Ancient Greece* (Berkeley: University of California Press, 1999).
50. Suetonius *Caligula* 59.
51. See, for example, Pliny the Younger *Letters* 7.27.
52. Suetonius *Vespasian* 19.
53. Cassius Dio *Histories* 75.4.2–5.5.
54. S. Price, "From Noble Funerals to Divine Cult: The Consecration of Roman Emperors," in *Rituals of Royalty: Power and Ceremonial in Traditional Societies,* ed. D. Cannadine and S. Price, 56–105 (Cambridge: Cambridge University Press, 1987).
55. J. R. Patterson, "Patronage, Collegia and Burial in Imperial Rome," in *Death in Towns,* ed. S. Bassett, 15–27 (Leicester: Leicester University Press, 1992).
56. Bodel, "Graveyards and Groves," 41–42; Bodel, "Dealing with the Dead."
57. Varro *On the Latin Language* 5.25.

58. Horace *Satires* 1.8.8–22; K. Hopkins, *Death and Renewal* (Cambridge: Cambridge University Press, 1983), 208–9; Bodel, "Graveyards and Groves," 40; E.-J. Graham, *Death, Disposal and the Destitute: The Burial of the Urban Poor in Italy in the Late Roman Republic and Early Empire,* British Archaeological Reports International Series 1565 (Oxford: Archaeopress, 2007).
59. See Bodel, "Graveyards and Groves," 83; D. Kyle, *Spectacles of Death in Ancient Rome* (London: Routledge, 1998), 169–70.
60. C. Edwards, "Unspeakable Professions: Public Performance and Prostitution in Ancient Rome," in *Roman Sexualities,* ed. J.P. Hallet and M.B. Skinner, 66–95 (Princeton, NJ: Princeton University Press, 1997); Bodel, "Dealing with the Dead"; Bodel, "Funerary Trade."
61. V.M. Hope, "Contempt and Respect: The Treatment of the Corpse in Ancient Rome," in *Death and Disease in the Ancient City,* ed. V. M. Hope and E. Marshall, 104–27 (London: Routledge, 2000), 118.
62. A.J.L. Van Hooff, *From Autothanasia to Suicide: Self-Killing in Classical Antiquity* (London: Routledge, 1990), 67; Hope, "Contempt and Respect," 119.
63. K.M. Coleman, "Fatal Charades: Roman Executions Staged as Mythological Enactments," *Journal of Roman Studies* 80 (1990): 44–73.
64. Appian *Civil War* 1.120.
65. Petronius *Satyricon* 111.
66. Kyle, *Spectacles,* 53.
67. Kyle, *Spectacles.*
68. Tacitus *Annals* 6.29.1.
69. Suetonius *Vitellius* 17.
70. Cassius Dio *Histories* 11.1–6.
71. P. Stewart, *Statues in Roman Society: Representation and Response* (Oxford: Oxford University Press, 2003); E. Varner, "Execution in Effigy: Severed Heads and Decapitated Statues in Imperial Rome," in *Roman Bodies: Antiquity to the Eighteenth Century,* ed. A. Hopkins and M. Wyke, 66–82 (London: The British School at Rome, 2005).
72. H.I. Flower, *The Art of Forgetting: Disgrace and Oblivion in Roman Political Culture* (Chapel Hill: University of North Carolina Press, 2006); E.R. Varner, *Mutilation and Transformation: Damnatio Memoriae and Roman Imperial Portraiture* (Leiden: Brill, 2004).
73. N. Fields, "Headhunters in the Roman Army," in *Roman Bodies: Antiquity to the Eighteenth Century,* ed. A. Hopkins and M. Wyke, 55–65 (London: The British School at Rome, 2005).
74. J.L. Voisin, "Les Romains, chasseurs de têtes," in *Du Châtiment dans la Cité: Supplices Corporels et Peine de Mort dans le Monde Antique,* Collection de l'École Française de Rome 79, 241–93 (Rome: l'École Française de Rome, 1984); A. Richlin, "Cicero's Head," in *Constructions of the Classical Body,* ed. J. Porter, 190–211 (Ann Arbor: University of Michigan Press, 1999).
75. Plutarch *Antony* 20.2; Plutarch *Cicero* 48.4–49.2.
76. Suetonius *Galba* 20; Plutarch *Galba* 28.2–3.
77. E.-J. Graham, "Memory and Materiality: Re-embodying the Roman Funeral," in *Memory and Mourning: Studies on Roman Death,* ed. V.M. Hope and J.A.R. Huskinson (Oxford: Oxbow, forthcoming).
78. Pliny the Elder *Natural History* 16.60.139.

79. D. J. Ochs, *Consolatory Rhetoric: Grief, Symbol and Ritual in the Greco-Roman Era* (Columbia: University of South Carolina Press, 1993).
80. For listing of sources see A. Richlin, "Emotional Work: Lamenting the Roman Dead," in *Essays in Honor of Gordon Williams: Twenty-five Years at Yale*, ed. E. Tywalsky and C. Weiss, 229–48 (New Haven, CT: Schwab, 2001); Hope, *Roman Death*, 72, n. 10.
81. Paulus *Opinions* 1.21.2–5; see also Cicero *On the Laws* 2.23.59; Plutarch *Numa* 12.
82. Cicero *Letters to His Friends* 9.20.3; Seneca *Consolation to Marcia* 2.3.4; Plutarch *Consolation to His Wife* 4.
83. Richlin, "Emotional Work"; A. Corbeill, *Nature Embodied: Gesture in Ancient Rome* (Princeton, NJ: Princeton University Press, 2004), 67–106; K. Mustakallio, "Roman Funerals: Identity, Gender and Participation," in *Hoping for Continuity. Childhood, Education and Death in Antiquity and the Middle Ages*, ed. K. Mustakallio, J. Hansks, H.-L. Sanio, and V. Vuolanto (Rome: *Acti Instituti Romani Findlandiae* 33, 2005), 179–90; A. Suter, *Lament: Studies in the Ancient Mediterranean and Beyond* (Oxford: Oxford University Press, 2008).
84. Compare Stears, "Death Becomes Her."
85. Mustakallio, "Roman Funerals."
86. Cicero *On the Laws* 2.23.59.
87. Livy 22.53.3, 22.56.4–5; Valerius Maximus *Memorable Deeds and Sayings* 1.1.15.
88. W. Eck, A. Caballos, and F. Fernández, *Das Senatus Consultum de Cn. Pisone Patre* (Munich: C. H. Beck, 1996).
89. Mustakallio, "Roman Funerals"; Hope, *Roman Death*, 125–32.
90. Hope, *Roman Death*, 145–48.
91. Corbeill, *Nature Embodied*, 67–106.
92. D. Dutsch, "*Nenia:* Gender, Genre and Lament in Ancient Rome," in *Lament: Studies in the Ancient Mediterranean and Beyond*, ed. A. Suter, 258–79 (Oxford: Oxford University Press, 2008).
93. Dutsch, "*Nenia,*" 263.
94. For the body as a symbol of power and of embodied experience and as an interface, see, for example, L. Meskell, "Writing the Body in Archaeology," in *Reading the Body: Representations and Remains in the Archaeological Record*, ed. A. E. Rautman, 13–21 (University of Pennsylvania Press, 2000); Y. Hamilakis, M. Pluciennik, and S. Tarlow, eds., *Thinking through the Body. Archaeologies of Corporeality* (New York: Kluwer/Plenum, 2002), 1; S. Tarlow, "The Aesthetic Corpse in Nineteenth-Century Britain," in *Thinking through the Body: Archaeologies of Corporeality*, ed. Y. Hamilakis, M. Pluciennik, and S. Tarlow, 85–97 (New York: Kluwer/Plenum, 2002), 87.
95. Stears, "Death Becomes Her."

Chapter 2

1. The ancient Greek word denoting health is *hygieia*. J. Jouanna, *Hippocrates*, trans. M. B. DeBevoise (Baltimore: Johns Hopkins University Press, 1999), 468n1) provides an enlightening etymology: "The word denoting health in Greek (*hygieia*) is in fact a compound whose second element (*-gi-*) stems from the Indo-European

root signifying 'life,' like the initial element (*bi-*) of 'biology' (literally, 'science of life'), and whose first element (*hy-*, from the Indo-European *su-*) means 'well.' Thus, the Greek term that we translate by 'health' literally means 'the state of one who is well in life.'" Health was also a goddess for the Greeks, as can be seen from the first line of the Hippocratic *Oath,* cited in W.H.S. Jones, *Hippocrates,* vol. 2 (Cambridge, MA: Harvard University Press, 1923): "I swear by Apollo Physician, by Asclepius, by Health (*Hygeian*), by Panacea and by all the gods and goddesses, making them my witnesses, that I will carry out, according to my ability and judgment, this oath and this indenture." Generally speaking the ancient Greek word denoting "disease" is *nousos* (in Ionic, the dialect of ancient Greek in which the Hippocratic Corpus is composed; in Attic Greek the word is *nosos*), though of course disease can manifest itself in a number of particular conditions such as fever (*puretos*), consumption (*phthisis*), or epilepsy (often called the "sacred disease" [*hê hierê nousos*] by the ancients), to name only a few. Indeed, epilepsy was such a fearsome condition in antiquity that it merited its own treatise among the Hippocratic writings: *On the Sacred Disease,* in Jones, *Hippocrates,* vol. 2; cf. J. Laskaris, *The Art Is Long: On the Sacred Disease* (Leiden: Brill 2002). Besides *nousos* or *nosos,* there is the related noun to denote a sickness or an illness, *nosêma,* and other general words for sickness or illness such as *astheneia* and *arrôstia.* As a cautionary note I should add that many of the diseases of antiquity bear little resemblance to diseases as we have them today, though often we retain the ancient names.

2. Because the treatises of the Hippocratic Corpus are anonymous, and it's unclear whether any one of them were actually written by the historical Hippocrates of Cos (the so-called "Hippocratic Question"), I adopt the convention of placing the name of Hippocrates in brackets when referring to Hippocratic treatises.

3. For a provocative survey of the prominence of disease in ancient Greek literature, see G.E.R. Lloyd, *In the Grip of Disease* (Oxford: Oxford University Press, 2003); for general surveys of ancient medicine, cf. E.D. Phillips, *Greek Medicine* (London: Thames & Hudson, 1973), L. Conrad, M. Neve, V. Nutton, R. Porter, and A. Wear, *The Western Medical Tradition 800 B.C. to A.D. 1800* (Cambridge: Cambridge University Press, 1995), and most recently H. King, *Greek and Roman Medicine* (London: Bristol Classical Press, 2003); and V. Nutton, *Ancient Medicine* (London: Routledge, 2004). One should also consult the valuable papers in L. Edelstein, *Ancient Medicine: Selected Papers of Ludwig Edelstein,* ed. O. Temkin and C.L. Temkin (Baltimore: Johns Hopkins University Press, 1967); and P.J. van der Eijk, *Medicine and Philosophy in Classical Antiquity* (Cambridge: Cambridge University Press, 2005).

4. Plato *Republic* 2.372e.

5. Cf. the papers of R. Brock, "Sickness in the Body Politic: Medical Imagery in the Greek Polis," in *Death and Disease in the Ancient City,* ed. V.M. Hope and E. Marshall, 24–34 (London: Routledge, 2000); and J.C. Kosak, "*Polis Nosousa*: Greek Ideas about the City and Disease in the Fifth Century BC," in Hope and Marshall, *Death and Disease,* 35–54, for an extended discussion of the application of health and disease to the ancient *polis.*

6. This is an approximation of the words that the centurion of Capernaum speaks to Jesus, when Jesus comes to heal the illness of the centurion's slave; cf. Matthew 8:5–13, Luke 4:6–7, and John 4:46–53.

7. Plato has Timaeus remark that health (*hygies*) is a "consistent, uniform, and proper proportion (*ana logon*)" of the elements (earth, air, fire, and water) both within and

without the body, while a disproportion among them causes "a multiplicity of altered states (*alloiotêtas pampoikilas*), and an infinity of diseases and degenerations (*nosous phthoras te apeirous*)," *Timaeus* 82b, in J. M. Cooper, ed., *Plato: Complete Works* (Indianapolis: Hackett, 1997). One can almost hear the sigh in the voice of the Roman writer Pliny when, in the course of describing the medicinal properties of plants in his massive *Natural History* (25.7.23) he laments: "To reflect on this makes one pity the lot of man (*humanae*); besides chances and changes and strange happenings that every hour brings, there are thousands of diseases (*milia morborum*) that every mortal has to dread" (in W.H.S. Jones, *Pliny: Natural History*, vol. 7 [Cambridge, MA: Harvard University Press, 1956]).

8. The central nosological treatises of the Hippocratic Corpus are the series *Diseases 1, Diseases 2*, and *Diseases 3; Affections;* and *Internal Affections*; all can be found in P. Potter, *Hippocrates*, vols. 5 and 6 (Cambridge, MA: Harvard University Press). The treatise *Diseases 4* is often collected with *On Generation* and *On the Nature of the Child* because it shares some of its subject matter with these works; see I. M. Lonie, *The Hippocratic Treatises "On Generation," "On the Nature of the Child," "Diseases IV"* (Berlin: Walter de Gruyter, 1981). Many of the individual treatises of the corpus also treat specific pathological conditions, such as *Ulcers, Haemorrhoids,* and *Fistulas* (in P. Potter, *Hippocrates*, vol. 8 [Cambridge, MA: Harvard University Press, 1995]) and *On Sight*, which deals with eye conditions (Greek text with English translation, introduction, and commentary in E. M. Craik, *Two Hippocratic Treatises* On Sight *and* On Anatomy [Leiden: Brill, 2006]). One could also find diseases or pathological conditions spoken of in some of the more programmatic treatises of the corpus, such as *On Breaths* (in Jones, *Hippocrates*, vol. 2), whose author argues that air, both external (winds) and internal (gases from digestion), causes all diseases. In the *Epidemics* (*Epidemics* 1 and 3 can be found in Jones, *Hippocrates*, vol. 1; *Epidemics* 2, 4, and 5–7 in W. Smith, *Hippocrates*, vol. 7 [Cambridge, MA: Harvard University Press,1994]) we are presented with a number of individual cases where doctors record their struggles attempting to heal various illnesses. Only in a few treatises does health take center stage, e.g., *Regimen in Health* and *Regimen* (both in W.H.S. Jones, *Hippocrates*, vol. 4 [Cambridge, MA: Harvard University Press, 1931]), both concerned with the proper diet and exercise one requires in order to maintain health.

9. The Roman encyclopedic writer Celsus observes that "at first the science of healing (*medendi scientia*) was held to be part of philosophy (*sapientiae*), so that the curing of diseases (*morborum curatio*) and the contemplation of the nature of things (*rerum naturae contemplatio*) came in through the same authorities; clearly because it was needed especially by those whose bodily strength (*corporum suorum robora*) had become weakened (*minuerant*) by quiet thinking (*quieta cogitatione*) and watching by night (*nocturnaque vigilia*)" (*On Medicine*, Prooemium 6–7; in W. G. Spencer, *Celsus: De Medicina*, vol. 1 [Cambridge, MA: Harvard University Press,1935]). Ancient medical writers and philosophers sometimes disputed which discipline was more knowledgeable and influential, since both shared important subject matters. Perhaps the most polemical treatise in the Hippocratic Corpus is *On Ancient Medicine*, where the author argues for medicine's priority over and autonomy from philosophy, attacking Empedocles in chapter 20. Melissus, a follower of Parmenides, is attacked for his monism in another Hippocratic treatise, *On the Nature of Man* 1. Aristotle, however, argues that medicine and natural philosophy, and doctors and philosophers, have much in common (cf. Aristotle, *Sense*

and Sensibilia 1.436a17–436b1 and *On Respiration* 21.480b21–30, with P. J. van der Eijk, "Aristotle on 'Distinguished Physicians' and on the Medical Significance of Dreams," in *Ancient Medicine in its Socio-Cultural Context: Papers Read at the Congress Held at Leiden University 13–15 April 1992*, 2 vols., ed. P. J. van der Eijk, H.F.J. Horstmanshoff, and P. H. Schrijvers, 447–59 [Amsterdam: Rodopi, 1995]). For the close ties of medicine and philosophy in antiquity, see Edelstein, *Ancient Medicine: Selected Papers*, 349–66, M. Frede, *Essays in Ancient Philosophy* (Minneapolis: University of Minnesota Press, 1987), 225–42; Jouanna, *Hippocrates*, 259–85; J. Longrigg, *Greek Rational Medicine: Philosophy and Medicine from Alcmaeon to the Alexandrians* (London: Routledge, 1993); and more recently, van der Eijk, *Medicine and Philosophy*, 1–42; and P. Pellegrin, "Ancient Medicine and Its Contribution to the Philosophical Tradition," in *A Companion to Ancient Philosophy*, ed. M. L. Gill and P. Pellegrin (Malden, MA: Blackwell, 2006), 664–85.
10. Tertullian *De anima* 2.
11. P. J. van der Eijk, "The Role of Medicine in the Formation of Early Greek Thought," in *The Oxford Handbook of Presocratic Philosophy*, ed. P. Curd and D. W. Graham (New York: Oxford University Press, 2008), 385–412, provides a recent thorough overview of the relationship between pre-Socratic philosophy and the Hippocratic Corpus.
12. See Lloyd, *Grip of Disease*, 14–39. Homer's *Iliad* (1.2.8–26) begins with the famous plague that devastates the Greek forces at Troy, as a result of Agamemnon's refusal to return the captured daughter of Chryses, a priest of Apollo. Chryses then prays to Apollo to punish the Greeks for this outrage, and a plague descends upon them, infecting the animals first but eventually infecting the Greek soldiers. Hesiod famously speaks of disease as Zeus's punishment for Prometheus's theft of fire in *Works and Days* (2.70–105).
13. Cf., e.g., [Hippocrates], *Airs, Waters, Places* 22 (in Jones, *Hippocrates*, vol. 2): "I too think that these diseases (*ta pathea*) are divine (*theia*), and so are all others, no one being more divine (*theioteron*) or more human (*anthrôpinôteron*) than any other; all are alike, and all divine. Each of them has a nature (*phusin*) of its own, and none arises without its natural cause (*phusios*)." Van der Eijk (*Medicine and Philosophy*, 56–60) argues that this seemingly paradoxical construal of diseases being at once divine, human, and natural means that diseases have unchanging, imperishable, and constant *natures* (or *forms*) that perpetually reoccur, yet they are all also treatable and curable by *human* means.
14. For Empedocles as a healer: frag. 112 (in G. S. Kirk, J.E.R. Raven, and M. Schofield, eds., *The Presocratic Philosophers* [Cambridge: Cambridge University Press, 1983], 313); for his notion that bodily tissues were ratios of the four elements: frag. 96 and 98 (in Kirk et al., *The Presocratic Philosophers*, 302). Empedocles also wrote a book that is lost, titled *Purifications*, and claimed to be able to cure diseases and bring the dead back to life; cf. Lloyd, *Grip of Disease*, 24–27. Empedocles was even more specific about the ratios of elements that went into the construction of the human body. Bone is a mixture of two portions of earth, two portions of water, and four portions of fire. The sinews are a result of a mixture of fire, earth, and water in the ratio of 1:1:2. Blood and flesh are the result of a mixture of all four elements in the perfect ratio of 1:1:1:1. These ratios are discussed by F. Solmsen, "Tissues and the Soul: Philosophical Contributions to Physiology," *Philosophical Review* 59 (1950): 437. Solmsen also observes that Empedocles's numerical ratios might be evidence of Pythagorean influence.

15. Frag. 4 (Aëtius 5.30.1); quoted from Longrigg, *Greek Rational Medicine*, 52.
16. Alcmaeon and Empedocles by no means exhaust pre-Socratic interest in health and disease; the large majority of the pre-Socratics were interested in medical themes, including Anaxagoras of Clazomenae, Democritus of Abdera, and Diogenes of Apollonia (cf. Nutton, *Ancient Medicine*, 37–52). There are even some fragments from Parmendies that deal with the body (frag. 16) and embryology (frag. 17). Diogenes Laertius, in his *Lives of Eminent Philosophers* (9.3), tells the story of Heraclitus of Ephesus seeking to be healed from a condition that ultimately kills him: "Finally, he became a hater of his kind and wandered on the mountains, and there he continued to live, making his diet (*diêitato*) of grass and herbs. However, when this gave him dropsy (*hyderon*), he made his way back to the city and put this riddle to the physicians (*iatrôn*), whether they were competent to create a drought after heavy rain. They could make nothing of this, whereupon he buried himself in a cowshed, expecting that the noxious damp humour would be drawn out of him by the warmth of the manure. But, as even this was of no avail, he died at the age of sixty" (in R. D. Hicks, *Diogenes Laertius: Lives of Eminent Philosophers*, vol. 2 [Cambridge, MA: Harvard University Press, 1931]; cf. M. D. Grmek, *Diseases of the Ancient World*, trans. M. Muellner and L. Muellner [Baltimore: Johns Hopkins University Press, 1989], 41–42, for discussion). Notice how, according to Diogenes, Heraclitus couches the description of his illness in meteorological terms.
17. *Physics* 7.3.246b3–8.
18. [Hippocrates] *On the Nature of Man* 4 (in Jones, *Hippocrates*, vol. 4). The word "humor" (*chymoi*) in ancient Greek simply means "juice"; cf. P. Demont, "About Philosophy and Humoural Medicine," in *Hippocrates in Context: Papers Read at the Eleventh International Hippocrates Colloquium, University of Newcastle upon Tyne, 27–31 August 2002*, ed. P. J. van der Eijk, 271–86 (Leiden: Brill, 2005).
19. See W. Smith, *The Hippocratic Tradition* (Ithaca, NY: Cornell University Press, 1979).
20. [Hippocrates] *On the Nature of Man* 4.
21. [Hippocrates] *On the Nature of Man* 4 (in Jones, *Hippocrates*, vol. 4; translation modified).
22. We also find this conception in the Hippocratic treatise *On Ancient Medicine* 14: "For there is in man salt and bitter, sweet and acid, astringent and insipid, and a vast number of other things, possessing properties of all sorts, both in number and in strength. These, when mixed (*memigmena*) and compounded (*kekrêmena*) with one another are neither apparent nor do they hurt a man; but when one of them is separated off, and stands alone, then it is apparent and hurts a man" (in Jones, *Hippocrates*, vol. 1). Cf. M. J. Schiefsky, *Hippocrates* On Ancient Medicine (Leiden: Brill, 2005) for commentary and discussion.
23. Or perhaps from doctors viewing bilious diarrhea, as Nutton (*Ancient Medicine*, 80) suggests.
24. Phillips, *Greek Medicine*, 48–49.
25. [Hippocrates] *On Breaths* 15 (in Jones, *Hippocrates*, vol. 2). The Hippocratic writer of *Diseases iv* (in Lonie, *Hippocratic Treatises*) attributes disease to the presence of phlegm, blood, bile, and "aqueous moisture" (*hydrôps*) in the body, speaking of four "moistures" (*hygron*) instead of humors (*chymoi*). According to von Staden (H. von Staden, *Herophilus: The Art of Medicine in Early Alexandria* [Cambridge: Cambridge University Press, 1989], 246), these terms are often used interchangeably in ancient Greek medical works. The author of *Regimen* 1.32 writes: "The

finest water (*hydatos leptotaton*) and the rarest fire (*puros araiotaton*), on being blended (*sungkrêsin*) together in the human body, produce the most healthy condition (*hygieinotatên hexin*)," which indicates that this author views the mixture of elements (and only two: fire and water) as crucial to health and disease (in Jones, *Hippocrates*, vol. 4).

26. [Hippocrates] *Airs, Waters, Places* (in Jones, *Hippocrates*, vol. 1); see R. Sallares, *The Ecology of the Ancient Greek World* (Ithaca, NY: Cornell University Press, 1991), 221–93. "The idea that diseases are influenced by the weather and its changes must have been old. Herodotus, for example, contends that the Egyptians are healthy because their seasons do not change" (V. Langholf, "L'air (*pneuma*) et les maladies," in *La maladie et les maladies dans la collection hippocratique*, ed. P. Potter, G. Maloney, and J. Desautels [Quebec: Éditions du Sphinx, 1990], 165, referring to Herodotus, *Histories* 2.73; see generally Langholf, "L'air (*pneuma*) et les maladies," 164–79, for meteorological themes in ancient medical thought). Connected with meteorological medicine is the theme of viewing the human body as a microcosm having a structure analogous to the macrocosm of the whole universe; cf. F. Le Blay, "Microcosm and Macrocosm: The Dual Direction of Analogy in Hippocratic Thought and the Meteorological Tradition," in van der Eijk, *Hippocrates in Context*, 251–69.

27. [Hippocrates] *On Breaths* 3 (in Jones, *Hippocrates*, vol. 2).

28. The singular emphasis that this author places on air perhaps connects him with certain pre-Socratic philosophers who also speak much about air in their natural philosophies, such as Anaximenes and Diogenes of Apollonia.

29. [Hippocrates] *On Breaths* 4.

30. "So whenever the air has been infected with such pollutions (*miasmasin*) as are hostile to the human race (*têi anthrôpeiêi phusei*), the men fall sick (*anthrôpoi noseousin*), but when the air has become ill-adapted to some other species of animals, then these fall sick" (*On Breaths* 6; in Jones, *Hippocrates*, vol. 2).

31. [Hippocrates] *On Breaths* 8 (headache); 12–14 (dropsy, apoplexy, and epilepsy).

32. Common exercises included walking, running, stretching the limbs, massage, and athletic exercises such as wrestling; see *Regimen* 2.61–66 (in Jones, *Hippocrates*, vol. 4). The author interestingly sees these exercises as involving some violence (*biê*), so that care should be taken in performing them. He also says that there are natural (*kata phusin*) exercises, and these are ones that involve the soul only, such as seeing, hearing, or thinking, and those that involve the voice: speaking, reading, or singing. These natural exercises move the soul in such a way that it becomes warm and dry, consuming the moisture of the body, and in doing so "it empties the flesh and makes a man thin" (*Regimen* 2.61). Just how the soul is able to have this effect on the body is unclear, but see B. Gundert, "Soma and Psyche in Hippocratic Medicine," in *Psyche and Soma: Physicians and Metaphysicians on the Mind-Body Problem from Antiquity to Enlightenment*, ed. J. P. Wright and P. Potter, 13–35 (Oxford: Clarendon Press, 2000).

33. For Hippocratic regimen in general, see Edelstein, *Ancient Medicine: Selected Papers*, 303–16; E.M. Craik, "Hippokratic Diaita," in *Food in Antiquity*, ed. J. Wilkins, D. Harvey, and M. Dobson, 343–50 (Exeter: University of Exeter Press, 1995); and E.M. Craik, "Diet, Diaita, and Dietetics," in *The Greek World*, ed. A. Powell, 387–402 (London: Routledge, 1995). Regimen also comprised other hygienic activities such as bathing (cf. *Regimen* 2.57), sleep (cf. *Prognostic* 10), and even sexual intercourse (almost like a sort of exercise; cf. *Regimen* 2.58). Jouanna (*Hippocrates*, 166–70) provides a good discussion of these elements of Hippocratic regimen.

34. [Hippocrates] *On Ancient Medicine* 3 (in Jones, *Hippocrates*, vol. 1).
35. Thomas Hobbes, *Leviathan* (1651), part 1, chapter 13 (in R. Tuck, ed., *Thomas Hobbes: Leviathan* [Cambridge: Cambridge University Press, 1996], 89).
36. [Hippocrates] *On Ancient Medicine* 3.
37. [Hippocrates] *On Ancient Medicine* 3.
38. Cf. [Hippocrates] *Regimen* 2.39–56, a thorough discussion of the various properties of foods and drinks (including barley, wheat, honey, water, beans, chickpeas, domesticated animals, wild animals, fish, birds, eggs, vegetables, and fruits), how they should be prepared, and when they should be consumed. More analysis of the powers of foods (both their healthy and unhealthy properties) can be found in *Affections* 47–61 (in Potter, *Hippocrates*, vol. 5).
39. [Hippocrates] *Regimen in Health* 1 (in Jones, *Hippocrates*, vol. 4).
40. [Hippocrates] *Regimen in Health* 1. This, of course, is an instance of the common Hippocratic doctrine that opposites cure opposites (allopathy). For formulations of this doctrine in other treatises, cf. *On Breaths* 1, *Epidemics* 6.2.1, *Aphorisms* 2.22, and *On the Nature of Man* 9. The author of *On Ancient Medicine* (chapter 13) argues that this method of treatment is imprecise sophistry. *Places in Man* 42 is interesting in that its author argues that opposites cure opposites, and "likes cure likes" (homeopathy), depending on the nature of the disease itself (cf. E.M. Craik, *Hippocrates: Places in Man* [Oxford: Clarendon Press, 1998], 204). Another renowned Hippocratic sentiment is that nature itself is the best doctor (*natura medicatrix*): "Natures (*phusies*) are physicians (*iêtroi*) of diseases (*nousôn*). Nature (*phusis*) finds the way for herself, not from thought (*dianoiês*). For example, blinking, and the tongue offers its assistance, and all similar things. Well trained (*eupaideutos*), readily and without instruction (*ou mathousa*), nature does what is needed (*ta deonta*)" (*Epidemics* 6.5.1, in Smith, *Hippocrates*, vol. 7, translation modfied; cf. Jouanna, *Hippocrates*, 346–47 for discussion). Other therapeutic measures (besides regimen) for the Hippocratic writers included incision and cautery (cf. Jouanna, *Hippocrates*, 155–61), and drugs (cf. L.M.V. Totelin, *Hippocratic Recipes: Oral and Written Transmission of Pharmacological Knowledge in Fifth- and Fourth-Century Greece* [Leiden: Brill, 2009]).
41. [Hippocrates] *Aphorisms* 1.14: "Growing creatures have most innate heat (*to emphuton thermon*), and it is for this reason that they need most food, deprived of which their body pines away. Old men have little innate heat, and for this reason they need but little fuel; much fuel puts it out. For this reason too the fevers of old men are less acute than others, for the body is cold" (in Jones, *Hippocrates*, vol. 4). E. Mendelsohn, *Heat and Life: The Development of the Theory of Animal Heat* (Cambridge, MA: Harvard University Press, 1964) provides a helpful survey of the concept of innate or vital heat; the concept is an ancient one, employed by many philosophers and medical writers (see Freudenthal, *Aristotle's Theory*, 74–105).
42. [Hippocrates] *On Ancient Medicine* 18–19, *Diseases* 4.32–53, and *Humors* 11.
43. [Hippocrates] *On Breaths* 7.
44. The aggressive, combative, or agonistic stance of the doctor toward the disease is a theme that occurs in a number of Hippocratic treatises: *On the Art* 11, *On the Nature of Man* 13, and *On Breaths* 1; cf. Jouanna, *Hippocrates*, 141 and 342–43.
45. Lloyd, *Grip of Disease*, 1. The most famous plague in antiquity was the great plague of Athens, which the Greek historian Thucydides describes in magnificent detail in his *History of the Peloponnesian War* 2.47–54. See Lloyd, *Grip of Disease*, 120–27; E.M. Craik, "Thucydides on the Plague: Physiology of Flux and Fixation,"

Classical Quarterly 51 (2001): 102–8; and R. Thomas, "Thucydides' Intellectual Milieu and the Plague," in *Brill's Companion to Thucydides*, ed. A. Rengakos and A. Tsakmakis, 87–108 (Leiden: Brill, 2006) for recent discussions. Studies of disease and illness in the ancient world abound; cf. D. Brothwell and A. T. Sandison, *Diseases in Antiquity* (Springfield, IL: C. C. Thomas, 1967); G. Majno, *The Healing Hand: Man and Wound in the Ancient World* (Cambridge, MA: Harvard University Press, 1975); Grmek, *Diseases*; the papers in D. Gourevitch, ed., *Maladie et maladies: Historie et conceptualization (mélanges en l'honneur de Mirko Grmek)* (Geneva: Librairie Droz S.A, 1992); and K. Kiple, ed., *The Cambridge World History of Human Disease* (Cambridge: Cambridge University Press, 1993). For disease in the Hippocratic Corpus, cf. the papers in P. Potter, G. Maloney, and J. Desautels, eds., *La maladie et les maladies dans la collection hippocratique* (Quebec: Éditions du Sphinx, 1990); and A. Thivel and A. Zucker, eds., *Le normal et la pathologique dans la collection hippocratique* (Nice: Université de Nice, 2002).

46. Cf. *On the Nature of Man* 9, where the author argues that whenever disease strikes many people at the same time, its cause is bad air, "some unhealthy exhalation (*nosêrên tina apokrisin*)" that we breathe (*anapneomen*), but when diseases of all sorts occur at one and the same time, the cause is the particular regimen (*diatêmata*) of each person.

47. This is largely the view of the Hippocratic writer of *On Breaths*; but cf. *On the Art* 10, where the author argues that in health (*hygiainon*) the body's cavities (*koilon*) are filled with air (*pneumatos*), while in disease (*asthenêsan*) they are filled with juice (*ichôros*). In much the same spirit, the author of *On the Sacred Disease*, in chapters 6–10, attributes epileptic seizure and all of its symptoms to phlegm blocking the conduits (vessels) that transmit respired air (*pneuma*) throughout the body. For the role of *pneuma* in Hippocratic nosology, see V. Langholf, "L'air (*pneuma*) et les maladies," in Potter et al., *La maladie*, 339–59.

48. There is a treatise in the Hippocratic Corpus devoted solely to wounds (*On Wounds in the Head*) and treatises that speak of mending fractures and dislocations (*Fractures, Joints*, and *Leverage*). All of these treatises are collected in E. T. Withington, *Hippocrates*, vol. 3 (Cambridge, MA: Harvard University Press, 1928). For war wounds in classical antiquity, see C. F. Salazar, *The Treatment of War Wounds in Graeco-Roman Antiquity* (Leiden: Brill, 2000).

49. [Hippocrates] *Diseases* 1.2 (in Potter, *Hippocrates*, vol. 5). Similarly, the Hippocratic writer of *Affections* 1 attributes all diseases to bile and phlegm: "The bile and phlegm produce diseases when, inside the body, one of them becomes too moist, too dry, too hot, or too cold; they become this way from foods and drinks, from exertions and wounds, from smell, sound, sight, and venery, and from heat and cold; this happens when any of the things mentioned are applied to the body at the wrong time, against custom, in too great amount and too strong, or in insufficient amount and too weak." (in Potter, *Hippocrates*, vol. 5).

50. [Hippocrates] *Nutriment* 25 (in Jones, *Hippocrates*, vol. 1). Besides *Nutriment*, other treatises of the Hippocratic Corpus are composed of aphorisms, such as *Aphorisms* and *Dentition*. The aphoristic style of these treatises might show the influence of the pre-Socratic philosopher Heraclitus of Ephesus.

51. Notice in this passage the telescopic movement from the inner depths of the human body to its outer surface (the skin). This telescopic movement is untypical of most Hippocratic descriptions of disease that go *a capite ad calcem* (from the head to the feet). A classification of diseases based on this model can be seen in the Hippocratic

treatise *Diseases* 2.12–75, which Langholf (*Medical Theories in Hippocrates*, 46) regards as one of the oldest nosological accounts in the entire Hippocratic Corpus.

52. [Hippocrates] *Epidemics* 1.23 (in Jones, *Hippocrates*, vol. 1).
53. The secondary literature on Hippocratic gynecology is happily large, but cf. the following with their bibliographies: A. E. Hanson, *Studies in the Textual Tradition and the Transmission of the Gynecological Treatises of the Hippocratic Corpus*, PhD diss., University of Pennsylvania, 1971; A. E. Hanson, "Diseases of Women I," *Signs* 1 (1975): 567–84; G.E.R. Lloyd, *Science, Folklore, and Ideology* (Cambridge: Cambridge University Press, 1983), 58–111; A. E. Hanson, "Diseases of Women in the *Epidemics*," in *Die Hippokratischen Epidemien: Theorie-Praxis-Tradition, Verhandlungen des Ve Colloque International Hippocratique, Berlin 10–15.9.1984, Sudhoffs Archiv*, ed. G. Baader and R. Winau, Heft 27 (Stuttgart: Franz Steiner, 1989), 38–51; L. Dean-Jones, *Women's Bodies in Classical Greek Science* (Oxford: Clarendon Press, 1994); N. Demand, *Birth, Death, and Motherhood in Classical Greece* (Baltimore: Johns Hopkins University Press, 1994); R. Flemming and A. Hanson, "Hippocrates' *Peri parthenîon (Diseases of Young Girls)*: Text and Translation," *Early Science and Medicine* 3 (1998): 241–52; H. King, *Hippocrates' Woman* (London: Routledge, 1998); Jouanna, *Hippocrates*, 171–76; H. King, *The Disease of Virgins: Green Sickness, Chlorosis and the Problems of Puberty* (London: Routledge, 2004); H. King, "Women's Health and Recovery in the Hippocratic Corpus," in *Health in Antiquity*, ed. H. King, 150–61 (London: Routledge, 2005); and Totelin, *Hippocratic Recipes*, 197–224. Totelin's bibliography is excellent, particularly regarding the works of A. E. Hanson, which are extensive (I have listed only a few here). For gynecology in Roman medicine, see R. Flemming, *Medicine and the Making of Roman Women* (Oxford: Oxford University Press, 2000). D. Gourevitch, "La gynécologie et l'obstétrique," *Aufstieg und Niedergang der römischen Welt*, 2nd ser., 37, part 3 (1996): 2083–2146, provides a comprehensive survey (with copious bibliography) of gynecology and obstetrics in antiquity. Under this category I might also mention Hippocratic reflection on the health and disease of children, but the only treatise of the corpus to deal with them specifically is the aphoristic *Dentition*; cf. J. Bertier, "Enfants malades et maladies des enfants dans le *corpus hippocratique*," in Potter et al., *La maladie*, 209–20. For other studies on infants and infant maladies in antiquity, cf. J. Bertier, "La médecine des enfants à l'époque impériale," *Aufstieg und Niedergang der römischen Welt*, 2nd ser., 37, part 3 (1996): 2147–2227; A. E. Hanson, "Your Mother Nursed You with Bile": Anger in Babies and Small Children," in *Ancient Anger: Perspectives from Homer to Galen*, Yale Classical Studies, vol. 32, ed. S. Braund and G. Most, 185–207 (Cambridge: Cambridge University Press, 2003); and the papers in V. Dasen, ed., *Naissance et petite enfance dans l'antiquité. Actes du colloque de Fribourg, 28 novembre–1er décembre 2001* (Fribourg: Academic Press and Vandenhoeck & Ruprecht, 2004).
54. The author of *On the Nature of Man* 7 ties each of the four humors to a particular season of the year to explain why certain diseases predominate depending on the weather. Thus, colds appear in winter (which is associated with cold and wet) since phlegm (also cold and wet) is in ascendency. Fevers appear in summer (hot and dry) since yellow bile (hot and dry) is in ascendency. The ancients typically thought that fever was a condition in its own right and not just a symptom of disease. Fevers in the Mediterranean were probably the result of malaria (cf. W.H.S. Jones, *Malaria: A Neglected Factor in the History of Greece and Rome* [Cambridge: Macmillan

and Bowes, 1907], and more recently P.F. Burke, "Malaria in the Greco-Roman World: A Historical and Epidemiological Survey," *Aufstieg und Niedergang der römischen Welt,* 2nd ser., 37, part 3 [1996]: 2252–81; and Sallares, *Ecology*).

55. Jouanna (*Hippocrates*, 145–54) helpfully discusses various Hippocratic attempts to classify disease. While some Hippocratic writers spoke of acute diseases (such as pleurisy, pneumonia, phrenitis, and fever; cf. *Regimen in Acute Diseases* 5), that is, lethal diseases often having a "crisis" (literally, a "decision," where the patient either recovered or died from the disease), the classification between acute and chronic diseases was first made by Aretaeus of Cappadocia (ca. 50 C.E.) and found its fullest expression in the massive treatise *On Acute and Chronic Diseases* by the Methodist writer Caelius Aurelianus (ca. 400 C.E.); see I. Drabkin, *Caelius Aurelianus: On Acute Diseases and On Chronic Diseases* (Chicago: University of Chicago Press, 1950).

56. Potter ("Some Principles of Hippocratic Nosology," in Potter et al., *La maladie,* 237–53) estimates that there are over two hundred diseases "named and/or described" in the Hippocratic Corpus; cf. Potter, *Hippocrates,* vol. 6, 333–43, for a helpful index of symptoms and diseases found in the central nosological treatises of the Hippocratic Corpus.

57. Plato, *Republic* 2.372b–d (in Cooper, *Plato*).

58. Cf. the anecdote Plutarch (*Advice about Keeping Well* 127a–b, in F.C. Babbitt, *Plutarch: Moralia,* vol. 2 [Cambridge, MA: Harvard University Press, 1928]) tells about Plato's own dining habits: "The remark Timotheus made, the day after he had dined with Plato at the Academy on the simple (*liton*) fare of the scholar, is in point here: 'Those who dine with Plato,' he said, 'get on pleasantly the next day also.'" Whether or not Socrates advocates a vegetarian diet in this passage of the *Republic,* many other ancients did, sometimes for reasons of health. Plutarch wrote a treatise arguing against eating meat, *On the Eating of Flesh (De esu carnium),* in H. Cherniss and W.C. Helmbold, *Plutarch: Moralia,* vol. 12 [Cambridge, MA: Harvard University Press, 1957]), which may have been influenced by older Pythagorean prohibitions against consuming animals. S.T. Newmyer (*Animals, Rights, and Reason* [London: Routledge, 2006], 85–102) argues that Plutarch advocated vegetarianism because he believed that animals shared in reason. For the view that vegetarianism is more conducive to human health, cf. D. Tsekourakis, "Pythagoreanism or Platonism and Ancient Medicine? The Reasons for Vegetarianism in Plutarch's 'Moralia,'" *Aufstieg und Niedergang der römischen Welt,* 2nd ser., 36, part 1 (1987): 366–93. The Neoplatonic philosopher Porphyry also argued against consuming animal flesh in his treatise *On Abstinence from Killing Animals*; cf. G. Clark, *Porphyry: On Abstinence from Killing Animals* (Ithaca, NY: Cornell University Press, 2000).

59. Pliny makes the same connection between desire for luxury, poor diet, and illness: "The greatest part however of man's trouble (*homini negotii*) is caused by the belly (*alvus*), the gratification of which is the life's work of the majority of mankind. For at one time it does not allow food to pass, at another it will not retain it, at another it does not take it, at another it does not digest it; and so much have our customs (*mores*) degenerated that it is chiefly through his food (*cibo*) that a man dies (*pereat*). This, the most troublesome organ (*pessimum vas*) in the body, presses as does a creditor, making its demands several times a day. It is for the belly's sake especially that avarice is so acquisitive; for its sake luxury uses spices, voyages are made to the Phasis, and the bottom of the ocean is explored. Nobody, again, is led

to consider how base an organ it is by the foulness of its completed work. Therefore the tasks (*opera*) of medicine concerned with the belly are very numerous (*numerosissima*)" (*Natural History* 26.28.43–44; in Jones, *Pliny: Natural History*).

60. Socrates famously mentions Hippocrates and his method in the *Phaedrus* at 270c–e (cf. M. L. Gill, "Plato's *Phaedrus* and the Method of Hippocrates," *Modern Schoolman* 80 [2003]: 295–314). In the *Symposium*, Plato's dialogue on love, a physician, Eryximachus, appears as one the participants who offers his own speech on love, within the context of ancient medical ideas: "The physician's task is to effect a reconciliation and establish mutual love between the most basic bodily elements. Which are those elements? They are, of course, those that are most opposed to one another, as hot is to cold, bitter to sweet, wet to dry, cases like those. In fact, our ancestor Asclepius first established medicine as a profession (*technên*) when he learned how to produce concord (*homonoian*) and love (*erôta*) between such opposites—that is what those poet fellows say, and—this time—I concur with them" (186d–e, in Cooper, *Plato*); cf. Edelstein, *Ancient Medicine: Selected Papers*, 153–71; and more recently E. M. Craik, "Plato and Medical Texts: *Symposium* 185c–193d," *Classical Quarterly* 51 (2001): 109–14; and M. McPherran, "Medicine, Magic, and Religion in Plato's *Symposium*," in *Plato's Symposium: Issues in Interpretation and Reception*, ed. J. Lesher, D. Nails, and F. Sheffield, 71–95 (Cambridge, MA: Harvard University Press, 2006). For medicine and philosophy in the *Gorgias*, see J. P. Anton, "Dialectic and Health in Plato's *Gorgias*: Presuppositions and Implications," *Ancient Philosophy* 1 (1980): 49–60; in the *Charmides*, see F. P. Coolidge, "The Relation of Philosophy to Σωφροσύνη: Zalmoxian Medicine in Plato's *Charmides*," *Ancient Philosophy* 13 (1993): 23–36.

61. In the discussion that follows, I offer just a brief outline of the nosology in the *Timaeus*; for a more detailed treatment, see H. W. Miller, "The Aetiology of Disease in Plato's Timaeus," *Transactions of the American Philological Association* 93 (1962): 175–87; and F. M. Cornford, *Plato's Cosmology* (London: Routledge & Kegan Paul, 1937).

62. But Plato still has Timaeus say that ignorance (*amathian*) is "the gravest disease (*megistên noson*) of all" (*Timaeus* 88b, in Cooper, *Plato*). Timaeus holds that other psychic (mental) illness seem to have a bodily basis; cf. *Timaeus* 86b–87b, with B. Simon, *Mind and Madness in Ancient Greece* (Ithaca, NY: Cornell University Press, 1978), 157–212.

63. As the dialogue opens, Socrates is counting the participants and notices that instead of four there are only three present. When he asks Timaeus where the fourth is, Timaeus answers: "Some illness (*astheneia tis*) befell him, Socrates. He wouldn't have missed our meeting willingly" (*Timaeus*, 17a; in Cooper, *Plato*, translation modified). P. Kalkavage (*Plato's Timaeus* [Newburyport, MA: Focus, 2001], 119n156) rightly calls attention to the significance of Timaeus's first words.

64. Plato *Timaeus* 84c–d.

65. Plato *Timaeus* 82a–b.

66. Timaeus also seems to adopt the politically charged language of Alcmaeon in this passage. Note the use of the term *pleonexia*, which appears all throughout the *Republic*. Cf. also *Timaeus* 85e–86a, where Timaeus explains that excess bile in the body is forcibly expelled from the belly "like an exile (*phugas*) from a city (*ek poleôs*) in civil strife (*stasiasasês*), so bringing on diarrhea, dysentery, and every disease (*nosêmata*) of that kind."

67. Plato *Timaeus* 82c.

68. Plato *Timaeus* 82e.
69. Plato *Timaeus* 83a.
70. Whether this is the air that is naturally in the body because of respiration or fumes from disintegrating tissue is unclear, but cf. *Timaeus* 84e: "And often, when flesh disintegrates inside the body, air (*pneuma*) is produced there, but is unable to get out. This air then causes just as much excruciating pain as the air that comes in from outside."
71. Plato *Timaeus* 83d.
72. Plato *Timaeus* 83a.
73. Plato *Timaeus* 84b.
74. According to Timaeus, microscopic triangles ultimately compose the element of fire (59a–b). Because triangles are small and sharp, this explains the cutting and piercing power of fire. However, as human beings grow old, their fire triangles get dull, eventually making it impossible for them to digest and assimilate food and drink. In effect, old age and death result from undigested food eventually overpowering the body's fire; our food begins to eat us (81b–e).
75. Plato *Timaeus* 84d.
76. Plato basically holds health to be proportion (*summetron*) among the constituents of the body (*Timaeus* 87c–88b), particularly the good proportion or balance of body and soul: "There is in fact one way to preserve (*sôtêria*) oneself, and that is not to exercise (*kinein*) the soul without exercising the body, nor the body without the soul, so that each may be balanced (*isorropô*) by the other and so be sound (*hygiê*)" (*Timaeus* 88b). The Roman satirist Juvenal similarly writes, "Let us pray for a healthy mind in a healthy body" (*orandum est ut sit mens sana in corpore sano*), *Satires* x 356 (cited in Edelstein, *Ancient Medicine: Selected Papers*, 356).
77. Heraclitus frag. 123 (in Kirk et al., *The Presocratic Philosophers*, 192).
78. P. Hadot, *The Veil of Isis: An Essay on the History of the Idea of Nature* (Cambridge, MA: Harvard University Press, 2006), 7; see generally 7–14.
79. The *Phaedo* opens with Socrates "sitting up" (*anakathizomenos*) on his prison bed to welcome his friends into his cell (60b). J. Burnet (*Platonis Opera: Phaedo* [Oxford: Clarendon Press, 1911], 13) observes that this verb is also used in some Hippocratic treatises. Socrates also makes a playful allusion (77e–78a) to magical medicine when he urges Simmias and Cebes to find someone to charm (*epaidein*) away their fear of death, if he is unable to convince them rationally through argumentation that the soul is immortal. Burnet (64) notes that Socrates speaks at length of singing charms in his discussion of Zalmoxian medicine in the *Charmides* at 155ff., and that he also mentions singing charms in connection with his maieutic method at *Theaetetus* 149d.
80. See especially 67a: "While we live, we shall be closest to knowledge if we refrain as much as possible from association with the body and do not join with it more than we must, if we are not infected (*anapimplômetha*) with its nature (*phuseôs*) but purify (*kathareuômen*) ourselves from it until the god himself frees us. In this way we shall escape the contamination of the body's folly; we shall be likely to be in the company of people of the same kind, and by our own efforts we shall know all that is pure, which is presumably the truth, for it is not permitted to the impure to attain the pure" (in Cooper, *Plato*). Burnet (*Platonis Opera*, 37) notes that the verb "infected" (*anapimplômetha*) is also used by the historian Thucydides in his account of the great plague of Athens in 430 B.C.E.

81. Cf. 66b–c: "The body keeps us busy in a thousand ways because of its need for nurture. Moreover, if certain diseases (*nosoi*) befall it, they impede our search for the truth" (in Cooper, *Plato*); cf. 67c–d, 81b–c, 82c, and 83d–e.
82. Plato, *Phaedo* 117a–c. The "poison" or "drug" that Socrates drinks to kill himself is traditionally thought to be hemlock, though there is a controversy over whether this is actually the drug that killed Socrates. For a recent discussion, see E. Bloch, "Hemlock Poisoning and the Death of Socrates: Did Plato Tell the Truth?" in *The Trial and Execution of Socrates: Sources and Controversies,* ed. T. Brickhouse and N. Smith (New York: Oxford University Press, 2002), 255–78.
83. Plato, *Phaedo* 118a (in Cooper, *Plato*).
84. M. McPherran, "Socrates, Crito, and Their Debt to Asclepius," *Ancient Philosophy* 23 (2003): 71–92, offers a comprehensive survey and analysis of the many interpretations of Socrates's last words. For the genealogy of Hippocrates, see Jouanna, *Hippocrates*, 10–17.
85. Friedrich Nietzsche, *Twilight of the Idols* (1888), "The Problem of Socrates" 1 (in W. Kaufmann, *The Portable Nietzsche* [New York: Penguin, 1976], 473).
86. Medicine along with litigiousness; see Plato *Republic* 3.405a–d.
87. For regimen in the *Republic*, cf. 3.403dff. Of course, Socrates underlines the fact that proper regimen should be balanced with a virtuous education. Timaeus argues that human beings should imitate the natural motion of the cosmos by thinking and exercising (89a); these sorts of motions are best because they are self-caused motions. Opposed to these self-caused motions are the harmful, forced motions that medical drugs cause (purges and emetics): "Now diseases have a similar makeup, so that when you try to wipe them out with drugs (*pharmakeiais*) before they have run their due course, the mild (*smikrôn*) diseases (*nosêmata*) are liable to get severe (*megala*), and the occasional (*oligôn*) ones frequent (*polla*). That is why you need to cater to all such diseases by taking care of yourself (*diaitais*) to the extent you are free and have the time to do that. What you should not do is aggravate a stubborn irritation with drugs (*pharmakeuonta*)" (*Timaeus* 89c–d; in Cooper, *Plato*). The theme of "letting nature take its course" that Plato presents here echoes *Epidemics* 6.5.1 (see note 40).
88. Xenophon, *Memorabilia* 4.7.9 (in E.C. Marchant and O.J. Todd, *Xenophon*, vol. 4 [Cambridge, MA: Harvard University Press, 1923]; translation modified).
89. Cf. Plato, *Alcibiades* 1.129a–135e.
90. Plato, *Phaedo* 59b.
91. Diogenes Laertius, *Lives of the Eminent Philosophers* 5.1. Scholars only recently have begun to debate the precise extent to which Aristotle could be classed as a medical writer or medical philosopher (Cf. F. Solmsen, "Greek Philosophy and the Discovery of the Nerves," *Museum Helveticum* 18 [1961]: 150–97; T. Tracy, *Physiological Theory and the Doctrine of the Mean in Plato and Aristotle* [Chicago: Loyola University Press, 1969]; V.P. Vizgin, "Hippocratic Medicine as a Historical Source for Aristotle's Theory of the Dynameis," *Studies in the History of Medicine* 4 [1980]: 1–12; J. Longrigg, "Medicine and the Lyceum," in *Ancient Medicine*, ed. van der Eijk et al., vol. 2, 431–45; D. Manetti, "'Aristotle' and the Role of Doxography in the Anonymous Londiniensis [PBRLIBR INV. 137]," in *Ancient Histories of Medicine: Essays in Medical Doxography and Historiography in Classical Antiquity,* ed. P.J. van der Eijk, 95–141 [Leiden: Brill, 1999]; D.M. Tress, "Aristotle against the Hippocratics on Sexual Generation: A Reply to Coles," *Phronesis* 44 [1999]: 228–41; P.J. van der Eijk, "Aristotle's Psycho-Physiological Account of

the Soul-Body Relationship," in *Psyche and Soma,* ed. Wright and Potter, 57–77; R. Polansky, "Is Medicine Art, Science, or Practical Wisdom? Ancient and Contemporary Reflections," in *Bioethics: Ancient Themes in Contemporary Issues,* ed. M. G. Kuczewski and R. Polansky [Cambridge, MA: MIT Press, 2000], 31–56; C. Oser-Grote, *Aristoteles und das Corpus Hippocraticum* [Stuttgart: Steiner, 2004]; and van der Eijk, *Medicine and Philosophy,* 139–237 and 259–75). There is certain agreement that he was a great philosopher of biology, if such an anachronistic term could be employed (cf. A. Gotthelf, ed., *Aristotle on Nature and Living Things* [Pittsburgh, PA, and Bristol, UK: Mathesis and Bristol Classical Press, 1985]; A. Gotthelf and J. G. Lennox, ed., *Philosophical Issues in Aristotle's Biology* [Cambridge: Cambridge University Press, 1987]; J. G. Lennox, *Aristotle's Philosophy of Biology: Studies in the Origins of Life Science* [Cambridge: Cambridge University Press, 2001]; and J. G. Lennox, *Aristotle: On the Parts of Animals I–IV* [Oxford: Clarendon Press, 2001]). However, if biology is taken to be a subdiscipline of medicine, then perhaps in some loose sense Aristotle could be taken as a medical philosopher as well.

92. Aristotle, *Physics* 7.3.246b3–8 (in J. Barnes, *The Complete Works of Aristotle,* 2 vols. [Princeton, NJ: Princeton University Press, 1995]). Important here is that Aristotle regards health as something relative, admitting of degrees. This is a common theme in ancient Greek medical thought; cf. F. Kudlien, "The Old Greek Concept of 'Relative' Health," *Journal of the History of the Behavioral Sciences* 9 (1973): 53–59.

93. Aristotle, *Parts of Animals* 2.7.652b26–653a10 (in Barnes, *Complete Works of Aristotle*; translation modified slightly). If Aristotle is referring to a work of his own called *Principles of Diseases,* we do not have it. He may have written a number of treatises on medical themes that are lost, including *On Health and Disease* and *On Remedies*; cf. van der Eijk, *Medicine and Philosophy,* 263–64. This may explain why Aristotle has little to say about regimen. He recommends that wet nurses dilute their wine with water, so that they do not harm infants whom they're nursing; cf. *On Sleep* 3.457a14–18: "This explains why wines (*hoi oinoi*) are not good for infants (*paidiois*) or for wet nurses (*titthais*) (for it makes no difference, doubtless, whether the infants themselves, or their nurses, drink them), but such persons should drink them diluted with water (*hydarê*) and in small quantity (*oligon*). For wine is spirituous (*pneumatôdes*), and of all wines the dark (*ho melas*) more so than any other" (in Barnes, *Complete Works of Aristotle*; for pathological *pneuma,* see below, n113). With respect to exercise, in the context of a discussion of the phrase "in vain" (*to matên*) in the *Physics,* Aristotle writes, "Taking a walk (*to badisai*) is for the sake of evacuation of the bowels (*lapaxeôs*); if this does not follow after walking, we say that we have walked in vain and that the walking was vain" (Aristotle, *Physics* 2.6.197b24–25). Aristotle apparently believed that walking helps digestion (cf. Plutarch, *Advice about Keeping Well* 21, 133F). In the *Nicomachean Ethics* (3.10.1118b5–8), Aristotle observes that full-body (*pan to sôma*) massage (*tripseôs*) is an appropriate pleasure, while touching only particular parts (*tina merê*) of the body (i.e., the genitals) is self-indulgent (*akolastou*) because it discourages temperance, appealing rather to our brutish animal natures. Finally, perhaps echoing the Hippocratic writer of *Regimen,* who thought that natural soul motions like seeing and speaking were healthy, Aristotle argues that smelling is healthy, but this is because, he argues, odors are light and contain heat, and help to moderate the coldness of the brain; cf. *On Sense and Sensibilia*

5.444a19–444b6, with Plutarch, *Isis and Osiris* 383D (in F.C. Babbitt, *Plutarch: Moralia*, vol. 5 [Cambridge, MA: Harvard University Press, 1936]).

94. For the importance of digestion in Aristotle, see M. Boylan, "The Digestive and 'Circulatory' Systems in Aristotle's Biology," *Journal of the History of Biology* 15 (1982): 89–118; and G.E.R. Lloyd, *Aristotelian Explorations* (Cambridge: Cambridge University Press, 1996), 83–103.
95. See generally Aristotle, *On Sleep* and *On Dreams*. Notably, at *On Sleep* 3.458a1–6, Aristotle describes vapor from digestion that is "nutritious" (*trophimos*) and not "pathological" (*nosôdês*) as properly cooling the heat in the heart, while "waste vapor" (*perittômatikê anathumiasis*) is what transforms into phlegm in the head, causing catarrhs (*katarroi*). Cf. also *Sense and Sensibilia* 5.444a10–15 for more on catarrhs.
96. Aristotle, *On Sleep* 3.458a10.
97. Aristotle, *Parts of Animals* 2.7.653b5: "disease (*nosous*), madness (*paranoias*), and death (*thanatous*)."
98. Aristotle indicates that digestion, despite being a normal physiological process of the body, also serves as a model to explain the exhaustion or debilitating effects characteristic of certain pathological phenomena such as lethargy, fatigue, and fever: "And it is after meals (*ta sitia*) especially that sleep comes on like this, for the exhalation (*anathumiasis*) from the foods eaten is then copious (*pollê*). It also follows certain forms of fatigue (*kopôn*); for fatigue (*kopos*) operates as a solvent (*suntêktikon*), and the dissolved matter (*suntêgma*) acts, if not cold, like food (*trophê*) prior to digestion (*apeptos*). Moreover, some kinds of illness (*nosoi*) have this same effect; those arising from moist (*hygrou*) and hot (*thermou*) secretions (*perittômatos*), as happens with fever-patients (*purettousi*) and in cases of lethargy (*lêthargois*)," *On Sleep* 3.456b32–457a4 (in Barnes, *Complete Works of Aristotle*).
99. Aristotle *On Dreams* 3.461a23–24. On melancholy in Aristotle, see van der Eijk, *Medicine and Philosophy*, 139–68.
100. Aristotle *On Sleep* 3.457a8.
101. Aristotle *Meteorology* 2.8.366b15–30; cf. *History of Animals* 8.24.604b4.
102. Aristotle *Generation of Animals* 4.3.768b34–38 with 5.4.784a23, 5.6.786a6–21, and *History of Animals* 3.11.518a12–13.
103. Aristotle *On Respiration* 20.479b32.
104. Aristotle *Parts of Animals* 3.9.672b5.
105. Aristotle *Generation of Animals* 4.7–10; cf. Preus 1983. Aristotle's views on women are controversial, but for fair assessments, see R. Mayhew, *The Female in Aristotle's Biology* (Chicago: University of Chicago Press, 2004); and D. Henry, "How Sexist Is Aristotle's Developmental Biology?" *Phronesis* 52 (2007): 251–69.
106. P. Macfarlane and R. Polansky ("Disability in Earlier Greek Philosophers," *Skepsis* 15 [2004]: 25–41) argue that Aristotle may have developed his important notion of ability or potentiality (*dunamis*) from consideration of the inability (incapacity) of various parts of the body to perform their function or characteristic activity (*ergon, energeia*), e.g., blindness in the eyes.
107. Aristotle *History of Animals* 10; cf. van der Eijk, *Medicine and Philosophy*, 259–75.
108. Aristotle *Generation of Animals* 5.4.784b32–33. Old age could be considered a natural disease because, Aristotle writes, "In sickness (*en têi arrôstiai*) the whole

body is deficient (*en endeiai*) in natural heat (*phusikês thermotêtos*) and so the parts besides, even the very small ones, participate (*metechei*) in this weakness (*arrôstias*)," *Generation of Animals* 5.4.784b25–27 (in Barnes, *Complete Works of Aristotle*). Because Aristotle holds that old age is generally caused by the decline in the power of our vital heat (*On Youth, Old Age, Life and Death* 4.469b5–20), having any disease is analogous to "acquiring old age" (*gêras epiktêton*), *Generation of Animals* 5.4.784b32. Consequently, recovering one's health is like becoming young again: "But when men have recovered health (*hygianantes*) and strength (*ischusantes*) again they change (*metaballousi*), becoming as it were young (*neoi*) again instead of old (*ek gerontôn*)," *Generation of Animals* 5.4.784b30–31 (in Barnes, *Complete Works of Aristotle*).

109. *Parts of Animals, Generation of Animals,* and *History of Animals.*
110. Aristotle *On Respiration* 20.479b27.
111. Aristotle *On Sleep* 3.457a9. Philip van der Eijk was the first to draw attention to the peculiarity of Aristotle explaining normal physiological functions by referring to pathological conditions. He calls specific attention to the case of sleep and epilepsy (cf. van der Eijk, *Medicine and Philosophy*, 169–205).
112. Cf. *History of Animals* 1.1.487a6–7, *History of Animals* 3.2.511b9–10, and *Parts of Animals* 2.7.653b9–15.
113. Aristotle refers to melancholy, fever, and drunkenness as "pneumatic afflictions" (*pathê pneumatôdê*) in *On Dreams* 3.461a23. Many of the other pathological conditions he refers to involve *pneuma* as well (e.g., leprosy, gangrene, and tetanus). Aristotle has two sorts of *pneuma* in his natural philosophy, respired *pneuma* and connate (*sumphuton*) *pneuma*. Respired *pneuma* is the atmospheric air that animals breathe in to help cool the vital heat of their hearts (see Aristotle *On Respiration*). Connate *pneuma* is generated in the body of the animal by digestion or other concoctive processes that occur throughout the body (in the heart, vessels, and kidneys); it is a key factor in Aristotle's account of sense perception, animal reproduction, and animal motion (see, e.g., Aristotle *Parts of Animals* 2.16.659b17–20, *Generation of Animals* 2.2.736a2, *Generation of Animals* 2.6.744a2–5, *Generation of Animals* 5.2.781a21–781b6, *De anima* 2.8.420a12, *Motion of Animals* 7 and 10.703a10, *On Sleep* 2.456a11–20, and *On Respiration* 9.475a1–21). Much has been written on the topic of *pneuma* in Aristotle; see the discussion and bibliography in G. Freudenthal, *Aristotle's Theory of Material Substance: Heat and Pneuma, Form and Soul* (Oxford: Clarendon Press, 1995); more recently, S. Berryman, "Aristotle on *Pneuma* and Animal Self-Motion," *Oxford Studies in Ancient Philosophy* 23 (2002): 85–97; A. P. Bos, "*Pneuma* and Ether in Aristotle's Philosophy of Living Nature," *Modern Schoolman* 79 (2002): 255–76; A. P. Bos, *The Soul and Its Instrumental Body: A Reinterpretation of Aristotle's Philosophy of Living Nature* (Leiden: Brill, 2003); van der Eijk, *Medicine and Philosophy*, 129; and R. W. Sharples, "Common to Body and Soul: Peripatetic Approaches After Aristotle," in *Common to Body and Soul: Philosophical Approaches to Explaining Living Behaviour in Greco-Roman Antiquity,* ed. R.A.H. King, 165–86 (Berlin: Walter de Gruyter, 2006). Aristotle may have written the treatise *De spiritu* (*On Pneuma*), but many scholars believe it is spurious. Recently, however, the authenticity of this work has been defended; see the helpful and detailed commentary of A. P. Bos and R. Ferwerda, *Aristotle, On the Life-Bearing Spirit* (De Spiritu) (Leiden: Brill, 2008). Hardly anyone, however, speaks of the pathological dimension of connate *pneuma*, though A. Roselli, *[Aristotele]*

de spiritu (Pisa: ETS Editrice, 1992), traces the medical themes in *On Pneuma*; cf. P. Macfarlane, "A Philosophical Commentary on Aristotle's *De spiritu*," PhD diss., Duquesne University, 2007, 238–62.

114. For example, he says that melancholics are addicted to sleep (*On Sleep* 3.457a27) and speaks of phlegmatic and bilious people (*Metaphysics* 1.1.981a11); what remains unclear is whether Aristotle based, as some Hippocratics did, a general theory of health and disease on the humors (van der Eijk, *Medicine and Philosophy*, 153, thinks it's "unlikely"). Van der Eijk observes (152) that Aristotle "remarkably claims that not all people possess bile and that, contrary to popular belief, bile is not the cause of acute diseases." Aristotle makes these claims in *Parts of Animals* 4.2.676b31–677a19.
115. Aristotle *History of Animals* 3.2.511b25–29.
116. Aristotle *History of Animals* 3.2.511b30–512b10.
117. Aristotle *History of Animals* 3.3.512b11–513a8. Polybus was reputedly the son-in-law of Hippocrates (cf. C.R.S. Harris, *The Heart and the Vascular System in Ancient Greek Medicine* [Oxford: Clarendon Press, 1973], 33). Because Polybus's account in this passage repeats the vascular scheme found in the Hippocratic treatise *On the Nature of Man*, it may serve as evidence that Aristotle read some Hippocratic treatises.
118. Aristotle *History of Animals* 3.3.513a9–6.516a7. Aristotle is even revealing of the cause for his predecessors' mistakes regarding the arrangement of animal vascular anatomy. Those who dissected *dead* animals had difficulty because the vessels collapse after the animal is slaughtered and all of its blood pours out. Those who attempted to study vascular anatomy by observing *living* animals had trouble inspecting these parts since by "their very nature they are internal" (*History of Animals* 3.2.511b19–20). Thus they were forced to reach their conclusions by studying the bodies of men "reduced to extreme attenuation" so that they arrive at their conclusions "from the manifestations then visible externally" (*History of Animals* 3.2.511b21–23). Rather than studying slaughtered animals or starved animals, Aristotle argues for a middle course: "The investigation of such a subject, as has been remarked, is one fraught with difficulties; but, if any one is keenly interested in the matter, he will get an adequate grasp of it only if he studies strangled animals which have been previously emaciated" (*History of Animals* 3.3.513a12–15). Strangling the animal keeps the blood in its vessels, thus allowing them to retain their shape, and by starving the animal, the dissector can get a map of the vascular terrain before he begins cutting. For more extensive studies of anatomy in Aristotle, cf. the older, though still useful T. E. Lones, *Aristotle's Researches in Natural Science* (London: West, Newman, 1912); Harris, *Heart and Vascular System*; and more recently C. E. Cosans, "Aristotle's Anatomical Philosophy of Nature," *Biology and Philosophy* 13 (1998): 311–39.
119. For the heart as the origin of the vessels (*phlebes*) in Aristotle, see *Parts of Animals* 3.4.665b33, *History of Animals* 3.3.513a21, and *On Youth, Old Age, Life and Death* 3.468b32.
120. Van der Eijk (*Medicine and Philosophy*, 207) uses the phrase "psycho-physiology" (which he acknowledges was first used by Charles Kahn) to capture the robust sense of Aristotelian hylomorphism: "Psychology, in Aristotle's view, amounts to psycho-*physiology*, an analysis of both the formal and the material (i.e. bodily) aspects of psychic functions. The fact that in *On the Soul* itself we hear relatively little of

these bodily aspects might then be explained as a result of a deliberate distribution and arrangement of information over *On the Soul* and the *Parva Naturalia*, which should be seen as complementary parts of a continuous psycho-physiological account which is in its turn complementary to the zoological works." For a good work on the final treatises of the *Parva Naturalia* that shares van der Eijk's sentiment, see R.A.H. King, *Aristotle on Life and Death* (London: Duckworth, 2001).
121. Aristotle *On Youth, Old Age, Life and Death* 3.469a5–7.
122. Aristotle *De anima* 2.4.416b14; cf. *On Youth, Old Age, Life and Death* 4.469b15–16 for Aristotle's remarks about the heart containing the vital heat, and the soul being "as it were, set aglow with fire (*empepureumenês*) in this part" (in Barnes, *Complete Works of Aristotle*); cf. P. Studtmann, "Living Capacities and Vital Heat in Aristotle," *Ancient Philosophy* 24 (2004): 365–79.
123. Aristotle *On Youth, Old Age, Life and Death* 3.469a10–12. As evidence for this view, Aristotle observes that two senses, taste and touch, "can be clearly seen to extend to the heart, and hence the others also must lead to it, for in it the other organs may possibly initiate changes, whereas in the upper region of the body taste and touch have no connexion. Apart from these considerations, if the life is always located in this part, evidently the principle of sensation must be situated there too, for it is *qua* animal that a body is said to be a living thing, and it is called animal because endowed with sensation. Elsewhere in other works we have stated the reasons why some of the sense-organs are, as is evident, connected with the heart, while others are situated in the head. (It is this fact that causes some people to think that it is in virtue of the brain that the function of perception belongs to animals)," *On Youth, Old Age, Life and Death* 3.469a12–22 (in Barnes, *Complete Works of Aristotle*). For the importance of the heart as the seat of the common sense organ, cf. the recent study of P. Gregoric, *Aristotle on the Common Sense* (Oxford: Clarendon Press, 2007).
124. Aristotle *Generation of Animals* 2.6.744a2–5 (in A.L. Peck, *Aristotle: Generation of Animals* [Cambridge, MA: Harvard University Press, 1942]): "[The sense organ of] smell and hearing are passages full of connate *pneuma*, connecting with the outer air and terminating at the small blood-vessels around the brain which extend thither from the heart." Cf. *Parts of Animals* 2.10.656b17–18: "From the eyes passages (*poroi*) run to the vessels (*phlebas*) around the brain."
125. Aristotle *Motion of Animals* 10; cf. M.F. Frampton, "Aristotle's Cardiocentric Model of Animal Locomotion," *Journal of the History of Biology* 24 (1991): 291–330.
126. Aristotle *Parts of Animals* 3.7.670a26–27. In fact, Aristotle compares the activity of the heart with the activity of the physician: "It is the part of the dominating (*kurion*) organ to achieve (*diatelein*) the final result (*to hou heneka*), as of the physician's (*iatros*) efforts to be directed toward health (*hygieian*), and not to be occupied with subordinate offices" (*On Youth, Old Age, Life and Death* 3.469a7–9, in Barnes, *Complete Works of Aristotle*). What appears at first glance as simply an analogy turns out to be the truth, since the heart, in all its various duties, constantly serves as the index of health in the body.
127. Cf. Aristotle *Parts of Animals* 3.4.667a32–667b14 (in Barnes, *Complete Works of Aristotle*): "The heart (*hê kardia*) again is the only one of the viscera (*splagchnôn*), and indeed the only part of the body, that is unable to tolerate any serious affection (*chalepon pathos*). This is but what might reasonably be expected. For, if the primary part be diseased (*phtheiromenês*), there is nothing from which the

other parts which depend upon it can derive succor (*boêtheia*). A proof (*sêmeion*) that the heart is thus unable to tolerate any serious affection (*pathos*) is furnished by the fact that in no sacrificial victim (*tôn thuomenôn hiereiôn*) has it ever been seen to be affected with those diseases (*toiouton pathos*) that are observable in the other viscera. For the kidneys are frequently found to be full of stones (*lithôn*), growths (*phumatôn*), and small abscesses (*dothiênôn*), as also are the liver, the lung, and more than all the spleen. There are also many other conditions (*pathêmata*) which are seen to occur in these parts, those which are least liable to such being the portion of the lung which is close to the windpipe, and the portion of the liver which lies about the junction with the great blood-vessel. This again admits of rational explanation (*eulogôs*). For it is in these parts that they are most closely in communion with the heart. On the other hand, when animals die not by sacrifice but from disease (*noson*), and from affections (*pathê*) such as are mentioned above, they are found on dissection to have morbid affections (*nosôdê pathê*) of the heart." Cf. *Parts of Animals* 4.2.677b4–10.
128. Traditional Aristotelian hylomorphism refers to the soul being the form of the visible body with organs (*De anima* 2.1.412b5–6; see R. Polansky, *Aristotle's De anima* [Cambridge: Cambridge University Press, 2007] for a comprehensive treatment). Bos, *The Soul*, challenges the traditional interpretation of Aristotelian hylomorphism, arguing that the body Aristotle refers to in his definition of the soul is the *pneumatic* body, since Aristotle often refers to *pneuma* as an organic body (cf., e.g., *Generation of Animals* 5.8.789bff.).
129. Lloyd, *Grip of Disease*, 202.
130. Sadly, only fragments of the works of these authors survive. For Mnesthius and Dieuches, see J. Bertier, *Mnésithée et Dieuchès* (Leiden: Brill, 1972); for Diocles, P. J. van der Eijk, *Diocles of Carystus*, 2 vols. (Leiden: Brill, 2000–2001); for Praxagoras, F. Steckerl, *The Fragments of Praxagoras of Cos and His School* (Leiden: Brill, 1958). In Aristotle's own school, interest in medical themes continued as well, especially with Theophrastus (see W. W. Fortenbaugh, R. W. Sharples, and M. G. Sollenberger, eds., *Theophrastus of Eresus: On Sweat, On Dizziness, and On Fatigue* [Leiden: Brill, 2003]) and Meno, whose history of medicine survives in fragmentary form, known as the *Anonymus Londinensis* (see W.H.S. Jones, *The Medical Writings of Anonymus Londinensis* [Cambridge: Cambridge University Press, 1947] with Manetti, "Aristotle").
131. Fragment 3 (van der Eijk, *Diocles of Carystus*, vol. 2, 7), however, enjoins caution about exactly what being a "follower" of Hippocrates is supposed to mean. Cf. also frag. 4, where Pliny says that Diocles was "second in age and fame" (*secundus aetate famaque*) to Hippocrates.
132. Van der Eijk, *Diocles of Carystus*, frags. 51b: "Diocles [says] that most diseases (*nosôn*) occur through an imbalance (*anômalian*) of the elements (*stoicheiôn*) in the body and of the constitution (*katastêmatos*) [of the weather]," and 51c: "Diocles [says] that most causes of disease (*nosôn*) occur through an imbalance (*anômalian*) of the elements (*stoicheiôn*) in the body and of the constitution (*katastêmatos*) of the air (*aeros*)." As van der Eijk notes (vol. 2, 112), frags. 51b and 51c both echo Alcmaeon and certain Hippocratic writers.
133. Van der Eijk, *Diocles of Carystus*, frag. 102 (Steckerl, *Fragments*, frag. 75).
134. See van der Eijk, *Medicine and Philosophy*, 119–35, for different views among the ancients on the location of the soul's mental, perceptive, and locomotive faculties (sometimes referred to as the *hegemonikon*, though the term is Stoic in origin and

if applied to Aristotle, Diocles, and Praxagoras is anachronistic and hence potentially misleading).
135. Van der Eijk, *Diocles of Carystus*, frag. 182.
136. Van der Eijk, *Diocles of Carystus*, frag. 182 (and cf. his vol. 2, 349, for discussion); for a survey of ancient dentistry, see W. Hoffmann-Axthelm, *History of Dentistry*, trans. H. M. Koehler (Chicago: Quintessence, 1981). Pennyroyal (*Mentha pulegium*) is from the mint family.
137. Steckerl, *Fragments*, frags. 21 and 22. For the renown of Praxagoras in antiquity, see Nutton, *Ancient Medicine*, 127.
138. Steckerl, *Fragments*, frag. 18 (Galen, *On the Natural Faculties* 2.8, in A. J. Brock, *Galen: On the Natural Faculties* [Cambridge, MA: Harvard University Press, 1916]).
139. Steckerl, *Fragments*, frags. 60 (fevers) and 69 (melancholy). Praxagoras is also famous for discovering that the pulse could be used for diagnostic purposes (frags. 26 and 27).
140. Herophilus is possibly connected through his teacher, Praxagoras of Cos. Nutton (*Ancient Medicine*, 358n86) observes that "his authorship of a treatise on *Foreign Diseases* (fr. 63 Steckerl) suggests that he spent some time away from Cos, either as a student or, later, as a travelling physician"; however, while Nutton thinks that Praxagoras was influenced by Aristotle's ideas, "There is no evidence that he ever studied in Athens at the Lyceum" (Nutton, *Ancient Medicine*, 126). Erasistratus was possibly associated with Theophrastus or other Peripatetics; cf. V. Nutton, "Erasistratus," in *Brill's New Pauly Encyclopedia of the Ancient World*, ed. H. Cancik and H. Schneider, vol. 5, cols. 13–15 (Leiden: Brill, 2004). Both men were venerated in antiquity for having been the first to dissect the human body in Ptolemaic Alexandria; for more on their dissections and anatomical work, see Holmes's contributions to this volume.
141. Von Staden, *Herophilus*, frag. 133.
142. I. Garofalo, *Erasistrati Fragmenta* (Pisa: Giardini, 1988), frag. 169 (cited in J. Longrigg, *Greek Medicine from the Heroic to the Hellenistic Age: A Source Book* [New York: Routledge, 1998], 116).
143. Cf. Nutton, *Ancient Medicine*, 135–36.
144. For Herophilus, von Staden, *Herophilus*, is essential; for Erasistratus, Garofalo, *Erasistrati Fragmenta*.
145. See K. Algra, J. Barnes, J. Mansfeld, and M. Schofield, eds., *The Cambridge History of Hellenistic Philosophy* (Cambridge: Cambridge University Press, 1999); and R. W. Sharples, *Stoics, Epicureans, and Sceptics: An Introduction to Hellenistic Philosophy* (London: Routledge, 1996).
146. The label "Rationalist" is an umbrella that includes many subsects, e.g., Hippocratics, Herophileans, Pneumatists, i.e., any medical sect that had a theoretical basis (cf. Nutton, *Ancient Medicine*, 202–15).
147. See Nutton, *Ancient Medicine*, 140–70, 187–215.
148. Edelstein (*Ancient Medicine: Selected Papers*, 173–91, 195–203) argues that philosophical skepticism influenced both the Empiricist and Methodist sects; cf. R. Walzer and M. Frede, eds., *Galen: Three Treatises on the Nature of Science* (Indianapolis: Hackett, 1985), xx–xxxi; Frede, *Essays in Ancient Philosophy*, 243–60; J. Allen, "Pyrrhonism and Medical Empiricism: Sextus Empiricus on Evidence and Inference," *Aufstieg und Niedergang der römischen Welt*, 2nd ser., 37, part 1 (1993): 646–90; J. Allen, *Inference from Signs: Ancient Debates about the*

Nature of Evidence (Oxford: Clarendon Press, 2001), 87–146; and R. J. Hankinson, *Cause and Explanation in Ancient Greek Thought* (Oxford: Clarendon Press, 1998), 306–18. Schiefsky (*Hippocrates* On Ancient Medicine, 345–59), noting the latent empirical tendencies in the Hippocratic treatise *On Ancient Medicine*, traces medical empiricism back to Plato and Aristotle. Methodism, a "Roman parvenu" (Nutton, *Ancient Medicine*, 187), was a sort of "*via media*" (R. J. Hankinson, *Galen: On the Therapeutic Method* [Oxford: Clarendon Press, 1991], xxxi) between Rationalism, which grounded its nosology and therapeutics on a theoretical basis, and Empiricism, which "argued that it was less important to know why any particular treatment worked, than *that* it worked" (Lloyd, *Grip of Disease*, 205). See Holmes's contributions to this volume for more references to the Methodist sect. The story of Roman medicine is largely one of Greek importation; see T. C. Allbutt, *Greek Medicine in Rome* (1921; repr., New York: Benjamin Blom, 1970); J. Scarborough, *Roman Medicine* (Ithaca, NY: Cornell University Press, 1969); R. Jackson, *Doctors and Diseases in the Roman Empire* (London: University of Oklahoma Press, 1988); and Nutton, *Ancient Medicine*, 157–86. Early Roman response to this importation was negative, with authors such as the elder Cato and the elder Pliny taking rather conservative views of medicine itself, similar to Plato. Roman encyclopedic works, such as Pliny's *Natural History* and Celsus's *On Medicine*, are valuable sources of information about the various doctors and sects that appeared in the Hellenistic era.

149. The secondary literature on Galen is large, but a good place to begin is Nutton, *Ancient Medicine*, 216–47, and the papers in R.J. Hankinson, ed., *The Cambridge Companion to Galen* (Cambridge: Cambridge University Press, 2008). For Galen as a philosopher, see R. J. Hankinson, "Galen's Philosophical Eclecticism," *Aufstieg und Niedergang der römischen Welt* 2nd ser., 36, part 5 (1992): 3505–22; and the papers in J. Barnes, V. Barras, and J. Jouanna, eds., *Galien et la philosophie* (Vandoeuvres-Genève: Fondation Hardt, 2003).

150. Galen, *Quod optimus medicus sit quoque philosophus* (1.53–63 K, in P. N. Singer, *Galen: Selected Works* [New York: Oxford University Press, 1997], 30–34). In the references to Galen made in this section, the K designations refer to the volume and pages of the standard edition of the Galenic Corpus assembled and edited by Kühn, 1819–1833. See Hankinson, *Cambridge Companion to Galen*, 391–403, for extremely helpful appendices to the correspondence between the Galenic treatises in Kühn and modern editions of these treatises, as well as to English translations of these treatises.

151. Galen, *The Best Constitution of Our Bodies* 2 (*De optima corporis nostri constitutione*) in Singer, *Galen: Selected Works*, 292; 4.741–742 K; translation modified.

152. Cf. generally, Aristotle *On Generation and Corruption* 2.3 and 8, and *Parts of Animals* 2.1–2 and 10.

153. Aristotle *Parts of Animals* 1.1.640a33–640b4, 640b18–29, and 1.5.645b15–20.

154. Cf. Galen, *On the Therapeutic Method* (*De methodo medendi*) 1.7.14–15 (10.58–60 K), 2.6.2–3 (10.115–117 K), in Hankinson, *Galen: Therapeutic Method*, with commentary (148–49 and 194); *Thrasybulus* (*Thrasybulus sive utrum medicinae sit an gymnasticae hygiene*) 11 (5.822–824 K), in Singer, *Galen: Selected Works*, 60–61; *On the Natural Faculties* [*De naturalibus facultatibus*] 2.8 (2.117–119 and 121 K) and 2.9 (2.125–126 K), in Brock, *Galen: Natural Faculties*.

155. For Erasistratus, see von Staden 2000. Asclepiades of Bythinia, "regularly cited as the last representative of the so-called Dogmatic physicians before Galen" (Nutton, *Ancient Medicine,* 168), is a hugely influential figure in ancient medicine. His atomistic and mechanistic explanations of health and disease are supposed to have inspired the Methodist sect; cf. J. Vallance, *The Lost Theory of Asclepiades of Bithynia* (Oxford: Clarendon Press, 1990); J. Vallance, "The Medical System of Asclepiades of Bithynia," *Aufstieg und Niedergang der römischen Welt,* 2nd ser., 37, part 1 (1993): 693–727; and Nutton, *Ancient Medicine,* 168–70. He may have been influenced by the Epicureanism of Lucretius's philosophical poem *On the Nature of Things,* which ends with a meditation on the rampant death caused by the great Plague of Athens (see Lloyd, *Grip of Disease,* 218–19, for a brief though excellent discussion).

156. Galen, *The Faculties of the Soul Follow the Mixtures of the Body* 9 (*Quod animi mores corporis temperamenta sequuntur*) (4.804–808 K, in Singer, *Galen: Selected Works,* 167–69, translation modified).

157. Lloyd, *Grip of Disease,* 211.

158. [Hippocrates] *On Ancient Medicine* 20 (in Jones, *Hippocrates,* vol. 1; translation modified).

159. R. W. Sharples, "Philosophy for Life," in *The Cambridge Companion to the Hellenistic World,* ed. G. R. Bugh, 223–40 (Cambridge: Cambridge University Press, 2006); cf. M. C. Nussbaum, *The Therapy of Desire: Theory and Practice in Hellenistic Ethics* (Princeton, NJ: Princeton University Press, 1994); and J. E. Annas, "Philosophical Therapy, Ancient and Modern," in Kuczewski and Polansky, *Bioethics,* 109–127.

160. Democritus, frag. 31 (cited in Annas, "Philosophical Therapy," 116).

161. [Hippocrates] *Decorum* 5 (in Jones, *Hippocrates,* vol. 2). The writer of *Decorum* goes on in this passage to list a number of qualities that wisdom and medicine share, including lack of avarice (*aphilarguriê*), sound opinion (*doxa*), judgment (*krisis*), purity (*kathariotês*), knowledge (*eidêsis*) of the things good (*chêstôn*) and necessary (*anagkaiôn*) for life (*pros bion*), freedom from superstition (*adeisidaimoniê*), and divine superiority (*hyperochê theiê*). However, the writer also includes some characteristics that might indicate his sophistry, such as selling of that which cleanses (*katharsios apempolêsis*) and sententious speech (*gnômologiê*); cf. Nutton, *Ancient Medicine,* 155–56.

162. Cf. Plato *Charmides* 156e–157b, where Socrates relates the methodology of the Zalmoxian physician from Thrace to Charmides: "'Just as one should not attempt to cure the eyes apart from the head, nor the head apart from the body, so one should not attempt to cure the body apart from the soul. And this, he says, is the very reason why most diseases are beyond the Greek doctors, that they do not pay attention to the whole as they ought to do, since if the whole is not in good condition, it is impossible that the part should be. Because,' he said, 'the soul is the source both of bodily health and bodily disease for the whole man, and these flow from the soul in the same way that the eyes are affected by the head. So it is necessary first and foremost to cure the soul if the parts of the head and of the rest of the body are to be healthy. And the soul,' he said, 'my dear friend, is cured by means of certain charms, and these charms consist of beautiful words. It is a result of such words that temperance arises in the soul, and when the soul acquires and possesses temperance, it is easy to provide health both for the head and the rest of the body'" (in Cooper, *Plato*). For discussion of this passage, see Annas,

"Philosophical Therapy," 112. By charms consisting of beautiful words, Socrates means philosophy; for a similar conception of medicine as treating both soul and body, cf. *Phaedrus* 270c–e.

163. R. Polansky ("The Unity of Plato's *Crito*," *Scholia* 6 [1997]: 49–67, 50n4) notes the etymological connection between the name of Crito (*Kritôn*) and the medical concept of *krisis*.

164. Plato *Crito* 48b.

165. Cicero *Tusculan Disputations* 3.2–3 (in J. E. King, *Cicero: Tusculan Disputations* [Cambridge, MA: Harvard University Press, 1927]); cf. the recent studies by M. Graver, *Cicero on the Emotions: Tusculan Disputations 3 and 4* (Chicago: University of Chicago Press, 2002); and B. Koch, *Philosophie als Medizin für die Seelen: Untersuchungen zur den Tusculanae Disputationes* (Stuttgart: Steiner, 2006).

166. Cf. Annas, "Philosophical Therapy"; Lloyd, *Grip of Disease,* 205–10; and J. Sellars, *Stoicism* (Berkeley: University of California Press, 2006), 114–20. Notice that the word for passion or emotion (*pathê*) is the same as the general Greek term for affection or illness (*pathê*); passions, emotions, and illness are similar in that they "befall" (*paschein*) human beings, and humans can "suffer" (*paschein*) from them. Aristotle (*On Respiration* 20.479b25–26) also noticed that certain strong feelings, such as fear (*phobon*) and potentially morbid emotion (*pathos nosêmatikon*), could actually result in death (*apothnêskein*), because the coldness that these emotions bring on in the heart would extinguish (*aposbennusthai*) the vital heat that resides there.

167. Similar notions can be found among the Epicureans: "One must not pretend to philosophize, but philosophize in reality. For we do not need the semblance of health but true health" (*The Vatican Collection of Epicurean Sayings*, 54, in B. Inwood and L. P. Gerson, eds., *Hellenistic Philosophy: Introductory Readings* [Indianapolis: Hackett, 1997], 38); and "Empty is the argument of the philosopher by which no human disease is healed; for just as there is no benefit in medicine if it does not drive out bodily diseases, so there is no benefit in philosophy if it does not drive out the disease of the soul" (Porphyry, *To Marcella* 31, in Inwood and Gerson, *Hellenistic Philosophy,* 97).

168. Luke 4:23. L. T. Johnson (*The Gospel of Luke* [Collegeville, MN: Liturgical Press, 1991], 80) notes, "The term *parabolê*, which elsewhere will be translated 'parable,' is here being used in the sense of *mashal*, or proverb, as in the LXX 1 Sam 10:12. Variations of the proverb itself are found in both Greek and Jewish writings; cf. e.g., 'Physician, physician, heal thine own limp!' in *Genesis Rabbah* 23:4."

169. Tertullian *Prescriptions against Heretics* 7. Interestingly at the beginning of this treatise, Tertullian likens heresies to fevers (*Prescriptions against Heretics* 2).

170. Despite his vehemence against any admixture of Greek philosophy with Christianity, Tertullian himself was not opposed to medical thought; in the passage from his *De anima* that I alluded to earlier, he claims to have studied both medicine and philosophy (cf. D. W. Amundsen, *Medicine, Society, and Faith in the Ancient and Medieval Worlds* [Baltimore: Johns Hopkins University Press, 1996], 145–46).

171. Paul Colossians 4:14. For a recent study of the possible Hippocratic antecedents to the cases of health and disease in Luke's Gospel, see A. Weissenrieder, *Images of Illness in the Gospel of Luke* (Tübingen: Mohr Siebeck, 2003). Building on earlier work, such as W. K. Hobart, *The Medical Language of St. Luke* (Piscataway, NJ: Gorgias Press, 2004; reprint of 1954 edition), Weissenrieder concludes: "One

thing, we can say with certainty: The author of the Gospel of Luke had some knowledge of ancient medicine" (335).
172. Luke 5:12–16 and 17:11–19; cf. Matthew 8:2–4, 9:27, 15:22, and Mark 1:40–45.
173. Luke 5:17–26; cf. Matthew 9:1–8 and Mark 2:1–12.
174. Luke 8:43–48; cf. Matthew 9:18–26 and Mark 5:21–43.
175. Luke 13:10–17.
176. Luke 14:1–6.
177. Luke 4:38–39.
178. Matthew 8:14–15 and Mark 1:29–31 both describe the condition simply as a fever (*puretos*), while Luke adds that it was a "severe" (*megalôi*) fever (*puretôi*).
179. Luke 22:44; there is some dispute, however, about the authenticity of this passage.
180. Cf., e.g., Aristotle *Parts of Animals* 3.5.668b6ff: "Instances, indeed, are not unknown of persons who in consequence of a bad general condition (*kachexian*) have secreted sweat (*hidrôsai*) that resembled blood (*haimatôdei*)" (in Barnes, *Complete Works of Aristotle*). In this passage, Aristotle attributes bloodlike sweat to insufficient concoction (*apepsian*), that is, weakness (*adunatousês*) and scantiness (*oligotêta*) of heat (*thermotêtos*) in the small blood vessels (*phlebiois*) near the surface of the skin. As we know from other passages in Aristotle (*On Respiration* 20.479b21–26), intense fear chills the heart, and the cold runs throughout the body, cooling the blood in the blood vessels, since all the vessels are ultimately connected to the heart (see note 119).
181. D.W. Amundsen and G.B. Ferngren, "The Perception of Disease and Disease Causality in the New Testament," *Aufstieg und Niedergang der römischen Welt*, 2nd ser., 37, part 3 (1996): 2948.
182. Cf. John 9:1–4, 11:4, and 11:40. In this regard, we should carefully interpret demonic etiologies in the New Testament. The common view that all the illnesses that Christ (and later in *Acts*, the Apostles themselves) heals are caused by demons is perhaps mistaken. Amundsen and Ferngren ("Perception of Disease," 2950–51) observe: "In most of the healings performed by Jesus, not only is there no mention of demonic involvement, but the symptoms are clearly distinguished from those of demonic possession. . . . Hence the attribution of sickness to demons is, in the New Testament, minimal, to say the least. There are only three cases in which it is clearly stated that a sickness was caused by demonic possession. In the other episodes of demon possession the erratic behavior *is* the demon possession. Most cases of disease and physical dysfunction in the New Testament are described in such a manner as to preclude even a hint of demonic activity."
183. Matthew 8:16–17. Notice in this passage that Matthew clearly distinguishes between those who are possessed by demons and those who are merely ill.
184. Not all illnesses are cured in the New Testament, however. Perhaps most conspicuous among those left to suffer from chronic illness is the Apostle Paul himself (2 Corinthians 12:1–10).
185. Cf., e.g., *Epidemics* 1, Case 9: the wife of Dromeades (in Jones, *Hippocrates*, vol. 1).
186. St. Augustine *Sermon* 87.11.13 (PL 38.537); cited in R. Arbesmann, "The Concept of 'Christus medicus' in St. Augustine," *Traditio* 10 (1954): 25.
187. Cf. Hebrews 12:7–13 and 1 Peter 4:12–19, and Ferngren and Amundsen, "Perception of Disease," 2962. Plato seems to foreshadow the Christian notion of the redemptive value of suffering. This occurs at the end of the *Republic*, when

Socrates relates the story of Er of Pamphylia (Pamphylia in Greek means "all tribes," i.e., Er stands as a symbol for all humankind), who had seen what occurs in the afterlife after being knocked out during a battle. Er speaks about how all the souls in the afterlife must choose their next life in a sort of lottery. Notably, the soul of Odysseus chooses last. We are told by Er that "it chanced that the soul of Odysseus got to make its choice last of all, and since memory of its former sufferings (*ponôn*) had relieved its love of honor, it went around for a long time, looking for the life of a private individual (*andros idiôtou*) who did his own work (*apragmonos*), and with difficulty it found one lying off somewhere neglected by the others. He chose it gladly and said that he'd have made the same choice even if he'd been first" (*Republic* 10.620c–d; in Cooper, *Plato*). Notice that the life that Odysseus chooses, that of "a private individual who did his own work," nicely embodies the definition that Socrates gives of justice earlier in the *Republic*—each part of the city and the soul minding its own business and doing its own work (cf. *Republic* 4.433a–434d and 441c–444c).

188. Mark 7:14–23 (cf. Matthew 15:1–20): "He summoned the crowd again and said to them, 'Hear me, all of you, and understand. Nothing that enters one from outside can defile that person; but the things that come out from within are what defile.' When he got home away from the crowd his disciples questioned him about the parable. He said to them, 'Are even you likewise without understanding? Do you not realize that everything that goes into a person from outside cannot defile, since it enters not the heart but the stomach and passes out into the latrine?' (Thus he declared all foods clean.) 'But what comes out of a person, that is what defiles. From within people, from their hearts, come evil thoughts, unchastity, theft, murder, adultery, greed, malice, deceit, licentiousness, envy, blasphemy, arrogance, folly. All these evils come from within and they defile.'"

189. John 6:48 and 54–61.

190. Controversy about how to square the doctrine of the Eucharist with the mundane facts about human digestion and metabolism continued up into the thirteenth century; see P. L. Reynolds, *Food and the Body: Some Peculiar Questions in High Medieval Theology* (Leiden: Brill, 1999).

191. St. Ignatius of Antioch (died 107 C.E.), *Letter to the Ephesians* 20 (cited in J. T. O'Connor, *The Hidden Manna: A Theology of the Eucharist* [San Francisco: Ignatius Press, 1988], 17).

192. Cf. Paul Romans 12:4–5, 1 Corinthians 12:12–31, and Colossians 1:18.

193. Cf. Revelation 21:5.

194. I have left out of consideration the ancient Greek healing cult of Asclepius, later transferred to Rome. See Holmes's contributions to this volume, and for more, see B. L. Wickkiser, *Asklepios, Medicine, and the Politics of Healing in Fifth-Century Greece* (Baltimore: Johns Hopkins University Press, 2008). Under Christianity, pagan devotion to Asclepius was eventually (though not easily) dislodged by the popularity of Christ the Physician (*Christus Medicus*); see Arbesmann, "Concept of 'Christus medicus.'"

Chapter 3

1. Here I summarize a few of the conclusions of my book *Sexuality in Greek and Roman Culture* (© 2005 by Marilyn B. Skinner. All rights reserved). Fuller treatment

of all the topics discussed below will be found there. I am grateful to the publisher, Blackwell Publishing Ltd., for permission to incorporate into this essay certain material, including translations of Greek and Latin passages, first published in that volume.

2. A. Richlin, "Towards a History of Body History," in *Inventing Ancient Culture: Historicism, Periodization, and the Ancient World*, ed. M. Golden and P. Toohey (London: Routledge, 1997), 16–35, traces in a concise and lucid manner the intersections between the writing of body history, feminism, and classics. She would disagree, however, with my perception of radical discontinuity between ancient and modern systems of sex and gender.
3. D. H. Garrison, *Sexual Culture in Ancient Greece* (Norman: University of Oklahoma Press, 2000), 10–18.
4. P. duBois, *Sowing the Body: Psychoanalysis and Ancient Representations of Women* (Chicago: University of Chicago Press, 1988), 39–85.
5. An alternative theory of generation, in which the child is the product of both male and female sperm and the respective strength of each seed determines the sex, is found in one treatise of the Hippocratic Corpus (*Genit.* 4–7) and probably reflects the belief of most trained Hippocratic physicians (L. Dean-Jones, *Women's Bodies in Classical Greek Science* [Oxford: Clarendon Press, 1994], 148–60).
6. G. Ferrari, *Figures of Speech: Men and Maidens in Ancient Greece* (Chicago: University of Chicago Press, 2002), 194–209.
7. S. B. Pomeroy, "Women's Identity and the Family in the Classical *Polis*," in *Women in Antiquity: New Assessments*, ed. R. Hawley and B. Levick, 111–21 (London: Routledge, 1995), 119.
8. The speaker of Demosthenes 57 points out that his father had married a sister (*adelphên*) not of the same mother (*ouch homomêtrian*, 20); Plutarch (*Them.* 32), who records a marriage between Themistocles's son and daughter, tells us they were children of his first and second wives, respectively.
9. R. Just, *Women in Athenian Law and Life* (London: Routledge, 1989), 78.
10. *Gen. an.* 716a.5–716b.10.
11. T. Laqueur, *Making Sex: Body and Gender from the Greeks to Freud* (Cambridge, MA: Harvard University Press, 1990), 4–8.
12. According to Aristotle (*Part. an.* 648a.28–31) this position was maintained by Parmenides and others, while Empedocles said the opposite. The Pythagoreans may also have viewed women as the colder sex (Dean-Jones, *Women's Bodies*, 44). It is worth observing that such claims were apparently based on *a priori* reasoning, not empirical measurement.
13. During the wet, cold winter, according to [Arist.] *Pr.* 4.28, women become sluggish, but men's heat and moisture are just sufficient to quicken the seed and arouse desire. At extremes of temperature, then, one sex will always be energetic, the other incapacitated, due to contrasting physiological makeup.
14. M. Foucault, *The Use of Pleasure*, trans. R. Hurley, *The History of Sexuality*, vol. 2 (New York: Vintage 1986), 125–39.
15. Alexander the Great is reported to have said that sleep and sexual intercourse most reminded him of his own mortality. Plutarch, who preserves the remark (*Alex.* 22.3), interprets it to mean that, in Alexander's view, the necessity of rest, on the one hand, and the experience of sexual languor, on the other, arose from the same natural limitation upon the body.
16. [Hippoc.] *Genit.* 4; cf. *Mul.* 1.2.

17. A. Carson, "Putting Her in Her Place: Woman, Dirt, and Desire," in *Before Sexuality: The Construction of Erotic Experience in the Ancient Greek World,* ed. D.M. Halperin, J.J. Winkler, and F.I. Zeitlin, 135–69 (Princeton, NJ: Princeton University Press, 1990), 137–45.
18. Xen. *Mem.* 2.2.5, *Oec.* 7.24; Arist. *Eth. Nic.* 1159a.26–34.
19. Xen. *Oec.* 7.25.
20. Pl. *Leg.* 781c.
21. *Pol.* 1260a.12–14.
22. M.B. Skinner, "*Alexander* and Ancient Greek Sexuality: Some Theoretical Considerations," in *Responses to Oliver Stone's Alexander: Film, History, and Cultural Studies,* eds. P. Cartledge and F. Greenland (Madison: The University of Wisconsin Press, 2010), pp. 119–34.
23. K.J. Dover, *Greek Homosexuality* (London: Duckworth, 1978), 68–73.
24. That last sentence requires some qualification, for Greek sources by no means exclude the possibility of mutual desire and pleasure. It is implicit, for example, in several Homeric accounts of heterosexual lovemaking (e.g., Paris and Helen, *Il.* 3.441–48; Odysseus and Circe, *Od.* 10.333–35; Odysseus and Penelope, *Od.* 23.295–96, 300). At a much later date, the lengthy prose fictions conventionally known as "Greek romances" also center on a heterosexual love relationship entered into by mutual consent and characterized by a remarkable "symmetry" of feelings and behaviors (D. Konstan, *Sexual Symmetry: Love in the Ancient Novel and Related Genres* [Princeton, NJ: Princeton University Press, 1994], 7–11). Scholars generally believe that the model of sexual relations between women in Sappho's poetry is an egalitarian one, although that assumption has been questioned (A. Giacomelli, "The Justice of Aphrodite in Sappho fr. 1," *Transactions of the American Philological Association* 110 [1980]: 135–42). However, the dominant erotic paradigm is one in which power relations are asymmetrical—particularly in verse, like that of Theognis, composed for the all-male symposium or drinking party, where pederasty is the controlling theme.
25. D.M. Halperin, *One Hundred Years of Homosexuality and Other Essays on Greek Love* (New York: Routledge, 1990), 95–96; J. Walters, "Invading the Roman Body: Manliness and Impenetrability in Roman Thought," in *Roman Sexualities,* ed. J.P. Hallett and M.B. Skinner, 29–43 (Princeton, NJ: Princeton University Press, 1997), 36–41.
26. Dover, *Greek Homosexuality,* 100–109.
27. Halperin, *One Hundred Years,* 30–31.
28. J.N. Davidson, *Courtesans and Fishcakes: The Consuming Passions of Classical Athens* (New York: HarperPerennial, 1997); J.N. Davidson, "Dover, Foucault and Greek Homosexuality: Penetration and the Truth of Sex," *Past and Present* 170 (2001): 3–51; T.K. Hubbard, "Popular Perceptions of Elite Homosexuality in Classical Athens," *Arion* ser. 3, 6.1 (1998): 48–78; T.K. Hubbard, "Pederasty and Democracy: The Marginalization of a Social Practice," in *Greek Love Reconsidered,* ed. T.K. Hubbard, 1–11 (New York: Wallace Hamilton Press, 2000).
29. Archaic Sparta and Crete are two other societies in which pederasty was institutionalized (Pl. *Leg.* 636b.3–d.1) and for which we have evidence of how the practice was conducted. Space restrictions, however, make it necessary to concentrate upon Athenian custom.
30. The most coherent justification of pederasty as an educational practice is put in the mouth of Pausanias, one of the speakers in Plato's *Symposium.* In this speech, presented as an encomium of the god Eros, Pausanias first distinguishes between

sacred and profane love and then explains the criteria that define the behavior of each partner as honorable (180c–185c).
31. M. Golden, "Slavery and Homosexuality at Athens," *Phoenix* 38 (1984): 312–16.
32. M. B. Skinner, *Sexuality in Greek and Roman Culture* (Oxford: Blackwell, 2005), 118–24.
33. *Symp.* 8.21.
34. M. Gleason, "The Semiotics of Gender: Physiognomy and Self-Fashioning in the Second Century C.E.," in Halperin, Winkler, and Zeitlin, *Before Sexuality,* 391.
35. For discussion see Dover, *Greek Homosexuality,* 169–70; J. J. Winkler, *The Constraints of Desire: The Anthropology of Sex and Gender in Ancient Greece* (New York: Routledge, 1990), 67–69.
36. 1148b.27–30.
37. Ibid., 31–32.
38. H. A. Shapiro, "Courtship Scenes in Attic Vase Painting," *American Journal of Archaeology* 85 (1981): 133–43.
39. R. F. Sutton Jr., "Pornography and Persuasion on Attic Pottery," in *Pornography and Representation in Greece and Rome,* ed. A. Richlin, 3–35 (New York: Oxford University Press, 1992), 24.
40. See, e.g., *Anth. Pal.* 5.145 (Asclep.); Callim. *Epigr.* 28.
41. B. Cohen, "Divesting the Female Breast of Clothes in Classical Sculpture," in *Naked Truths: Women, Sexuality, and Gender in Classical Art and Archaeology,* ed. A. O. Koloski-Ostrow and C. L. Lyons, 66–92 (London: Routledge, 1997).
42. Outside Athens, as the editor of this volume reminds me, the tradition of partial and even complete female nudity had a longer pedigree: witness the Aphrodite and the flute girl on the early fifth-century Ludovisi Throne from Locri and the near-contemporary Lapith woman with bared breast on the west pediment of the Temple of Zeus at Olympia. Illustrations of all monuments discussed may be found in Garrison, *Sexual Culture,* 199–204.
43. Isoc. 4.133, 167–69; 5.96, 120–22; 8.24.
44. S. B. Pomeroy, *Families in Classical and Hellenistic Greece: Representations and Realities* (Oxford: Clarendon Press, 1997), 108–14, discusses the appearance of funerary foundations beginning in the late fourth century. As described in inscriptions, the wealthy set up funds to insure that offerings for the dead would continue to be made for their ancestors and themselves. This had not been the practice earlier because parents could assume that their children would perform due rites.
45. Garrison, *Sexual Culture,* 233–36.
46. S. B. Pomeroy, "TECHNIKAI KAI MOUSIKAI: The Education of Women in the Fourth Century and in the Hellenistic Period," *American Journal of Ancient History* 2 (1977): 51–68.
47. U. Kron, "Priesthoods, Dedications and Euergetism: What Part Did Religion Play in the Political and Social Status of Greek Women?" in *Religion and Power in the Ancient Greek World: Proceedings of the Uppsala Symposium 1993,* ed. P. Hellström and B. Alroth, 139–82 (Uppsala: Acta Universitatis Upsalensis, 1996).
48. L. Koenen, "The Ptolemaic King as a Religious Figure," in *Images and Ideologies: Self-Definition in the Hellenistic World,* ed. A. Bulloch, E. S. Gruen, A. A. Long, and A. Stewart, 25–115 (Berkeley, CA: University of California Press, 1993), 61–62; J. D. Reed, "Arsinoe's Adonis and the Poetics of Ptolemaic Imperialism," *Transactions of the American Philological Association* 130 (2000): 336–38.

49. Solon, the leading Athenian statesman of his time (chief magistrate in 594/3 B.C.E.), heard his nephew play a song of Sappho at a party and asked the boy to teach it to him, "so that," he said, "I may learn it and die" (Ael. *ap.* Stob. 3.29.58).
50. *Symp.* 191e.
51. Dover, *Greek Homosexuality*, 174.
52. P. Oxy. 1800 frag. 1.
53. *Leg.* 636c.
54. K.J. Dover, "Two Women of Samos," in *The Sleep of Reason: Erotic Experience and Sexual Ethics in Ancient Greece and Rome*, ed. M.C. Nussbaum and J. Sihvola, 222–28 (Chicago: University of Chicago Press, 2002), 226.
55. Suet. *Calig.* 36.1.
56. *Ep.* 95.21.
57. See, e.g., Hor. *Sat.* 2.5; Juv. 1 and 3; cf. Trimalchio in Petronius's *Satyricon*.
58. The speaker in Juvenal's third satire observes that the cruelest thing about wretched poverty itself is that "it makes human beings ridiculous" (152–53).
59. J.P. Hallett, "Women as Same and Other in Classical Roman Elite," *Helios* 16 (1989): 59–78.
60. J.H.W.G. Liebeschuetz, *Continuity and Change in Roman Religion* (Oxford: Clarendon Press, 1979), 45–47, with reference to Catull. 68.41–160.
61. Hor. *Carm.* 3.6.17–44.
62. C. Edwards, *The Politics of Immorality in Ancient Rome* (Cambridge: Cambridge University Press, 1993), 52–53.
63. It is reaffirmed in the emperor Justinian's sixth-century C.E. compilation of Roman law, the *Institutes* (4.18.4), where capital punishment is stipulated for homoeroticism and adultery, and confiscation of property or corporeal punishment and banishment (the penalty depending on rank) for seduction of an unmarried girl or widow.
64. *Dig.* 48.51.6 (Papinian); 50.16.101. pr. (Modestinus).
65. *Dig.* 48.5.13 (Ulpian).
66. C.A. Williams, *Roman Homosexuality: Ideologies of Masculinity in Classical Antiquity* (New York: Oxford University Press, 1999), 96; cf. E. Fantham, "*Stuprum*: Public Attitudes and Penalties for Sexual Offences in Republican Rome," *Échos du Monde Classique/Classical Views* 35 n.s. 10 (1991): 267–91.
67. *Dig.* 47.10.15.15–23 (Ulpian).
68. *Dig.* 47.11.1.2 (Paulus).
69. *Cael.* 9–10.
70. Williams, *Roman Homosexuality*, 113–15.
71. Skinner, *Sexuality in Greek and Roman Culture*, 200.
72. Tac. *Ann.* 1.54.
73. C. Edwards, "Unspeakable Professions: Public Performance and Prostitution in Ancient Rome," in *Roman Sexualities*, ed. Hallett and Skinner, 66–95.

Chapter 4

1. Recent research has demonstrated just how powerful the placebo effect may be. For one attempt to explain the effect from the perspective of evolutionary biology, see N. Humphrey, "Great Expectations: The Evolutionary Psychology of Faith Healing and the Placebo Effect," in *The Mind Made Flesh: Essays from the Frontiers of*

Psychology and Evolution (Oxford: Oxford University Press, 2002), 255–85, with n. 2 for further bibliography.

2. For a more detailed account of how the physical body emerges as a "conceptual object" in the Greek world and a more in-depth discussion of Greek medicine in the classical period, see B. Holmes, *The Symptom and the Subject: The Emergence of the Physical Body in Ancient Greece* (Princeton, NJ: Princeton University Press, 2010).

3. We cannot, however, assume that medicine exercised the same kind of authority over popular ideas about the body in the ancient Greco-Roman world as it does today. See especially the remarks on "naturalization" in R. Flemming, *Medicine and the Making of Roman Women* (Oxford: Clarendon Press, 2000), 3–27.

4. In recognizing the virtual absence of divine and demonic causes from the learned medical tradition, I am making no claims about the writers' views of the gods *tout court*.

5. Vivian Nutton, "Healers in the Medical Marketplace: Towards a Social History of Graeco-Roman Medicine," in *Medicine and Society: Historical Essays*, ed. A. Wear (Cambridge: Cambridge University Press, 1992), 15–58. See also Vivian Nutton, "The Medical Meeting Place," in *Ancient Medicine in Its Socio-Cultural Context*, 2 vols., ed. P. van der Eijk, H.F.J. Horstmanshoff, and P. H. Schrijvers (Amsterdam: Rodopi, 1995), 3–25.

6. For a recent account of the dating of individual treatises, see J. Jouanna, *Hippocrates*, trans. M. DeBevoise (Baltimore: Johns Hopkins University Press, 1999), 373–416. On the formation of the Corpus, see W. D. Smith, *The Hippocratic Tradition* (Ithaca, NY: Cornell University Press, 1979), 177–246.

7. On the "Hippocratic Question" (i.e., which texts were written by the historical Hippocrates) and its demise, see G.E.R. Lloyd, "The Hippocratic Question," *Classical Quarterly* 25 (1975): 171–92, reprinted in G.E.R. Lloyd, *Methods and Problems in Greek Science* (Cambridge: Cambridge University Press, 1991), 194–223. In addition to the extant Hippocratic writings, our other major source for medical views on the body and disease in the classical period is the Meno Papyrus (English translation in W.H.S. Jones, *The Medical Writings of Anonymous Londinensis* [Cambridge: Cambridge University Press, 1947]). The text is probably based on a compilation of medical doctrines by one of Aristotle's students.

8. On the methods and pitfalls of medical history and historiography in the ancient Greco-Roman world, see the overview in P. van der Eijk, "Historical Awareness, Historiography, and Doxography in Greek and Roman Medicine," in *Ancient Histories of Medicine: Essays in Medical Doxography and Historiography in Classical Antiquity*, ed. P. van der Eijk (Leiden: Brill, 1999), 1–31.

9. Vivian Nutton, "Ancient Medicine: Asclepius Transformed," in *Science and Mathematics in Ancient Greek Culture*, ed. C.J. Tuplin and T.E. Rihll (Oxford: Oxford University Press, 2002), 48–49.

10. The claim that Homer does not have a concept of the unified body was made controversially by Bruno Snell in the first chapter of *The Discovery of the Mind: The Greek Origins of European Thought*, trans. T. Rosenmeyer (Oxford: Blackwell, 1953), 1–22. While Snell's argument has been challenged weakly on philological grounds, it has gained widespread acceptance, particularly as the revised claim that Homer recognizes multiple modes of embodiment. See M. Clarke, *Flesh and Spirit in the Songs of Homer* (Oxford: Clarendon Press, 1999), 115–19, and Holmes, *Symptom and Subject*.

11. Homer's anatomy, evident in his wound descriptions, has long impressed medical historians. For a contemporary look at the wounding scenes, see K. B. Saunders, "The Wounds in *Iliad* 13–16," *Classical Quarterly* 49 (1999): 345–63. On the symbolic dimension of wounding, see C. F. Salazar, *The Treatment of War Wounds in Graeco-Roman Antiquity* (Leiden: Brill, 2000), 127–58; B. Holmes, "The *Iliad*'s Economy of Pain," *Transactions of the American Philological Association* 137 (2007): 45–84.
12. Homer *Iliad* 3.23, 7.79, 18.161, 22.342, 23.169; *Odyssey* 11.53, 12.67, 24.187. But compare Hesiod *Works and Days* 539–40, where the *sôma* is indisputably alive. Note: I have used Loeb Classical Texts when available.
13. Homer *Odyssey* 17.383–85. On the healer as a craftsman, see O. Temkin, "Greek Medicine as Science and Craft," *Isis* 44 (1953): 213–25; H.F.J. Horstmanshoff, "The Ancient Physician: Craftsman or Scientist?" *Journal of the History of Medicine and Allied Sciences* 45 (1990): 176–97. On the gradual democratization of the medical craft, which opened up apprenticeship to those outside the family, see L. Dean-Jones, "Literacy and the Charlatan in Ancient Greek Medicine," in *Written Texts and the Rise of Literate Culture in Ancient Greece*, ed. H. Yunis (Cambridge: Cambridge University Press, 2003), 97–121. Although doctors continued to be itinerant, as the Hippocratic *Epidemics* attest, in the classical period it appears that cities would pay physicians a fee to practice in the city, thereby ensuring not free medical care for the citizens, but the on-call presence of a reputable physician (L. Cohn-Haft, *The Public Physicians of Ancient Greece* [Northampton, MA: Department of History, Smith College, 1956]).
14. Homer *Odyssey* 19.457, *Iliad* 11.623–39. See also Homer *Odyssey* 4.230–31: every Egyptian has experience with all kinds of healing drugs. Warriors, too, appear to have had basic knowledge of wound-care, as at *Iliad* 11.841–48, where Patroclus treats the arrow-wound of Eurypylus: see C. Mackie, "The Earliest Jason: What's in a Name?" *Greece and Rome* 48 (2001): 1–17.
15. See Homer *Iliad* 11.515, where the *iatros* is valued because he cuts out arrows and dresses wounds. See also Pindar *Pythian* 3.40–54, where the healer is credited with the ability to administer drugs, undertake surgery, and sing charms: the celebrated Indo-European linguist Émile Benveniste believed he had discovered here the tripartite medical doctrine of the Indo-Europeans (*Révue de l'histoire des religions* 130 [1945]: 5–12).
16. Homer *Iliad* 1.62–64.
17. Celsus *On Medicine* Proem 3-4.
18. Calchas anticipates the shadowy "healer-seers" (*iatromanteis*), who we begin to hear about in the seventh and sixth centuries B.C.: see Robert Parker, *Miasma: Pollution and Purification in Early Greek Religion* (Oxford: Clarendon Press, 1983), 209–12.
19. For Babylonian diagnosis, see N. Heeßel, "Diagnosis, Divination and Disease: Towards an Understanding of the *Rationale* behind the Babylonian *Diagnostic Handbook*," in *Magic and Rationality in Ancient Near Eastern and Graeco-Roman Medicine*, ed. H.F.J. Horstmanshoff and M. Stol (Leiden: Brill, 2004), 97–116, especially 108–10. For the correlation between gods and symptoms in the classical period, see *On the Sacred Disease* 4, in Émile Littré, *Œuvres complètes d'Hippocrate*, 10 vols. (Paris: Baillière, 1839–1861), vol. 6, 466–70; translation in W.H.S. Jones, trans., *Hippocrates*, vol. 2 (Cambridge, MA: Harvard University Press, 1923).

20. Disease as demonic or nebulously personified: e.g., Homer *Odyssey* 5.394–97; Hesiod *Works and Days* 100–105; [Hippocrates] *On the Sacred Disease* 4 (Littré 6.360–62), in Jones, *Hippocrates*, vol. 2.
21. See *The Sack of Ilion*, frag. 2, in M.L. West, *Greek Epic Fragments* (Cambridge, MA: Harvard University Press, 2003), 149. On the dating of the fragment, see M. Davies, *Epicorum Graecorum fragmenta* (Göttingen: Vandenhoek & Ruprecht, 1988), 3–6, 65, 77.
22. Compare [Hippocrates] *On Breaths* 1 (Littré 6.90), in Jones, *Hippocrates*, vol. 2, where laypersons can grasp matters of the body, but only the physician is an expert in matters of understanding. The manual training for surgery is classed as a matter of the body, while the knowledge of causes is presented as more difficult and prestigious.
23. [Hippocrates] *On Places in a Human Being* 2 (Littré 6.278), in P. Potter, trans., *Hippocrates*, vol. 8 (Cambridge, MA: Harvard University Press, 1995).
24. [Hippocrates] *On the Sacred Disease* 1–4, 14 (Littré 6.352-364, 380-82), in Jones, *Hippocrates*, vol. 2. See also [Hippocrates] *Airs, Waters, Places* 22 (Littré 2.76-82), in W.H.S. Jones, trans. *Hippocrates*, vol. 1 (Cambridge, MA: Harvard University Press, 1923).
25. On the intellectual climate of the latter half of the fifth century B.C., including the role of medicine within it, see Rosalind Thomas, *Herodotus in Context: Ethnography, Science and the Art of Persuasion* (Cambridge: Cambridge University Press, 2000), 1–27. For the geographical scope of the works in the Hippocratic Corpus, see Jouanna, *Hippocrates*, 25–36.
26. On the circulation of medical treatises: Xenophon *Memorabilia* 4.2.10.
27. See Holmes, *Symptom and Subject*, chap. 2.
28. For the triad chance, necessity, and nature: Plato *Laws* 10.889b–c.
29. Although many of the physicists took an active interest in biology and indeed in human nature, we can see medicine trying to stake out the ground of a materialist anthropology in *On Ancient Medicine* 1–2 (Littré 1.570-74), in Jones, *Hippocrates*, vol. 1.
30. Alcmaeon frag. 310 in G.S. Kirk, J.E. Raven, and M. Schofield, eds., *The Presocratic Philosophers: A Critical History with a Selection of Texts*, 2nd ed. (Cambridge: Cambridge University Press, 1983). For his work on "medical things," see Diogenes Laertius *The Lives of the Philosophers* 8.83.
31. [Hippocrates] *On Ancient Medicine* 14 (Littré 1.602), in Jones, *Hippocrates*, vol. 1.
32. [Hippocrates] *On the Art* 11 (Littré 6.20), in Jones, *Hippocrates*, vol. 1.
33. Anaxagoras frag. 510 in Kirk, Raven, and Schofield, *The Presocratic Philosophers*. The dictum, handed down by Sextus Empiricus (*Against the Mathematicians*, 7.140), was also sometimes attributed to Democritus.
34. An imaginative look at the inner body can be found in [Hippocrates] *On the Art* 10 (Littré 6.16-18), in Jones, *Hippocrates*, vol. 2.
35. [Hippocrates] *On Regimen in Acute Diseases* 17 (Littré 2.260-62), in Jones, *Hippocrates*, vol. 2.
36. [Hippocrates] *On Ancient Medicine* 14 (Littré 1.600-604), in Jones, *Hippocrates*, vol. 1.
37. [Hippocrates] *On the Nature of a Human Being* 2 (Littré 6.34), in W.H.S. Jones, trans. *Hippocrates*, vol. 1 (Cambridge, MA: Harvard University Press, 1931).
38. [Hippocrates] *On the Nature of a Human Being* 7 (Littré 6.46), in Jones, *Hippocrates*, vol. 4; [Hippocrates] *On the Sacred Disease* 14 (Littré 6.382), in Jones, *Hippocrates*, vol. 2.

39. [Hippocrates] *On Fractures* 42 (Littré 3.552), *On Joints* 19 (Littré 4.132), in E. T. Withington, trans., *Hippocrates,* vol. 3 (Cambridge, MA: Harvard University Press, 1928).
40. [Hippocrates] *Epidemics* V 64 (Littré 5.242), in W. D. Smith, trans., *Hippocrates,* vol. 7 (Cambridge, MA: Harvard University Press, 1994).
41. [Hippocrates] *On Regimen* I 2 (Littré 6.472), in Jones, *Hippocrates,* vol. 4.
42. For the "race" against the disease, see [Hippocrates] *On the Art* 11 (Littré 6.20-22), in Jones, *Hippocrates,* vol. 1.
43. V. Langholf, *Medical Theories in Hippocrates: Early Texts and the "Epidemics"* (Berlin: De Gruyter, 1990).
44. [Hippocrates] *Prognostic* 12 (Littré 2.42), in Jones, *Hippocrates,* vol. 2.
45. For evidence of the idea of inner or vital heat in the medical writers, see B. Gundert, "Soma and Psyche in Hippocratic Medicine," in *Psyche and Soma: Physicians and Metaphysicians on the Mind-Body Problem from Antiquity to Enlightenment,* ed. J. P. Wright and P. Potter (Oxford: Clarendon Press, 2000), 16–17. The equation of life and warmth is already formalized in pre-Socratics such as Parmenides and Empedocles. In the medical writers, as well as in Plato (*Timaeus* 78b–79a), heat is closely associated with the body's ability to "cook" or, more technically, "to concoct" humors: see Langholf, *Medical Theories,* 88–89; M. Schiefsky, *Hippocrates, On Ancient Medicine* (Leiden: Brill, 2005), 280–83.
46. See L. Edelstein, "Hippocratic Prognosis," in *Ancient Medicine: Selected Papers,* ed. O. Temkin and L. Temkin (Baltimore: Johns Hopkins University Press, 1967), 65–85; G. E. R. Lloyd, *Magic, Reason, and Experience: Studies in the Origin and Development of Greek Science* (Cambridge: Cambridge University Press, 1979), 86–125; P. van der Eijk, "Towards a Rhetoric of Ancient Scientific Discourse: Some Formal Characteristics of Greek Medical and Philosophical Texts (Hippocratic Corpus, Aristotle)," in *Grammar as Interpretation: Greek Literature in Its Linguistic Contexts,* ed. E. Bakker (Leiden: Brill, 1997), 77–129; J. Laskaris, *The Art Is Long: "On the Sacred Disease" and the Scientific Tradition* (Leiden: Brill, 2002).
47. [Hippocrates] *On Affections* 1 (Littré 6.208), in P. Potter, trans., *Hippocrates,* vol. 5 (Cambridge, MA: Harvard University Press, 1988). See also Plato *Laws* 4.720d–e.
48. On dietetics and the care of the self in this period, see M. Foucault, *History of Sexuality,* vol. 2, *The Use of Pleasure,* trans. R. Hurley (New York: Pantheon, 1985).
49. Greek athletics, of course, has strong roots in the archaic period, where it develops in a religious context. See T. Scanlon, *Eros and Greek Athletics* (New York: Oxford University Press, 2002). On the intersection of the sculpted body and classical Greek notions of health and the body, see S. Kuriyama, *The Expressiveness of the Body and the Divergence of Greek and Chinese Medicine* (New York: Zone, 1999), 134–43.
50. [Hippocrates] *On Regimen* I 1 (Littré 6.466), in Jones, *Hippocrates,* vol. 4.
51. [Hippocrates] *On Regimen* III 69 (Littré 6.604), in Jones, *Hippocrates,* vol. 4.
52. [Hippocrates] *On Regimen* I 35 (Littré 6.512-22), in Jones, *Hippocrates,* vol. 4.
53. Diocles of Carystus frag. 182 in P. van der Eijk, *Diocles of Carystus: A Collection of the Fragments with Translation and Commentary,* 2 vols. (Leiden: Brill, 2000–2001).
54. Plato *Republic* 3.405c–d.
55. Ibid., 3.407c. Translated by C. D. C. Reeve, *Plato, Republic* (Indianapolis: Hackett, 2004).

56. B. Holmes, "Body, Soul, and Medical Analogy in Plato," in *When Worlds Elide: Classics, Politics, Culture*, ed. J. P. Euben and K. Bassi (Lanham, MD: Rowman and Littlefield, forthcoming).
57. See, e.g., Plato *Alcibiades* I 130a–c; *Protagoras* 313a; *Phaedo* 114e.
58. On the close relationship of mind and body in the *Timaeus* and other later dialogues, see T. J. Tracy, *Physiological Theory and the Doctrine of the Mean in Plato and Aristotle* (The Hague: Mouton, 1969), 77–156; R. Sorabji, "The Mind-Body Relation in the Wake of Plato's *Timaeus*," in *Plato's "Timaeus" as Cultural Icon*, ed. G. J. Reydams-Schils (Notre Dame, IN: University of Notre Dame Press, 2003), 152–62; G. R. Carone, "Mind and Body in Late Plato," *Archiv für Geschichte der Philosophie* 87 (2005): 227–69.
59. Aristotle was the son of a physician and saw medicine as an integral part of natural philosophy: *On Sense and the Sensible* 436a17–21; *On Respiration* 480b28–30. For Aristotle's work on psycho-physiological phenomena, see P. van der Eijk, *Medicine and Philosophy in Classical Antiquity: Doctors and Philosophers on Nature, Soul, Health and Disease* (Cambridge: Cambridge University Press, 2005), chap. 5–9. For the medical analogy in his ethics: W. Jaeger, "Aristotle's Use of Medicine as Model of Method in His Ethics," *Journal of Hellenic Studies* 77 (1957): 54–61; Tracy, *Physiological Theory*, 157–333.
60. M. Nussbaum, *The Therapy of Desire: Theory and Practice in Hellenistic Ethics* (Princeton, NJ: Princeton University Press, 1994).
61. In his zoological treatises, Aristotle occasionally refers to a book called *Anatomai*, which featured diagrams of dissected animals (e.g., *Generation of Animals* 719a9–10). On the need to construct the human body analogically: Aristotle *Parts of Animals* 644b1-16. For fragmentary evidence of the anatomical interests of Diocles of Carystus, see frag. 17–24c (van der Eijk).
62. Prior to Aristotle, however, anatomy was used primarily to prove specific claims rather than to provide an organized model of the body (J. M. Annoni and V. Barras, "La découpe du corps humain et ses justifications dans l'Antiquité," *Canadian Bulletin for Medical History* 10 [1993]: 202). For empirical investigation more generally in the Hippocratic writings, see Lloyd, *Magic, Reason, and Experience*, 146–69.
63. Aristotle, however, also recognized the limits of teleological explanation: see H. von Staden, "Teleology and Mechanism: Aristotelian Biology and Early Hellenistic Medicine," in W. Kullmann and S. Föllinger, eds., *Aristotelische Biologie* (Stuttgart: Steiner Verlag, 1997), 183–85, with further bibliography in n. 4. For evidence of Diocles' teleology, see frags. 23d and 31 (van der Eijk).
64. Heat as the triumph over formlessness: Aristotle *Generation of Animals* 732b26–733b16. The presence of heat for Aristotle is a key factor in differentiating male and female bodies, the latter being colder and thus less articulated: see L. Dean-Jones, *Women's Bodies in Classical Greek Science* (Oxford: Clarendon Press, 1994), 60–61. Diocles and vital or innate heat: frags. 31, 109 (van der Eijk). For the delicate relationship between vital heat and the soul in Plato and Aristotle, see F. Solmsen, "The Vital Heat, the Inborn Pneuma and the Aether," *Journal of Hellenic Studies* 77 (1957): 119–23. The Stoics identify the hot with the soul.
65. Transmission of perception: Aristotle *Generation of Animals* 743b37–744a5; 781a20–b5. Voluntary motion: Aristotle *On the Motion of Animals* 703a4-27. See further discussion of these passages in F. Solmsen, "Greek Philosophy and the Discovery of the Nerves," *Museum Helveticum* 18 (1961): 150–97, esp. 174–78. Aristotle's decision to locate the ruling part in the heart supported a long tradition

of cardiocentrism that persisted despite the anatomists' arguments, based on their studies of the nerves, for seeing the brain as hegemonic.
66. For *pneuma* in Diocles, see frags. 78, 80, 101, 107 (van der Eijk).
67. F. Steckerl, *The Fragments of Praxagoras of Cos and His School* (Leiden: Brill, 1958), 17.
68. Praxagoras of Cos frag. 11 (Steckerl).
69. In fact, the Stoics will use Praxagoras to reject further anatomical work on nerves and *pneuma*: see Solmsen, "Greek Philosophy," 195.
70. Praxagoras frags. 26–28 (Steckerl). Kuriyama (*Expressiveness of the Body,* 23–32) argues for a relationship between anatomical inquiry and the birth of the pulse.
71. Celsus *On Medicine* Proem 23–24. See L. Edelstein, "The History of Anatomy in Antiquity," in *Ancient Medicine,* ed. Temkin and Temkin, 274–81; H. von Staden, "The Discovery of the Body: Human Dissection and Its Cultural Contexts in Ancient Greece," *Yale Journal of Biology and Medicine* 65 (1992): 223–41.
72. Kuriyama, *Expressiveness of the Body,* 111–51.
73. R. Flemming, "Empires of Knowledge: Medicine and Health in the Hellenistic World," in *A Companion to the Hellenistic World,* ed. A. Erskine (Malden, MA: Blackwell, 2003), 451–53; P. Lang, "Medical and Ethnic Identities in Hellenistic Egypt," in *Re-inventions: Essays on Hellenistic and Early Roman Science,* special issue of *Apeiron* 37 (2004): 107–31.
74. Herophilus T 90-92, 96–99, 124 in H. von Staden, *Herophilus: The Art of Medicine in Early Alexandria* (Cambridge: Cambridge University Press, 1989).
75. On the pulse and its relationship to life: Herophilus T 144, 145a, 155, 164 (von Staden).
76. On the brain: Herophilus T 137–39 (von Staden). For voluntary nerves: T 81, 141 (von Staden). It is clear that Herophilus held that the nerves contained a kind of *pneuma;* it is less clear how he understood the relationship between this *pneuma* and the *pneuma* of the soul—there is no direct testimony for *pneuma* in the "voluntary nerves"—or the *pneuma* in the arteries: for discussion, see Solmsen, "Greek Philosophy," 185–88; H. von Staden, "Body, Soul, and Nerves: Epicurus, Herophilus, Erasistratus, the Stoics, and Galen," in *Psyche and Soma: Physicians and Metaphysicians on the Mind-Body Problem from Antiquity to Enlightenment,* ed. J.P. Wright and P. Potter (Oxford: Clarendon Press, 2000), 87–91.
77. For the evidence on vivisection, see esp. Celsus, *On Medicine* Proem 23–26 (T 63a von Staden) and Tertullian *On the Soul* 10.4 (T 66 von Staden); further evidence at T 63b–64b (von Staden). The evidence is judiciously discussed by von Staden (*Herophilus,* 141–53), who concludes that it is highly likely that vivisection was practiced on humans in this context.
78. Celsus *On Medicine* Proem 23–24.
79. Erasistratus of Cos frags. 158, 161–62, 240 in I. Garofalo, *Erasistrati fragmenta* (Pisa: Giardini, 1988).
80. M. Vegetti, "L'épistémologie d'Érasistrate et la technologie hellénistique," in *Ancient Medicine,* ed. van der Eijk, Horstmanshoff, and Schrijvers, 461–71; M. Vegetti, "Between Knowledge and Practice: Hellenistic Medicine," in *Western Medical Thought from Antiquity to the Middle Ages,* ed. M.D. Grmek, trans. A. Shugaar (Cambridge, MA: Harvard University Press, 1999), 72–103.
81. The principle was called "the following into that which is being emptied" (*hê pros to kenoumenon akolouthia*) and is widely thought to have been influenced by the views of Erasistratus's older contemporary Strato of Lampsacus on void.

82. H. von Staden, "Teleology and Mechanism," 199–203. See also H. von Staden, "Body and Machine: Interactions between Medicine, Mechanics, and Philosophy in Early Alexandria," in *Alexandria and Alexandrianism* (Malibu, CA: J. Paul Getty Museum, 1996), 85–106.
83. See, e.g., Empedocles frag. 471 in Kirk, Raven, and Schofield, *The Presocratic Philosophers*; [Hippocrates] *On Ancient Medicine* 22 (Littré 1.626-630; Jones, *Hippocrates,* vol. 1), where the author offers the principle that one can learn about concealed body parts and processes "through visible things outside the body," such as cupping glasses.
84. Vegetti ("Between Knowledge and Practice," 93) points out the rejection of Aristotle's principle of innate life-principles (*pneuma,* heat) among Hellenistic physicians. Galen's opposition of his own teleological vitalism to mechanism can be seen throughout the treatise *On the Natural Faculties* (translation in A.J. Brock, *Galen, On the Natural Faculties* [Cambridge, MA: Harvard University Press, 1916]).
85. Von Staden, "Teleology and Mechanism."
86. Erasistratus frag. 86 (Garofalo). According to Erasistratus, each strand carries its own substance—psychic *pneuma* in the nerves, vital *pneuma* in the arteries, and nutrient-rich blood in the veins—and has its own point of origin (the brain, the heart, and the liver, respectively).
87. Vegetti, "L'épistémologie d'Érasistrate."
88. R.J. Hankinson, "The Growth of Medical Empiricism," in *Knowledge and the Scholarly Medical Traditions,* ed. D. Bates (Cambridge: Cambridge University Press, 1995), 63–64.
89. On Asclepiades of Bithynia: J. Vallance, *The Lost Theory of Asclepiades of Bithynia* (Oxford: Oxford University Press, 1990); J. Vallance, "The Medical System of Asclepiades of Bithynia," *Aufstieg und Niedergang der Römischen Welt,* Band 2.37.1, Berlin, 1993, 693–727. The idea of things visible only to reason first occurs in fifth-century B.C. medical treatises: [Hippocrates] *On the Art* 11 (Littré 6.20) and *On Breaths* 3 (Littré 6.94), in Jones, *Hippocrates,* vol. 2.
90. See Galen's treatise *An Outline of Empiricism,* in M. Frede and R. Walzer, *Galen: Three Treatises on the Nature of Science* (Indianapolis: Hackett, 1985), 21–46. See also M. Frede, "The Ancient Empiricists," in *Essays in Ancient Philosophy* (Minneapolis: University of Minnesota Press, 1987), 243–60; Hankinson, "The Growth of Medical Empiricism." On the medical sects (*haireseis iatrikai*), see H. von Staden, "Hairesis and Heresy: The Case of the *Haireseis Iatrikai,*" in *Jewish and Christian Self-Definition,* vol. 3, *Self-Definition in the Graeco-Roman World,* ed. B.F. Meyer and E.P. Sanders (London: S.C.M., 1982), 76–100.
91. L. Edelstein, "The Methodists," in Temkin and Temkin, *Ancient Medicine,* 173–91; M. Frede, "The Method of the So-Called Methodical School of Medicine," in *Science and Speculation: Studies in Hellenistic Theory and Practice,* ed. J. Barnes et al. (Cambridge: Cambridge University Press, 1982), 1–23, reprinted in *Essays in Ancient Philosophy,* 261–78; Vivian Nutton, *Ancient Medicine* (London: Routledge, 2004), 187–201. There is now an edition of the fragments of the Methodists with English translations: M. Tecusan, *The Fragments of the Methodists,* 2 vols. (Leiden: Brill, 2004).
92. Another sect, the "Pneumatists," which formed in the first century A.D., should be mentioned here. Its adherents develop the idea that *pneuma* is the major principle of life, a position that allies them with the Stoic school of philosophy. Yet they also borrow from the other sects, fostering the growth of the eclecticism that is

systematized by Galen. For a brief overview of the Pneumatists, see Nutton, *Ancient Medicine*, 202–8.

93. Later perceptions of the sects are shaped by Galen's often biased account of them. At the same time, he plays a key role in establishing their importance in the Greek medical tradition as it is transmitted in the Byzantine, Arabic, and Latin worlds: his work *On the Sects for Beginners* (translated in Frede and Walzer, *Three Treatises*) was the first book read by medical students in Alexandria through the eighth century A.D.
94. Celsus *On Medicine* Proem 40–44.
95. Apollonius of Rhodes *Argonautica* 3.762–63.
96. On why systematic human dissection was so short-lived, see von Staden, *Herophilus*, 148–51. For evidence of public reactions to anatomical inquiry and vivisection, see G. B. Ferngren, "Roman Lay Attitudes towards Medical Experimentation," *Bulletin of the History of Medicine* 59 (1985): 495–505. For lay attitudes more generally: Vivian Nutton, "Murders and Miracles: Lay Attitudes to Medicine in Classical Antiquity," in *Patients and Practitioners: Lay Perceptions of Medicine in Pre-Industrial Society*, ed. R. Porter (Cambridge: Cambridge University Press, 1985), 23–53.
97. For an overview of the arrival of Greek medicine in Rome, see Nutton, *Ancient Medicine*, 157–70; Vivian Nutton, "Roman Medicine: Tradition, Confrontation, Assimilation," *Aufstieg und Niedergang der Römischen Welt*, Band 2.37.1, Berlin: De Gruyter, 1993, 49–78.
98. Pliny *Natural History* 29.6.12–15. The Roman moralist Cato the Elder claimed, according to Pliny, that Greek doctors had sworn an oath to kill foreigners by means of medicine.
99. For dietetic medicine in the Hellenistic period, see J. Scarborough, "Diphilus of Siphnos and Hellenistic Medical Dietetics," *Journal of the History of Medicine and Allied Sciences* 25 (1970): 194–201; W. D. Smith, "Erasistratus's Dietetic Medicine," *Bulletin of the History of Medicine* 56 (1982): 398–409.
100. Indeed, it may be in part because the ethics of body-care were so central to the cultural identity of Greek medicine that Greek physicians resisted an idea of contagion, which downplays individual responsibility: see Vivian Nutton, "The Seeds of Disease: An Explanation of Contagion and Infection from the Greeks to the Renaissance," *Medical History* 27 (1983): 1–34.
101. O. Temkin, *Galenism: Rise and Decline of a Medical Philosophy* (Ithaca, NY: Cornell University Press, 1973), 39.
102. See, e.g., Seneca *Letters to Lucilius* 54, 78. See G. Bowersock, *Greek Sophists in the Roman Empire* (Oxford: Clarendon Press, 1969), 69–73. For Seneca, see C. Edwards, "The Suffering Body: Philosophy and Pain in Seneca's *Letters*," in *Constructions of the Classical Body*, ed. J. I. Porter (Ann Arbor: University of Michigan Press, 1999), 252–68.
103. Nussbaum, *The Therapy of Desire*. On the care of the self in this period, see M. Foucault, *History of Sexuality*, vol. 3, *The Care of the Self*, trans. R. Hurley (New York: Pantheon, 1986); P. Hadot, *Philosophy as a Way of Life: Spiritual Exercises from Socrates to Foucault*, trans. M. Chase (Cambridge, MA: Harvard University Press, 1995).
104. For a reading of the importance of the body in both pagan and Christian texts of the first centuries A.D., see J. Perkins, *The Suffering Self: Pain and Narrative Representation in the Early Christian Era* (London: Routledge, 1995).

105. See especially the account of a visit to Asclepius in Aristophanes's *Wealth* (633–747) and the late fourth-century B.C. testimonia from Epidaurus, T 423, in E. J. Edelstein and L. Edelstein, *Asclepius: Collection and Interpretation of the Testimonies*, 2 vols. (Baltimore: Johns Hopkins University Press, 1945). A fuller edition of the Epidaurian tablets may be found in L. R. LiDonnici, *The Epidaurian Miracle Inscriptions: Text, Translation, and Commentary* (Atlanta: Scholars Press, 1995).
106. *Inscriptiones Graecae* IV^2 1 no. 122 (T 423 Edelstein and Edelstein).
107. See H.F.J. Horstmanshoff, "Did the God Learn Medicine?: Asclepius and Temple Medicine in Aelius Aristides' *Sacred Tales*," in *Magic and Rationality in Ancient Near Eastern and Graeco-Roman Medicine*, ed. H.F.J. Horstmanshoff and M. Stol (Leiden: Brill, 2004), 325–42; M. E. Gorrini, "The Hippocratic Impact on Healing Cults: The Archaeological Evidence in Attica," in *Hippocrates in Context: Papers Read at the Eleventh International Hippocrates Colloquium, University of Newcastle upon Tyne, 27–31 August 2002*, ed. P. van der Eijk (Leiden: Brill, 2005), 135–56.
108. For Asclepius's integration into people's daily lives, see H. C. Kee, "Self-Definition in the Asclepius Cult," in *Jewish and Christian Self-Definition*, ed. B. F. Meyer and E. P. Sanders, vol. 3, *Self-Definition in the Graeco-Roman World* (London: S.C.M., 1982), 118–36.
109. C. A. Behr, *Aelius Aristides and the Sacred Tales* (Amsterdam: A. M. Hakkert, 1968).
110. B. Holmes, "Aelius Aristides' Illegible Body," in *Aelius Aristides between Greece, Rome, and the Gods*, ed. W. V. Harris and B. Holmes (Leiden: Brill, 2008), 77–109. On the concept of relative health: F. Kudlien, "The Old Greek Concept of 'Relative Health,'" *Journal of the History of Behavioral Sciences* 9 (1973): 53–59.
111. For Galen's Hippocratism: Smith, *The Hippocratic Tradition*, 62–176.
112. See Temkin, *Galenism*, 16–22.
113. See, e.g., Galen *Art of Medicine* 4, in C. G. Kühn, *Claudii Galeni opera omnia*, 20 vols., Leipzig: C. Cnobloch, 1821–1833, vol. 1, 314–15.
114. For Galen's views on women, see Flemming, *Medicine*, chaps. 5–6.
115. H. von Staden, "Lexicography in the Third Century B.C.: Bacchius of Tanagra, Erotian, and Hippocrates," in *Tratados hipocráticos: estudios acerca de su contenido, forma e influencia; actas del VIIe Colloque international hippocratique, Madrid, 24–29 de septiembre de 1990*, ed. J. A. López Férez (Madrid: Universidad Nacional de Educación a Distancia, 1992), 549–69.
116. For his public demonstrations: H. von Staden, "Anatomy as Rhetoric: Galen on Dissection and Persuasion," *Journal of the History of Medicine and Allied Sciences* 50 (1995): 47–66; A. Debru, "Les demonstrations médicales à Rome au temps de Galien," in van der Eijk, Horstmanshoff, and Schrijvers, *Ancient Medicine*, 69–81. On Galen's anatomy, see Nutton, *Ancient Medicine*, 230–33. For Galen's defense of the brain as the origin of the nerves (versus the heart, as in Aristotle), see *On the Doctrines of Hippocrates and Plato* 1.2, in P. de Lacy, *Galen, On the Doctrines of Hippocrates and Plato*, 3 vols., Corpus Medicorum Graecorum 5.4.1.2 (Berlin: Akademie-Verlag, 1978–1984).
117. Like Plato, he saw three major centers of life: the liver (appetitive), the heart (emotional), and the brain (mental and volitional). For Galen's physiology, see R. E. Siegel, *Galen's System of Physiology and Medicine* (Basel: Karger, 1968).

118. Anatomy "had given surgeons the ability to interpret or visualize the conditions they were tackling" (M. McVaugh, "Therapeutic Strategies: Surgery," in *Western Medical Thought from Antiquity to the Middle Ages,* ed. M.D. Grmek, trans. A. Shugaar [Cambridge, MA: Harvard University Press, 1999], 275). It must be noted, nevertheless, that Galen, like the early medical writers, preferred treatment through diet or drugs to surgical intervention, since he shared with them the idea that health belongs to the whole body.
119. Vivian Nutton, "Galen at the Bedside: The Methods of a Medical Detective," in *Medicine and the Five Senses,* ed. W.F. Bynum and R. Porter (Cambridge: Cambridge University Press, 1993), 7–16; T. Barton, *Power and Knowledge: Astrology, Physiognomics, and Medicine under the Roman Empire* (Ann Arbor: University of Michigan Press, 1994), 133–68; Perkins, *The Suffering Self,* 142–72.
120. On Galen's concept of experience, see P. van der Eijk, "Galen's Use of the Concept of 'Qualified Experience' in His Dietetic and Pharmacological Works," in *Galen on Pharmacology: Philosophy, History and Medicine,* ed. A. Debru (Leiden: Brill, 1997), 35–57; reprinted in van der Eijk, *Medicine and Philosophy,* 279–98.
121. See n. 63.
122. See, e.g., *On the Utility of the Parts* (Kühn xx). For discussion, see R.J. Hankinson, "Galen and the Best of All Possible Worlds," *Classical Quarterly* 39 (1989): 206–27.
123. Galen *On the Utility of the Parts* 11.14 (Kühn 2.158–60). See also R. Walzer, *Galen on Jews and Christians* (Oxford: Clarendon Press, 1949), 23–37; S. Gero, "Galen on the Christians: A Reappraisal of the Arabic Evidence," *Orientalia Christiana Periodica* 56 (1990): 371–411.
124. Two of Galen's treatises on the soul, *On the Passions of the Soul* and *On the Errors of the Soul,* are translated in P.W. Harkins, *Galen on the Passions and Errors of the Soul* (Columbus: Ohio State University Press, 1963). See also L. García Ballester, "Soul and Body, Disease of the Soul and Disease of the Body in Galen's Medical Thought" in *Le opere psicologiche di Galeno: atti del terzo colloquio Galenico internazionale, Pavia, 10–12 settembre 1986,* ed. P. Manuli and M. Vegetti (Naples: Bibliopolis, 1988), 117–52; R.J. Hankinson, "Galen's Anatomy of the Soul," *Phronesis* 36 (1991): 197–233.
125. He is present at the fictional symposium in Athenaeus's *Deipnosophists* (1.1e, 1.26c, 3.115c), written around 200 A.D., for example, and he was held in high esteem by the influential Aristotelian commentator Alexander of Aphrodisias, active at the end of the second century and the beginning of the third century A.D.
126. O. Temkin, "Studies on Late Alexandrian Medicine," *Bulletin of the History of Medicine* 3 (1935): 405–30; A. Cunningham, "The Theory/Practice Division of Medicine: Two Late-Alexandrian Legacies," in *History of Traditional Medicine,* ed. T. Ogawa (Osaka: Division of Medical History, the Taniguchi Foundation, Tokyo, 1986), 303–24, esp. 313–21.
127. Although Alexandria was still known for the expertise of its surgeons: O. Temkin, "Byzantine Medicine: Tradition and Empiricism," *Dumbarton Oaks Papers* 16 (1962): 101; Vivian Nutton, "From Galen to Alexander: Aspects of Medicine and Medical Practice in Late Antiquity," *Dumbarton Oaks Papers* 38 (1984): 5.
128. Eunapius, *Lives of the Sophists,* 497–98. Magnus's lectures, it should be noted, were not only on Galen, nor were they mere paraphrases of Galenic texts. His partially preserved work on the diagnostic analysis of urines, for example, greatly

expands on what was a small part of Hippocratic and Galenic semiology and perhaps founded the tradition of uroscopy in the Middle Ages.

129. See, e.g., the commentary by Stephanus of Athens, teaching in sixth-century A.D. Alexandria, on Hippocrates's *Aphorisms* in the edition of L. G. Westerink (Berlin: Akademie Verlag, 1985–1995). So entrenched do Galen's views on Hippocrates become that the Hippocratic writings extant in Arabic translation are simply extracts from Galen's commentaries (G. Strohmaier, "Reception and Tradition: Medicine in the Byzantine and Arab World," in *Western Medical Thought from Antiquity to the Middle Ages*, ed. M.D. Grmek, trans. A. Shugaar [Cambridge, MA: Harvard University Press, 1998], 144).

130. Nutton, *Ancient Medicine*, 296.

131. A. Ghersetti, "The Semiotic Paradigm: Physiognomy and Medicine in Islamic Culture," in *Seeing the Face, Seeing the Soul: Polemon's Physiognomy from Classical Antiquity to Medieval Islam*, ed. S. Swain (Oxford: Clarendon Press, 2007), 281–308.

132. See Temkin, *Galenism*, 73–80; Strohmaier, "Reception and Tradition," 156–62.

133. See, for example, Nemesius of Emesa *On the Nature of Man* 2.23-25 (in *Nemesius, On the Nature of Man*, trans. R. W. Sharples and P. van der Eijk [Liverpool: Liverpool University Press, 2008]); see also Temkin, *Galenism*, 81–92. Despite his unease with Galen's views on the soul, Nemesius drew extensively from Galen in creating a Christian anthropology.

134. Vivian Nutton, "God, Galen, and the Depaganisation of Ancient Medicine," in *Religion and Medicine in the Middle Ages*, ed. P. Biller and J. Ziegler (Woodbridge, UK: York Medieval Press, 2001), 15–32; Nutton, *Ancient Medicine*, 304–5.

135. Nutton, "From Galen to Alexander," 9.

136. See Philostorgius *Ecclesiastical History* 8.10, reporting that the late-fourth-century A.D. physician Posidonius was unusual insofar as he blamed mental illness on humoral rather than demonic causes.

137. For shifting ideas of the body in the early Christian period, see esp. P. Brown, *The Body and Society: Men, Women, and Sexual Renunciation in Early Christianity* (New York: Columbia University Press, 1988).

138. See, however, Nutton, *Ancient Medicine*, 248–71, on the panoply of people working as physicians in the Roman Empire.

139. Note, however, that the idea of demonic possession as the actual entry of demons into the body does not appear to exist in the archaic and classical Greek periods: W. D. Smith, "So-Called Possession in Pre-Christian Greece," *Transactions of the American Philological Association* 96 (1965): 403–26.

140. Aulus Gellius *Attic Nights* 18.10: it is shameful (*reprehensum*) to be ignorant of such matters, not only for a physician but also for a cultivated and educated man.

141. Vivian Nutton, "Medicine in Late Antiquity and the Early Middle Ages," in *The Western Medical Tradition: 800BC to AD1800*, ed. L.I. Conrad et al. (Cambridge: Cambridge University Press, 1995), 83–87.

142. J. Scarborough, "Theoretical Assumptions in Hippocratic Pharmacology," in *Formes de pensée dans la collection hippocratique: actes du IVe Colloque international hippocratique, Lausanne, 21–26 septembre 1981*, ed. F. Lasserre and P. Mudry (Geneva: Droz, 1983), 307–25. The basic theoretical principle of pharmacology in the early medical treatises is that opposites cure opposites, although there often appear to be sympathetic assumptions at work in their authors' therapeutic choices (H. von Staden, "Women and Dirt," *Helios* 19 [1992]: 7–30).

NOTES 299

Sympathy becomes an accepted mode of explanation with the rise of learned magic in the Hellenistic period. For Dioscorides, see J. Riddle, *Dioscorides on Pharmacy and Medicine* (Austin: University of Texas Press, 1985), 169–76 on Galen's pharmacology.

Chapter 5

1. *Characters* 19.
2. Michel Foucault, *History of Sexuality,* vol. 3, *The Care of the Self*, trans. R. Hurley (New York: Vintage, 1988).
3. Foucault, *History of Sexuality,* vol. 3, 56.
4. Foucault, *History of Sexuality,* vol. 3, 101.
5. Celsus *De Medicina* 1.2.45, 49.
6. "On the Genealogy of Ethics: An Overview of Work in Progress," in *The Foucault Reader*, ed. Paul Rabinow (New York: Pantheon, 1984), 340–72; passage cited, 357.
7. Aristotle, *Physiognomics,* in *Minor Works,* trans. W.S. Hett (Cambridge, MA: Harvard University Press and W. Heinemann, 1936 [Loeb Classical Library]), 83–137.
8. Aristotle, *Physiognomics* 807b.
9. Aristotle, *Physiognomics* 806b.
10. Aristotle, *Physiognomics* 808b.
11. Aristotle, *Physiognomics* 811b.
12. Aristotle, *Physiognomics* 812a.
13. Aristotle, *Physiognomics* 813b.
14. Aristotle, *Physiognomics* 813b.
15. Petronius, *Satyrica*, ed. and trans. R.B. Branham and D. Kinney (London: J.M. Dent, 1996), 62.
16. Ovid *Metamorphoses* 8.618–724.
17. Theophrastus, *Characters*; Herodas, Mimes, Sophron *and Other Mime Fragments,* trans. J. Rusten and I.C. Cunningham (Cambridge, MA: Harvard University Press [Loeb Classical Library], 2002), 87.12.2–4.
18. Aristotle, *Physiognomics* 809b.
19. Aristotle, *Physiognomics* 810a.
20. Aristotle, *Physiognomics* 813a.
21. Aristotle, *Physiognomics* 814a.
22. Giulia Sissa, *Greek Virginity*, trans. A. Goldhammer (Cambridge, MA: Harvard University Press, 1990).
23. Aristophanes, *Clouds, Wasps, Peace,* trans. Jeffrey Henderson (Cambridge, MA: Harvard University Press, 1998 [Loeb Classical Library]), lines 428–29.
24. Aristophanes, *Clouds, Wasps, Peace,* 450.
25. Aristophanes *Wasps,* lines 1292–96.
26. Aristophanes *Wasps,* lines 1342–44.
27. Aristophanes *Wasps,* lines 1372–77.
28. Aristophanes *Wasps,* lines 223–27.
29. See Jeffrey Henderson, *The Maculate Muse: Obscene Language in Attic Comedy* (New Haven, CT: Yale University Press, 1975); on allusions to the phallus, see 108–30.

30. Eva Keuls, *The Reign of the Phallus: Sexual Politics in Ancient Athens* (New York: Harper and Row, 1985).
31. Henderson, *Maculate Muse*, 130–48.
32. Aristophanes *Wasps*, lines 1075–90.
33. Aristophanes *Wasps*, lines 1107–11.
34. Amy Richlin, *The Garden of Priapus: Sexuality and Aggression in Roman Humor* (New Haven, CT: Yale University Press, 1983).
35. Richlin, *The Garden of Priapus*, 82.
36. Diehl 615.
37. Richlin, *The Garden of Priapus*, 83.
38. Richlin, *The Garden of Priapus*, 110.
39. Richlin, *The Garden of Priapus*, 112.
40. Richlin, *The Garden of Priapus*, 146.
41. Richlin, *The Garden of Priapus*, 119.
42. Aristotle, *Physiognomics* 812a.
43. Aristotle, *Physiognomics* 812a.
44. Aristotle, *Physiognomics* 812b.
45. Herodotus *The Histories* 2.34, trans. Robin Waterfield (Oxford: Oxford University Press, 1998).
46. Herodotus *The Histories* 2.36.
47. *Hippocratic Writings*, ed. G.E.R. Lloyd (New York: Penguin, 1978), 15.
48. *Hippocratic Writings*, 22.
49. Plato *Phaedo* 81c–d, *Collected Dialogues*, ed. E. Hamilton and H. Cairns (Princeton, NJ: Princeton University Press, 1961), 64.
50. Plautus *Mostellaria* 1.32.4.
51. Daniel Ogden, *Magic, Witchcraft, and Ghosts in the Greek and Roman Worlds: A Sourcebook* (Oxford: Oxford University Press, 2002), 162.
52. Suetonius *Nero* 34.
53. Pindar *Pythian* 3.47–53, in *Odes*, trans. J. Sandys (Cambridge, MA: Harvard University Press, 1937), translation modified.
54. Porphyry *Life of Plotinus* 10, in Ogden, *Magic*, 218.
55. *Moralia* Table Talk 7, in Ogden, *Magic*, 223.
56. On this and other finds, see Christopher Faraone, "Binding and Burying the Forces of Evil: The Defense Use of 'Voodoo' Dolls in Ancient Greece," *Classical Antiquity* 10 (1991): 165–205.
57. Ogden, *Magic*, 246.
58. *Brutus* 217, in Ogden, *Magic*, 212.
59. Aristophanes *Wasps* 450.
60. Cicero, *Selected Letters*, trans. D.R. Shackleton-Bailey (Harmondsworth: Penguin, 1982), 109, no. 39. (Leucas, November 7, 50 [B.C.E.])

Chapter 7

1. Herodes *Mimiambi* 5.77–79 (in I.C. Cunningham, ed., *Herodas, Mimiambi* [Oxford: Clarendon Press, 1971]).
2. Demosthenes *Against Androtion* 55. Translated by J.H. Vince, in *Demosthenes*, vol. 5 (Cambridge, MA: Harvard University Press, 1935). Note: I have used Loeb Classical Texts when available.

3. Cicero *On Behalf of Gaius Rabirius on a Charge of Treason* 12.
4. Livy 1.58.5–12; Seneca *On Anger* 3.18.3–4; Suetonius *Caligula* 27.3; Procopius *Secret History* 3.8–13, 4.7–12.
5. Galatians 3:27–28.
6. Women: Aristotle *Generation of Animals* 737a27–28; 775a15–16. Children: Aristotle *Parts of Animals* 686b12. Elderly: Aristotle *On the Longevity and Brevity of Life* 466a17–20.
7. Barbarians: Aristotle *Politics* 1252b6–9; 1285a19–21. Slaves: *Politics* 1254b20–28.
8. [Aristotle] *Problemata* 14, 909a27–32. Cf. Ptolemy *Tetrabiblos* 2.2.
9. Galen *On a Good Habitus*, in C. G. Kühn, ed., *Claudii Galeni opera omnia*, 20 vols. (Leipzig: C. Cnobloch, 1821–1833), vol. 4, 751-752.
10. Aristotle *Politics* 1252a34–b5.
11. Pliny the Younger *Letters* 8.18.9-10, cited in Robert Garland, *The Eye of the Beholder: Deformity and Disability in the Graeco-Roman World* (London: Duckworth, 1995), 20.
12. Aristotle *Politics* 1335b19–21; cf. Plato *Republic* 5.459d-e; Plutarch *Life of Lycurgus* 16.1.
13. Cato *On Agriculture* 2.7; Suetonius *Claudius* 25.
14. Aristotle *Generation of Animals* 773a16–20; [Hippocrates] *Diseases of Young Virgins*, in Émile Littré, ed., *Œuvres complètes d'Hippocrate*, 10 vols. (Paris: Baillière, 1839–1861), vol. 8, 466–470. Soranus, however, argued that menstruation was not necessarily healthy: see *Gynecology* 1.29 (I. Ilberg, ed., *Soranus*, Corpus Medicorum Graecorum 4 [Leipzig: Teubner, 1927]).
15. T. G. Laqueur, *Making Sex: Body and Gender from the Greeks to Freud* (Cambridge, MA: Harvard University Press, 1990), 28–62.
16. Aristotle *Generation of Animals* 726a30–727b33.
17. Paola Manuli, "Fisiologia e patologia del femminile negli scritti ippocratici dell'antica ginecologia greca," in *Hippocratica: Actes du Colloque hippocratique de Paris, 4–9 septembre 1978,* ed. M. D. Grmek (Paris: Éditions du Centre national de la recherche scientifique, 1980), 402; Lesley Dean-Jones, *Women's Bodies in Classical Greek Science* (Oxford: Clarendon Press, 1994), 124.
18. Seneca *Letters* 95.21. See Jonathan Walters, "Invading the Roman Body: Manliness and Impenetrability in Roman Thought," in *Roman Sexualities*, ed. J. P. Hallett and M. B. Skinner (Princeton, NJ: Princeton University Press, 1997), 30–43.
19. Soranus *Gynecology* 2.10 (Ilberg, 57–58).
20. Rufus *On Buying Slaves*, fragments in Franz Rosenthal, *The Classical Heritage in Islam* (trans. Emile and Jenny Marmorstein) (Berkeley: University of California Press, 1975), 204.
21. Ulpian *Digest* 21.1.12.1.
22. Aristotle *Politics* 1254b27–1255a3.
23. Aristotle *Politics* 1254b2–20.
24. Aristotle *History of Animals* 538b10.
25. Aristotle *Politics* 1254a37–b2.
26. Ann Ellis Hanson, "Continuity and Change: Three Case Studies in Hippocratic Gynecological Therapy and Theory," in *Women's History and Ancient History*, ed. S. B. Pomeroy (Chapel Hill: University of North Carolina Press, 1991), 82.
27. Pompeius Festus 272L. Cited and discussed in Amy Richlin, "Cicero's Head," in *Constructions of the Classical Body*, ed. J. I. Porter (Ann Arbor: University of Michigan Press, 1999), 194.

28. Homer *The Iliad* 2.216–19 (Thersites); 3.170 (Agamemnon).
29. Homer *The Iliad* 13.275–86.
30. Edith Hall, *Inventing the Barbarian: Greek Self-Definition through Tragedy* (Oxford: Clarendon Press, 1989), 79–83; Timothy McNiven, "Behaving Like an Other: Telltale Gestures in Athenian Vase Painting," in *Not the Classical Ideal: Athens and the Construction of the Other in Greek Art*, ed. Beth Cohen (Leiden: Brill, 2000), 83–94.
31. For the challenges posed by Odysseus to the Iliadic ideal, see Nancy Worman, *The Cast of Character: Style in Greek Literature* (Austin: University of Texas Press, 2002).
32. On non-ideal somatypes and *kalokagathia*, see Ingomar Weiler, "Inverted *Kalokagathia*," in *Representing the Body of the Slave*, ed. T.E.J. Wiedemann and J.F. Gardner, special issue of *Slavery and Abolition* 23 (2002): 9–28.
33. Victoria Wohl, *Love among the Ruins: The Erotics of Democracy in Classical Athens* (Princeton, NJ: Princeton University Press, 2002).
34. See, for example, D. Halperin, "The Democratic Body: Prostitution and Citizenship in Classical Athens," in *One Hundred Years of Homosexuality: And Other Essays on Greek Love* (London: Routledge, 1990), 88–112; V. Hunter, "Constructing the Body of the Citizen: Corporal Punishment in Classical Athens," *Échos du monde classique* 36 (1992): 271–91; Nicole Loraux, *The Children of Athena: Athenian Ideas about Citizenship and the Division between the Sexes* (trans. Caroline Levine) (Princeton, NJ: Princeton University Press, 1993); Ellen Davis, ed., *Representations of the "Other" in Athenian Art, c. 510–400BC*, New York: Ars Brevis Foundation, 1995; Cohen, *Not the Classical Ideal*; M.M. Sassi, *The Science of Man in Ancient Greece* (trans. P. Tucker) (Chicago: University of Chicago Press, 2001), 82–139.
35. Benjamin Isaac, *The Invention of Racism in Classical Antiquity* (Princeton, NJ: Princeton University Press, 2004), 55–109.
36. [Hippocrates] *Airs, Waters, Places* 15 (Littré 2.60-62), translation in W.H.S. Jones, trans., *Hippocrates,* vol. 1 (Cambridge, MA: Harvard University Press, 1923. On the sogginess of female flesh: [Hippocrates] *Diseases of Women* I 1 (Littré 8.12); *Glands* 16 (Littré 8.572), in P. Potter, trans., *Hippocrates*, vol. 8 (Cambridge, MA: Harvard University Press, 1995).
37. On the Scythians, see [Hippocrates] *Airs, Waters, Places* 17–22 (Littré 2.66-82), in Jones, *Hippocrates*, vol. 1. See also S. Kuriyama, *The Expressiveness of the Body and the Divergence of Greek and Chinese Medicine* (New York: Zone, 1999), 111–51.
38. Aristotle *Politics* 1327b18–33; Polybius *Histories* 4.21.1-7; Diodorus Siculus 3.34.6–8; Pliny *Natural History* 2.80.189–90; Vitruvius *On Architecture* 6.1.9–11; Galen *The Faculties of the Soul Follow the Mixtures of the Body* 9 (Kühn 4.805).
39. Hesiod *Works and Days* 235.
40. [Aristotle] *Physiognomica* 808b11–30. See Elizabeth C. Evans, *Physiognomics in the Ancient World* (Philadelphia: American Philosophical Society, 1969).
41. Maud Gleason, *Making Men: Sophists and Self-Presentation in Ancient Rome* (Princeton, NJ: Princeton University Press, 1995), 59.
42. Adamantius 2.1 (in R. Foerster, ed., *Scriptores physiognomonici Graeci et Latini*, 2 vols., Leipzig: Teubner, 1893, vol. 1, 349). Translated by Ian Repath in "The *Physiognomy* of Adamantius the Sophist," in *Seeing the Face, Seeing the Soul: Polemon's Physiognomy from Classical Antiquity to Medieval Islam*, ed. S. Swain

(Oxford: Clarendon Press, 2007), 517. Cf. [Aristotle] *Physiognomica* 814a6-8: "It is a good idea to refer all signs that have been mentioned to overall impression and to gender, male and female." On "overall impression," see also Tamsyn Barton, *Power and Knowledge: Astrology, Physiognomics, and Medicine under the Roman Empire* (Ann Arbor: University of Michigan Press, 1994), 109–10; Gleason, *Making Men*, 33–37.

43. Barton, *Power and Knowledge*, 113.
44. J. Davidson, "Dover, Foucault and Greek Homosexuality: Penetration and the Truth of Sex," *Past and Present* 170 (2001): 3–51. On female excesses, see also Anne Carson, "Putting Her in Her Place: Woman, Dirt, and Desire," in D. Halperin, J. J. Winkler, and F. I. Zeitlin (eds.), *Before Sexuality: The Construction of Erotic Experience in the Ancient Greek World* (Princeton, NJ: Princeton University Press, 1990), 135–69.
45. Aeschines *Against Timarchus* 26. See Giulia Sissa, "Sexual Bodybuilding: Aeschines against Timarchus," in Porter, *Constructions*, 147–68.
46. Adamantius 1.19 (Foerster 1.341). Translated by Repath, "The *Physiognomy*."
47. Hesiod *Theogony* 570–89; Ovid *Art of Love* 3.201–30; Tertullian *On Feminine Adornment* 2.5.2 (in Marie Turcan, trans., *Tertullien, De cultu feminarum*, Sources chrétiennes 173, Paris: Les Éditions du Cerf, 1971). See Froma I. Zeitlin, "Signifying Difference: The Case of Hesiod's Pandora," in *Playing the Other: Gender and Society in Classical Greek Literature* (Chicago: University of Chicago Press, 1996), 53–86; Maria Wyke, "Woman in the Mirror: The Rhetoric of Adornment in the Roman World," in *Women in Ancient Societies: An Illusion of the Night*, ed. Léonie J. Archer, Susan Fischler, and Maria Wyke (Basingstoke: Macmillan, 1994), 134–51; Amy Richlin, "Making Up a Woman: The Face of Roman Gender," in *Off with Her Head!: The Denial of Women's Identity in Myth, Religion, and Culture*, ed. H. Eilberg-Schwartz and Wendy Doniger (Berkeley: University of California Press, 1995), 185–213; Eric Downing, "Anti-Pygmalion: The *Praeceptor* in *Ars Amatoria*, Book 3," in Porter, *Constructions*, 235–51.
48. Diogenes Laertius *Lives of the Philosophers* 7.173.
49. McNiven, "Behaving Like an Other," 72–75.
50. Andrew Stewart and Celina Gray, "Confronting the Other: Childbirth, Aging, and Death on an Attic Tombstone at Harvard," in Cohen, *Not the Classical Ideal*, 262.
51. Plato *Protagoras* 326b-c. Translated by W.R.M. Lamb, in *Plato*, vol. 2 (Cambridge, MA: Harvard University Press, 1924).
52. [Hippocrates] *On Regimen* I 35–36 (Littré 6.512–524), in W.H.S. Jones, trans., *Hippocrates*, vol. 4 (Cambridge, MA: Harvard University Press, 1931).
53. Cicero *Orator* 18.59; Quintilian *Education of an Orator* 11.3.122–36.
54. [Aristotle] *Physiognomica* 806b26–27; Adamantius 2.48 (Foerster 1.413).
55. Quintilian *Education of an Orator* 11.3.19–29. See Erik Gunderson, *Staging Masculinity: The Rhetoric of Performance in the Roman World* (Ann Arbor: University of Michigan Press, 2000), 81–82.
56. *Rhetorica ad Herennium* 3.12.22. Translated by Harry Caplan, [*Cicero*], *Ad C. Herennium (Rhetorica ad Herennium)* (Cambridge, MA: Harvard University Press, 1954).
57. Quintilian *Education of an Orator* 11.3.85–121. On the orators' performance of masculinity, see also Joy Connolly, "Virile Tongues: Rhetoric and Masculinity," in *A Companion to Roman Rhetoric*, ed. William Dominik and Jon Hall (Malden, MA: Blackwell, 2007), 83–97.

58. Quintilian *Education of an Orator* 11.3.76, 83. On the characteristic gestures of comic slaves, see Fritz Graf, "Gestures and Conventions: The Gestures of Roman Actors and Orators," in *A Cultural History of Gesture*, ed. Jan Bremmer and Herman Roodenburg (Ithaca, NY: Cornell University Press, 1992), 36–58.
59. E.g., Cicero *On Oratory* 1.59.251; Quintilian *Education of an Orator* 11.3.88–89, 111–12. See Gunderson, *Staging Masculinity*, 111–48.
60. M. Foucault, *The History of Sexuality*, vol. 2, *The Use of Pleasure*, trans. R. Hurley (New York: Pantheon, 1985), 22–23. See also Amy Richlin, "Zeus and Metis: Foucault, Feminism, Classics," *Helios* 18 (1991): 160–80; Lesley Dean-Jones, "The Politics of Pleasure: Female Sexual Appetite in the Hippocratic Corpus," *Helios* 19 (1992): 72–91.
61. Foucault, *The Use of Pleasure*, 10. See also M. Foucault, *The History of Sexuality*, vol. 3, *The Care of the Self*, trans. R. Hurley (New York: Pantheon, 1986).
62. See especially Gunderson, *Staging Masculinity*, 22–25, where he uses Foucault's own earlier work in *Discipline and Punish* on the body inscribed by power to critique Foucault's later work on antiquity.
63. Soranus *Gynecology* 2.14–15, 32–35 (Ilberg, 60–62, 77–79); Plato *Laws* 7.789e.
64. Soranus *Gynecology* 2.34 (Ilberg, 79).
65. Circumcised barbarians: Aristophanes *Acharnians* 158; *Wealth* 267. Idealization of small penises: Aristophanes *Clouds* 1014.
66. Paul of Aegina *Medical Compendium in Seven Books* 6.46, in I. L. Heiberg, ed., *Paulus Aegineta*, 2 vols., Corpus Medicorum Graecorum 9.1–2, (Berlin: Teubner, 1921–1924), vol. 2, 86–87.
67. Celsus *On Medicine* 7.25.1; Paul of Aegina *Medical Compendium in Seven Books* 6.53 (Heiberg 2.94). The operation is explained in detail in Ralph Jackson, "Circumcision, De-circumcision and Self-Image: Celsus' 'Operations on the Penis,'" in *Roman Bodies: Antiquity to the Eighteenth Century*, ed. A. Hopkins and M. Wyke (London: The British School at Rome, 2005), 23–32.
68. Lactantius *The Workmanship of God* 12. Translated by Mary Francis McDonald, in *Lactantius, Minor Works* (Washington DC: Catholic University of America Press, 1965).
69. Aristotle *Generation of Animals* 727a26–30, 728a32–34.
70. [Hippocrates] *On Regimen* I 28–29 (Littré 6.500–504), in Jones, *Hippocrates*, vol. 4. On female seed: [Hippocrates] *On Generation/On the Nature of Child* 4 (Littré 7.474).
71. Aristotle *Generation of Animals* 770b9–17.
72. Aristotle *Generation of Animals* 766a18–24.
73. Colder: [Hippocrates] *Aphorisms* 1.14 (Littré 4.466), in Jones, *Hippocrates*, vol. 4; Aristotle *On the Longevity and Brevity of Life* 466a17–20. Dryer: Aristophanes *Wealth* 1054; [Hippocrates] *The Nature of Women* 1 (Littré 7.12); Galen *On the Preservation of Health* 5.3 (Kühn 6.319); Wetter: [Hippocrates] *On Regimen* I 33 (Littré 6.510–512), in Jones, *Hippocrates*, vol. 4; [Hippocrates] *On the Nature of a Human Being* 17 (Littré 6.74–76), in Jones, *Hippocrates*, vol. 4.
74. [Hippocrates] *On Diseases* I 22 (Littré 6.184), in P. Potter, trans., *Hippocrates*, vol. 5 (Cambridge, MA: Harvard University Press, 1988).
75. [Hippocrates] *On Generation/On the Nature of Child* 18 (Littré 7.498–500); *On the Eight-Month Fetus* 9 (Littré 7.450). Cf. Euripides *Aeolus* frag. 11, F. Jouan and H. van Looy, eds., *Euripide, Fragments*, 4 vols. (Paris: Les Belles Lettres, 1998), vol. 1, 31.

76. [Hippocrates] *Diseases of Women* I 2 (Littré 8.14-16).
77. [Hippocrates] *Epidemics* VI 8.32 (Littré 5.356), in W.D. Smith, trans., *Hippocrates*, vol. 7 (Cambridge, MA: Harvard University Press, 1994).
78. Seneca *Epistulae Morales* 95.16–21. See the Elder Seneca *Controversiae* 1, praef. 8–10 on men becoming women.
79. Galen *On Mixtures* 2.6 (Kühn 1.625-626); Clement of Alexandria *Paidagogos* 3.19.
80. Clement of Alexandria *Paidagogos* 3.15, cited and translated in Gleason, *Making Men*, 69. See also Quintilian *Education of an Orator* 8, praef. 19–20; Ovid *The Art of Love* 1.505–524; Tertullian *On Feminine Adornment* 2.8.2.
81. John Chrysostom *Against Those Who Keep Virgins from Outside the Family* (in J.-P. Migne, ed., *Patrologiae Cursus Completus, Series Graeca* [Paris: J.-P. Migne, 1857–1866]), vol. 47, col. 510).
82. John Chrysostom *Against Games and the Theatre* (*Patrologiae cursus completus, Series Graeca* 56.267); Tertullian *On the Veiling of Virgins* 7.2 (in Eva Schulz-Flügel, trans., *Tertullien, De virginibus velandis*, Sources chrétiennes 424, Paris: Les Éditions du Cerf, 1997).
83. Heliodorus *Ethiopian Story* 10.14.
84. Galen *On Mixtures* 2.2 (Kühn 1.578).
85. Plutarch *Table Talk* 680D; *The Education of Children* 3F–4A.
86. Aulus Gellius *Attic Nights* 12.1.17.
87. Quintilian *Education of an Orator* 1.11.2–3. See also Plato *Republic* 3.395d.
88. Plato *Republic* 3.394e–396e; *Laws* 7.816e.
89. Lucian *On the Dance* 1; Gregory of Nazianzus *Carmina* 2.1.88–90 (*Patrologiae Cursus Completus, Series Graeca* 37.1438).
90. Ovid *Metamorphoses* 9.786–91. See Shilpa Raval, "Cross-Dressing and 'Gender Trouble' in the Ovidian Corpus," *Helios* 29 (2002): 149–72.
91. A.G. Geddes, "Rags and Riches: The Costume of Athenian Men in the Fifth Century," *Classical Quarterly* 37 (1987): 307–31.
92. Tertullian *On the Pallium* 5, in J.-P. Migne, ed., *Patrologiae Cursus Completus, Series Latina* (Paris: J.-P. Migne, 1844–1864]), vol. 2, col. 1045B-1049A.
93. Romans 10:12.
94. Galatians 6:14–15. Cf. Acts 15:1–29; Romans 2:28–29; Colossians 2:11; Ephesians 2:11. See Gillian Clark, "'In the Foreskin of Your Flesh': The Pure Male Body in Late Antiquity," in Hopkins and Wyke, *Roman Bodies*, 43–53.
95. Athanasius, "On Virginity," in L.-Th. Lefort, "Saint Athanase: Sur la Virginité," *Le Muséon* 42 (1929): 246 (translation), cited in Peter Brown, *The Body and Society: Men, Women and Sexual Renunciation in Early Christianity* (New York: Columbia University Press, 1988), 274.
96. See Eusebius *Ecclesiastical History* 5.1.41-42, on the martyr Blandina. On the complex gendering of endurance as a virtue of martyrs, see Brent D. Shaw, "Body/Power/Identity: Passions of the Martyrs," *Journal of Early Christian Studies* 4 (1996): 269–312.
97. Corinthians I, 11:5–16.
98. See Garry Wills, *What Paul Meant* (New York: Viking, 2006), 89–104.
99. *Acts of Thecla* 25, 40.
100. J. Drescher, *Three Coptic Legends: Hilaria, Archellites, the Seven Sleepers* (Cairo: Institut français d'archéologie orientale, 1947), 75; T. Wilfong, "Reading the Disjointed Body in Coptic: From Physical Modification to Textual Fragmentation," in *Changing Bodies, Changing Meanings: Studies on the Human Body in Antiquity*,

ed. Dominic Montserrat (London: Routledge, 1998), 127–30. See also Stephen J. Davis, "Crossed Texts, Crossed Sex: Intertextuality and Gender in Early Christian Legends of Holy Women Disguised as Men," *Journal of Early Christian Studies* 10 (2002): 28–29.
101. *The Martyrdom of Perpetua and Felicitas* 6, 10.
102. Elizabeth Castelli, "'I Will Make Mary Male': Pieties of the Body and Gender Transformation of Christian Women in Late Antiquity," in *Body Guards: The Cultural Politics of Gender Ambiguity*, ed. J. Epstein and K. Straub (London: Routledge, 1991), 33. See also Daniel Boyarin, "Paul and the Genealogy of Gender," *Representations* 41 (1993): 1–33; Elizabeth A. Clark, "Holy Women, Holy Words: Early Christian Women, Social History, and the 'Linguistic Turn,'" *Journal of Early Christian Studies* 6 (1998): 413–17.
103. Patricia Cox Miller, "Is There a Harlot in This Text?: Hagiography and the Grotesque," *Journal of Medieval and Early Modern Studies* 33 (2003): 423.
104. Tertullian *On the Veiling of Virgins* 7.1.
105. *Codex Theodosianus* 16.2.27.1.
106. D. Dexheimer, "Portrait Figures on Funerary Altars of Roman *Liberti* in Northern Italy: Romanization or the Assimilation of Attributes Characterising Higher Social Status?," in *Burial, Society and Context in the Roman World*, ed. J. Pearce, M. Millett, and M. Struck (Oxford: Oxbow Books, 2000), 78–84.
107. Judith Lynn Sebesta, "Women's Costume and Feminine Civic Morality in Augustan Rome," in M. Wyke (ed.), *Gender and the Body in the Ancient Mediterranean*, Malden, MA: Blackwell, 1998, 105–17.
108. Appian *The Civil Wars* 2.120; Martial *Epigrams* 4.66; Juvenal *Satires* 3.171–72. See S. Stone, "The Toga: From National to Ceremonial Costume," in *The World of Roman Costume*, ed. Judith Lynn Sebesta and Larisa Bonfante (Madison: University of Wisconsin Press, 1994), 13–45.
109. Suetonius *Life of Augustus* 40.5.
110. Persius *Satires* 5.30–31.
111. Mary Harlow, "Clothes Maketh the Man: Power Dressing and Elite Masculinity in the Later Roman World," in *Gender in the Early Medieval World: East and West, 300–900*, ed. Leslie Brubaker and Julia M. H. Smith (Cambridge: Cambridge University Press, 2004), 48. See esp. Aulus Gellius *Attic Nights* 6.12; on trousers, see also Polybius 2.28.7; Diodorus Siculus 5.30.1.
112. *Codex Theodosianus* 14.10.2.
113. E.g., Sappho frag. 98 (in E. Lobel and D. L. Page, *Poetarum Lesbiorum fragmenta* [Oxford: Clarendon Press, 1955]).
114. Bonnie Effros, "Dressing Conservatively: Women's Brooches as Markers of Ethnic Identity?" in Brubaker and Smith, *Gender*, 165–84.
115. Quintilian *Education of an Orator* 11.3.137–149. Wearing the imperial toga well is comparable to being able to tie an impeccable bow tie: Glenys Davies, "What Made the Roman Toga *Virilis*?" in *The Clothed Body in the Ancient World*, ed. Liza Cleland, Mary Harlow, and Lloyd Llewellyn-Jones (Oxford: Oxbow Books, 2005), 127.
116. Hans van Wees, "Trailing Tunics and Sheepskin Coats: Dress and Status in Early Greece," in Cleland, Harlow, and Llewellyn-Jones, *The Clothed Body*, 45–46.
117. Gillian Clark, *Women in Late Antiquity: Pagan and Christian Life-Styles* (Oxford: Clarendon Press, 1993), 113–116.

118. Donatus *On Comedy* 8.6–7 (in P. Wessner, ed., *Aeli Donati quod fertur Commentum Terenti*, 3 vols. [Stuttgart: Teubner, 1902–1908], vol. 1, 29).
119. Anacreon frag. 388 (in D. L. Page, ed., *Poetae melici graeci* [Oxford: Clarendon Press, 1962]).
120. Dio Cassius 47.10.2–4; Appian *The Civil Wars* 4.6.44; Seneca *On Benefits* 3.25.1.
121. Aristophanes *Frogs* 494–673.
122. Seneca *Letter* 66.23.
123. *Codex Theodosianus* 15.7.11–12.
124. Pliny *Natural History* 13.5.25.
125. Xenophon *Symposium* 2.4.
126. Pliny *Natural History* 13.5.25. For Pliny's attacks on perfume see *Natural History* 13.1.3, 13.4.20–21, together with David S. Potter, "Odor and Power in the Roman Empire," in Porter, *Constructions*, 175–79.
127. Seneca *Letters* 86.12. Cf. 108.16, Martial *Epigrams* 6.55.
128. As Ralph Jackson points out in *Doctors and Diseases in the Roman Empire* (London: British Museum Publications, 1987), 55.
129. Pliny *Natural History* 7.30.117; Suetonius *Augustus* 44.
130. Matthew B. Roller, *Dining Posture in Ancient Rome: Bodies, Values, and Status* (Princeton, NJ: Princeton University Press, 2006). Cf. John D'Arms, "The Roman *Convivium* and the Idea of Equality," in *Sympotica: A Symposium on the Symposium*, ed. Oswyn Murray (Oxford: Clarendon Press, 1990), 308–20; Carlin Barton, *The Sorrows of the Ancient Romans: The Gladiator and the Monster* (Princeton, NJ: Princeton University Press, 1993), 109–12.
131. Suetonius *Caligula* 26.2; Philo of Alexandria *The Contemplative Life* 72.
132. [Xenophon] *Constitution of Athens* 1.10–12.
133. Aristotle *Politics* 1254b32–1255a1.
134. P. Oxy LI 3617, quoted and translated in Dominic Montserrat, "Experiencing the Male Body in Roman Egypt," in *When Men Were Men: Masculinity, Power and Identity in Classical Antiquity*, ed. Lin Foxhall and John Salmon (London: Routledge, 1998), 158.
135. Michele George, "Slave Disguise in Ancient Rome," in *Representing the Body of the Slave*, ed. T.E.J. Wiedemann and J.F. Gardner, special issue of *Slavery and Abolition* 23 (2002): 50.
136. Myron of Priene *FGrH*106 F2, cited in van Wees, "Trailing Tunics," 49.
137. Seneca *On Clemency* 1.24.1.
138. Suetonius *Caligula* 47.
139. Virgil *Aeneid* 8.722–23.
140. I. M. Ferris, *Enemies of Rome: Barbarians through Roman Eyes* (Stroud, UK: Sutton, 2000).
141. W. Pohl, "Gender and Ethnicity in the Early Middle Ages," in Brubaker and Smith, *Gender*, 36 n. 61.
142. W. Pohl, "Telling the Difference: Signs of Ethnic Identity," in *Strategies of Distinction: The Construction of Ethnic Communities, 300–800*, ed. W. Pohl and H. Reimitz (Leiden: Brill, 1998), 17–69.
143. Effros, "Dressing Conservatively," 166–68. See also Dawn Hadley, "Negotiating Gender, Family and Status in Anglo-Saxon Burial Practices, c. 600–950," in Brubaker and Smith, *Gender*, 301–23.

144. Pliny *Natural History* 6.35.187–88; Pomponius Mela 1.48. See Barton, *Power and Knowledge*, 122–24; Rhiannon Evans, "Ethnography's Freak Show: The Grotesques at the Edges of the Roman Earth," *Ramus* 28 (1999): 54–73.
145. W. Pohl, "Introduction: Strategies of Distinction," in Pohl and Reimitz, *Strategies of Distinction*, 4.
146. Historia Augusta *Aurelianus* 34.
147. *Life of Aesop* (G) 1; Himerius *Orations* 13.4–5. Some possible representations of Aesop may be found in François Lissarrague, "Aesop, between Man and Beast: Ancient Portraits and Illustrations," in Cohen, *Not the Classical Ideal*, 132–49.
148. Alcibiades describes Socrates as a Silenus figure: Plato *Symposium* 215a–b. For Plato's appropriation of the aristocratic ideal for the ugly Socrates, see Nicole Loraux, "Socrates, Plato, Herakles: A Heroic Paradigm of the Philosopher," in *The Experiences of Tiresias: The Feminine and the Greek Man*, trans. P. Wissing (Princeton, NJ: Princeton University Press, 1993), 167–77.
149. Mikeal C. Parsons, *Body and Character in Luke and Acts: The Subversion of Physiognomy in Early Christianity* (Grand Rapids, MI: Baker Academic, 2006).
150. Plato *Phaedrus* 253e.
151. Ovid *Amores* 2.7.19–22.
152. Quintilian *Education of an Orator* 1.3.13–17, against corporal punishment; cf. Augustine *Confessions* 1.9.
153. Maud Gleason, "Truth Contests and Talking Corpses," in Porter, *Constructions*, 300–302.
154. Scholia *ad* Aeschines 2.83, cited in C. P. Jones, "Stigma: Tattooing and Branding in Graeco-Roman Antiquity," *Journal of Roman Studies* 77 (1987): 148.
155. Aeschylus *Agamemnon* 1326.
156. Suetonius *Caligula* 32; *Galba* 9.
157. Herodas *Mimiambi* 5.16–17.
158. Barton, *The Sorrows*.
159. Garland, *Eye of the Beholder*, 16–18. For hermaphrodites as prodigies, see Livy 27.11.4–5.
160. Pliny *Natural History* 7.16.75; Historia Augusta *Severus Alexander* 34.2–4. On the use of dwarves as personal attendants in Greece, see Véronique Dasen, *Dwarfs in Ancient Egypt and Greece* (Oxford: Clarendon Press, 1993), 226–30.
161. Plutarch *On Curiosity* 520C; Quintilian *Education of an Orator* 2.5.11. Translated by D. A. Russell in *Quintilian, The Orator's Education, Books 1-2* (Cambridge, MA: Harvard University Press, 2001).
162. *Rhetorica ad Herennium* 4.50.63; see also Theophrastus *Characters* 21.4.
163. On the analogous fascinating power of amulets, see Plutarch *Table Talk* 681F–682A.
164. On *curiosi* at the baths: Martial *Epigrams* 11.63; Seneca *Natural Questions* 1.16.3.
165. B. Fehr, "Entertainers at the *Symposium*: The *Akletoi* in the Archaic Period," in *Sympotica: A Symposium on the Symposium*, ed. Oswyn Murray (Oxford: Clarendon Press, 1990), 187.
166. Lucian *Lapithae* 18.
167. Pliny *Natural History* 7.3.34–35.
168. See especially Froma I. Zeitlin, "Playing the Other: Theater, Theatricality, and the Feminine in Greek Drama" and "Travesties of Gender and Genre in Aristo-

phanes' *Thesmophoriazousae*," in *Playing the Other: Gender and Society in Classical Greek Literature* (Chicago: University of Chicago Press, 1996), 341–416.
169. [Xenophon] *Constitution of the Athenians* 2.8.
170. Tertullian *On Spectacles* 23.
171. Barton, *The Sorrows*, 143–44.
172. Homer *The Iliad* 2.265–77.
173. See K. M. Coleman, "Fatal Charades: Roman Executions Staged as Mythological Enactments," *Journal of Roman Studies* 80 (1990): 44–73.
174. Prudentius *Peristephanon* 3.135–140, cited in Shaw, "Body/Power/Identity," 306. On the mimesis of Christ, see S. Brock and S. A. Harvey, *Holy Women of the Syrian Orient* (Berkeley: University of California Press, 1987), 14.
175. Shaw, "Body/Power/Identity," 304.
176. John 10:37–38; Acts 9:34–42; Mark 1:40–42.
177. Livio Pestilli, "Disabled Bodies: The (Mis)representation of the Lame in Antiquity and Their Reappearance in Early Christian and Medieval Art," in Hopkins and Wyke, *Roman Bodies*, 85–97.
178. Ian Morris, "Remaining Invisible: The Archaeology of the Excluded in Classical Athens," in *Women and Slaves in Greco-Roman Culture: Differential Equations*, ed. Sandra R. Joshel and Sheila Murnaghan (London: Routledge, 1998), 196–97.
179. For the Roman world, see the comments of Kathleen McCarthy, *Slaves, Masters, and the Art of Authority in Plautine Comedy* (Princeton, NJ: Princeton University Press, 2000), 18–19.
180. Galen *The Passions of the Soul* 1.8 (Kühn 5.40–41); Ovid *Art of Love* 3.239–42, *Amores* 1.14.13–22; Juvenal *Satires* 6.474–495; Procopius *Secret History* 5.8–12, 15.5, 16.22.
181. Achilles Tatius *The Story of Leucippe and Cleitophon* 6.18–22.
182. E.g., Horace *Epodes* 8, 12.
183. Susan B. Matheson, "The Elder Claudia: Older Women in Roman Art," in *I, Claudia II: Women in Roman Art and Society*, ed. Diana E. E. Kleiner and Susan B. Matheson (Austin: University of Texas Press, 2000), 135.
184. Janet Huskinson, "Representing Women on Roman Sarcophagi," in *The Material Culture of Sex, Procreation, and Marriage in Premodern Europe*, ed. Anne L. McClanan and Karen Rosoff Encarnación (New York: Palgrave, 2002), 11.
185. Catherine Edwards, "Archetypally Roman?: Representing Seneca's Ageing Body," in Hopkins and Wyke, *Roman Bodies*, 13.
186. Herodotus *Histories* 5.6. On the marked *Britanni*, Caesar *Gallic Wars* 5.14; Herodian 3.14.7.
187. Aristotle *Generation of Animals* 766a26–30; Clement of Alexandria *Stromata* 3.15; Gregory of Nazianus *In Praise of Athanasius* (*Patrologiae Cursus Completus, Series Graeca* 35.1105).
188. Adamantius 2.21 (Foerster 1.369–70).
189. Claudian *Against Eutropius* 1.423–24.
190. Liz James and Shaun Tougher, "Get Your Kit On! Some Issues in the Depiction of Clothing in Byzantium," in Cleland, Harlow, and Llewellyn-Jones, *The Clothed Body*, 154–61.
191. *Narration of the Miracles of the Mighty Archangel Michael*, cited in James and Tougher, "Get Your Kit On!," 160. Indeed, Christ was often represented as androgynous (Davis, "Crossed Texts," 35).

192. See Daniel F. Caner, "The Practice and Prohibition of Self-Castration in Early Christianity," *Vigiliae Christianae* 51 (1997): 396–415.
193. Ray Laurence, "Health and the Life Course at Herculaneum and Pompeii," in *Health in Antiquity*, ed. Helen King (London: Routledge, 2005), 90. Laurence describes and compares both skeletons noted above.
194. Lynn Meskell, "Archaeologies of Identity," in *Archaeological Theory Today*, ed. I. Hodder (Cambridge: Cambridge University Press, 2001), 187–213; Montserrat, "Experiencing the Male Body"; R. A. Joyce, "Archaeology of the Body," *Annual Review of Anthropology* 34 (2005): 139–58.

Chapter 8

1. Ovid *Metamorphoses* 1.1–4. Translations of Greek and Latin texts are my own, unless stated otherwise.
2. Jean-Pierre Vernant, "Dim Body, Dazzling Body," trans. Anne M. Wilson, in *Fragments for a History of the Human Body*, vol. 3, part 1, ed. Michel Feher, Ramona Naddaff, and Nadia Tazi, 18–47 (New York: Zone, 1989), 23.
3. Ibid.
4. *Met.* 3.316–38.
5. *Met.* 6.424–674.
6. Ovid provides a different reason: Juno (Hera), infuriated that Tiresias claimed that women enjoyed sex more than men, blinded him (although Jupiter [Zeus] counterbalanced the punishment by giving him prophetic vision); see *Metamorphoses* 3.330–36.
7. Ovid *Met.* 3.194–203.
8. *Menexenus* 237d.
9. Gary Steiner, *Anthropocentrism and Its Discontents: The Moral Status of Animals in the History of Western Philosophy* (Pittsburgh: University of Pittsburgh Press, 2005), 60.
10. A notable exception is the Cynic philosophical disposition; Diogenes, for example, regarded animals as superior to humans (see Kenneth S. Rothwell, *Nature, Culture, and the Origins of Greek Comedy: A Study of Animal Choruses* [Cambridge: Cambridge University Press, 2007]), 90.
11. *Republic* 441a.
12. *The History of Animals* 588b.
13. Galen *De placitis Hippocratis et Platonis* 5.5.4–4; in *Galen. De placitis Hippocratis et Platonis*, Corpus Medicorum Graecorum V.4.2.1–4, ed. and trans. P. H. de Lacy (Berlin: Akademie Verlag).
14. Semonides 7.1.
15. Semonides 7.2–5.
16. The word also denotes a female "mate" in relation to animals (see Aristotle *Politics* 1262a22).
17. Friedrich Nietzsche, "The Greek Woman," in *The Complete Works of Friedrich Nietzsche*, vol. 2, trans. M. A. Mügge (London: T. N. Foulis, 1911), 23.
18. The alternate order of constituent beings on the divine-to-animal continuum is discussed by Mark Golden, "Change or Continuity? Children and Childhood in Hellenic Historiography," in *Inventing Ancient Culture: Historicism, Periodization and the Ancient World*, ed. Mark Golden and Peter Toohey, 176–91 (London:

Routledge, 1977). In his analysis of Greek historians, for example, Golden writes: "Children are almost always mentioned before women in Herodotus (sixteen of nineteen times), in about two times in three (thirteen of nineteen) in Thucydides, and in both cases in *Hellenica*. In Polybius, however, the numbers are more nearly equal" (183). See also John Heath, *The Talking Greeks: Speech, Animals, and the Other in Homer, Aeschylus, and Plato* (Cambridge: Cambridge University Press, 2005), 172–73.

19. 716a5–23. For a comparison of women and children by Aristotle, see *On the Generation of Animals* 784a5–12. For further discussion of the earth as female body, see Page duBois, *Sowing the Body: Psychoanalysis and Ancient Representations of Women* (Chicago: University of Chicago Press, 1988), 57–63.
20. Iliad 16.156–66, in Richmond Lattimore, trans., *The Iliad of Homer* (Chicago: University of Chicago Press, 1951). All subsequent citations are to this edition.
21. *Il.* 22.345–54.
22. A persuasive counterview is offered by Heath, *Talking Greeks*, 136–37, in his discussion of the Greek words *krea* and *oma* translated as "meat" and "raw," respectively, at 22.347. As *krea* is consistently associated with animals in epic literature, never being used alone as a term for human flesh, Heath argues that "by insisting that it is uncooked *kera*, he reveals his primary wish to *become* an animal himself, a hungry beast who, like a lion, sees the dying figure in front of him as a meal of pulsating flesh" (137).
23. *Il.* 24.212–14.
24. Jonathan Gottschall, "Homer's Human Animal: Ritual Combat in the *Iliad*," *Philosophy and Literature* 25 (2001): 281.
25. *Republic* 571c.
26. Martha Craven Naussbaum, *The Fragility of Goodness: Luck and Ethics in Greek Tragedy and Philosophy* (Cambridge: Cambridge University Press, 1986), 204.
27. 589c–d.
28. Paul Cartledge, *The Greeks: A Portrait of Self and Others* (Oxford: Oxford University Press, 2002), 13.
29. For more on Keuls's views on women as pigs, see Eva Keuls, *The Reign of the Phallus: Sexual Politics in Ancient Athens* (Berkeley: University of California Press, 1985), 353–57. Keuls also discusses the related term *kapraina* (wild sow), in relation to women, pointing out that "to the Greek mind" it "was the equivalent of 'lewd' woman" (354). For an interesting association between the Greek male and the pig, see the transformation passage from Homer's *Odyssey*, in which the goddess-sorceress, Circe, turns Odysseus's men into swine. In view of the fears and anxieties inherent in stories of metamorphosis, and the latent meanings discussed previously in relation to the male-female transformation of Tiresias, the act may reveal a secondary form of punishment besides simple male-beast conversion, namely that the process may have an inherently female connection—thereby augmenting the men's humiliation.
30. See Jeffrey Henderson, *The Maculate Muse: Obscene Language in Attic Comedy*, 2nd rev. ed. (New York: Oxford University Press, 1991), 126–29 (for animals and male genitalia) and 131–33 (for female genitalia).
31. See J.N. Adams, *The Latin Sexual Vocabulary* (Baltimore: Johns Hopkins University Press, 1982), 29–34.
32. Kenneth S. Rothwell, *Nature, Culture, and the Origins of Greek Comedy: A Study of Animal Choruses* (Cambridge: Cambridge University Press, 2007), 36.

33. William F. Hansen, *Handbook of Classical Mythology* (Santa Barbara, CA: ABC-CLIO, 2004), 279–80. Hansen comments on the ambiguity regarding Satyrs in antiquity: "The question of whether they were mortals or immortals evidently was unclear to the ancients" (280).
34. Paul Murgatroyd, *Mythical Monsters in Classical Literature* (London: Duckworth, 2007), 2.
35. See François Lissarrague, "The Sexual Life of Satyrs," in *Before Sexuality: The Construction of Erotic Experience in the Ancient Greek World,* ed. David M. Halperin, John J. Winkler, and Froma I. Zeitlin, 53–81 (Princeton, NJ: Princeton University Press, 1990), 54, on the variability of the Satyrs' closeness to humanity based on individual artistic depictions that can accentuate or decrease their physical resemblance to men.
36. See Mark Griffith, "Slaves of Dionysos: Satyrs, Audience, and the Ends of the *Oresteia.*" *Classical Antiquity* 21, no. 2 (2002): 199.
37. Rothwell, *Nature,* 151.
38. Ibid., 85.
39. See E. O. Bassett, "Plato's Theory of Social Progress," *International Journal of Ethics* 38, no. 4 (1928): 467–77.
40. Also in tragedies such as Sophocles's *Ajax* and, more pronounced, in his *Antigone;* see Chiara Thumiger, "ἀνάγκης ζεύγματ' ἐμπεπτώκαμεν: Greek Tragedy between Human and Animal," *Leeds International Classical Studies* 7, no. 3 (2008): 6n16.
41. Hence the widespread ritual in both cultures of omens being performed through the interpretation of bird "signs." See also Walter Burkert, *Greek Religion: Archaic and Classical,* 2nd rev. ed., trans. John Raffan (London: Blackwell, 1987), 40. See also Thumiger, "Greek Tragedy": "Human emotions themselves can be animalised: especially (but not exclusively) in the image of the bird. We find states of violent mental or emotional affection, or even critical experiences represented as winged and feathered, and the subject of 'flying away'—an exchange that underlines the equation between the animalised subject and the animalised emotional affection" (7).
42. By intercepting sacrifices offered to the gods on earth, thereby literally starving them into submission, the rule of the Olympians will be ended. Of course, as the gods consume only ambrosia, the point of the starvation is somewhat strange and thereby in keeping with the absurdist nature of the comedy.
43. Discussed briefly in Edith Hall and Amanda Wrigley, eds., *Aristophanes in Performance, 421 BC–AD 2007: Peace, Birds and Frogs* (London: MHRA, 2007).
44. Martial *Liber Spectaculorum* 5.
45. Kathleen M. Coleman, *M. Valerii Martialis Liber Spectaculorum* (Oxford: Oxford University Press, 2006), 64.
46. See ibid., 64–65, and Coleman, "Fatal Charades: Roman Executions Staged as Mythological Enactments," *Journal of Roman Studies* 80 (1990): 63–64.
47. Coleman, *M. Valerii Martialis,* lxxv.
48. A.E.R. Boak, "The Theoretical Basis of the Deification of Rulers in Antiquity," *Classical Journal* 11, no. 5 (1916): 293. "Most of the great families of Greece traced their descent from some god or hero, just as the Macedonian royal house itself claimed Heracles as its ancestor" (293). See also A. B. Bosworth, *Conquest and Empire: The Reign of Alexander the Great* (Cambridge: Cambridge University Press, 1993), 278.
49. On the beginnings of Hellenistic kings instigating the minting of coins with their own portraits, see M. Bieber, *Alexander the Great in Greek and Roman Art* (Chicago: Argonaut, 1964), 52–53.

50. This particular Hellenistic stater is dated to approximately 286–281 B.C., some 37 years (at least) after his death. However, as Worthington (Ian Worthington, ed., *Alexander the Great: A Reader* [London: Routledge, 2003], 236) points out: "The road to Alexander's deification began in winter 332 when he visited the oracle of Zeus Ammon in the oasis of Siwah in Egypt, guided there in miraculous fashion . . . Here, the priests apparently told him (we have only Alexander's word for it, for he met with them in private) that he was a son of Zeus. This was the turning-point in his belief, and from then on he referred to himself in this way (for example, before the Battle of Gaugamela in 331)."
51. "The decree included a set of honours, namely, the building of a temple and the celebration of festivals. The priest in charge of the cult, Mark Antony, the first *flamen divi Iulii*, was finally inaugurated after the peace of Brundisium in 40" (Fernando Lozano, "*Divi Augusti* and *Theoi Sebastoi*: Roman Initiatives and Greek Answers," *Classical Quarterly* 57, no. 1 [2007]: 139).
52. Of course there are exceptions to the rule: according to Suetonius (*Domitian* 13.2), Domitian insisted on being addressed as *dominus et deus*. In contrast, Tiberius—another exception—rejected deification.
53. *Liber Spec.* 7.
54. Otto Weinrich, *Studien zu Martial* (Stuttgart: W. Kohlhammer, 1928), 35; U. Carratello, "Omnis Caesareo cedit labor amphitheatre! (not a Mart. *Spect. Lib.*)," *Giornale italiano di filogia* 18 (1965): 301; and G. Moretti, "L'arena, Cesare e il mito: appunti sul *De Spectaculis* di Marziale," *Maia*, 44 (1992): 57.
55. Coleman, *M. Valerii Martialis*, 70n2.
56. On this point, Coleman discusses the occasions in the Arena when animals act unpredictably or "contrary to their natures" as "proof of the emperor's divinity" within the interpretive confines of the *Liber Spectaculorum* (ibid., 244). On the argument contra Coleman on the possibility of the gladiators dressed as gods, one could also consider the role of the *editor* in introducing "divine" characters in the Arena.
57. In line with Durkheim's understanding of "social facts," which (as implied above) is an obliterated possibility, theoretically speaking, for interpretive analysis in the case of the divine human, especially when juxtaposed to the related concept of reality. A Foucauldian reading is clearly preferred above.
58. Bosworth discusses the intricate problems inherent in assessing Alexander's divinity:

> What is far from clear is the process of evolution in the course of the reign and the extent of Alexander's belief in his own divinity. At one level there is his deep consciousness of his heroic ancestry (as an Argead he took his lineage back to Heracles and ultimately to Dionysus), at another there is his conviction that he was in some sense the son of Zeus, the equal at least of Heracles, and finally there is the conception of himself as a god among men. These categories are fundamentally different and represent different aspects of Greek religious thought, but in Alexander's mind they must have been conflated. (Bosworth, *Conquest and Empire*, 278)

59. As Bosworth (ibid., 286) notes, Alexander, according to some commentators (Arrian 4.10.6–7 and Curtius 8.5.11–12), surpassed or "excelled" both Heracles and Dionysus in terms of action or "achievements" and thereby "fully deserved to be honoured as a god."

60. Tacitus *Annals* 15.44.
61. Exemplified by the pronouncement of Pope St. Pius (1566–72) that advised the person in need of the sanctity of a religious relic to acquire sand from the Colosseum, which, according to the Pontiff, was infused with the blood of martyrs.
62. The table of Pythagorean Principles is from Cartledge (*The Greeks,* 14).

unlimited	limited
even	odd
plurality	one
left	right
female	male
in motion	at rest
crooked	straight
darkness	light
evil	good
oblong	square

63. Hesiod *Works and Days* 60–68.
64. Rosi Braidotti, "Mothers, Monsters, and Machines," in *Writing on the Body: Female Embodiment and Feminist Theory,* 7th ed., ed. Katie Conboy, Nadia Medina, and Sarah Stanbury (New York: Columbia University Press, 1997), 64.
65. *Il.* 16.148; *Od.* 1.241, 14.371, 20.66ff; *Th.* 265ff.
66. Such a literary device recalls the depiction of Polyphemus in the *Odyssey* where the focus is more on the spectators' reactions *to* him rather than a description *of* him.
67. Hesiod has them as the offspring of Thaumas and Electra (*Th.* 265ff), a sea god and goddess of clouds, respectively.
68. Virgil *Aen.* 3.215–18.
69. *Aen.* 3.225–28.
70. Homer *Od.* 12. 39–46; Lattimore trans.
71. Murgatroyd, *Mythical Monsters,* 45.
72. *Od.* 12.184–90; Lattimore trans.
73. Helene Cixous, "The Laugh of the Medusa," trans. Keith and Paula Cohen, *Signs* 1 (1976): 885.
74. Ibid.
75. Ibid.
76. Ibid, 889.
77. *Met.* 4. 795–804.
78. See, for example, Strabo 11.5.3.
79. *History* 4.114. On Herodotus's complex evocation of Amazon/Scythian, see Cartledge, *The Greeks,* 94–95.
80. Edith Hall, "Asia Unmanned: Images of Victory in Classical Athens," in *War and Society in the Greek World,* ed. J. Rich and G. Shipley (London: Routledge, 1993), 114.
81. Gail Kern Paster, *The Body Embarrassed: Drama and the Disciplines of Shame in Early Modern England* (Ithaca, NY: Cornell University Press, 1993), 234.
82. Montrose in ibid., 234.
83. See Diodorus Siculus 2.46 and Lysias 2.4 (perhaps not to be taken literally); see also Apollonius Rhodius *Argonautica* 2.989ff. Diodorus also aligns the Amazons with the Gorgons geographically (3.52); on the Amazon/Gorgon connection, especially

with reference to Medusa, see W. M. Blake Tyrrell, *Amazons: A Study in Athenian Mythmaking* (Baltimore: Johns Hopkins University Press, 1984), 104–10.
84. Ellen D. Reeder, *Pandora: Women in Classical Greece* (Princeton, NJ: Princeton University Press, 1995), 381.
85. On the use of similes in Euripides's *Medea* see Herbert Musurillo, "Euripides' *Medea*: A Reconsideration," *American Journal of Philology* 87 (1966): 52–74.
86. Seneca *Medea* 910.
87. Horace *Epode* 5.15–31.
88. See Figure 8.4; Goya's witches reflect the bizarre creatures of Horace's imagination.
89. *Epode* 5.97–102.
90. Lucan *Civil War* 6.515–22.
91. Anne Carson, "Putting Her in Her Place: Woman, Dirt, and Desire," in *Before Sexuality: The Construction of Erotic Experience in the Ancient Greek World,* ed. David M. Halperin, John J. Winkler, and Froma I. Zeitlin (Princeton, NJ: Princeton University Press, 1990), 136.
92. Ibid., 158.
93. Lucan *Civil War* 6.554–59.
94. Lucan *Civil War* 6.560–68.

Chapter 9

1. On nudity and the formation of the body in Greek art, see J. Tanner, "Nature, Culture and the Body in Classical Greek Religious Art," *World Archaeology* 33 (2001): 257–76.
2. *LIMC* (*Lexicon Iconographicum Mythologiae Classicae* 1974, vol. 4 [Zurich: Artemis]; subsequent references to this work are designated *LIMC,* followed by the volume number) 4 Herakles 728 (J. Boardman): Herakles's name means Hera's Glory. On Hera nursing Herakles, see Diod. 4.9.4. On the relationship of Herakles and Hera, see M. W. Padilla, *The Myths of Herakles in Ancient Greece* (Lanham, MD: University Press of America, 1998), 2.
3. *LIMC* 4 Herakles 728–31 (J. Boardman); *LIMC* V Herakles 5–16 (J. Boardman); F. Brommer, *Heracles: The Twelve Labors of the Hero in Ancient Art and Literature,* trans. S. J. Schwarz (New York: A. D. Caratzas, 1986), 1–6; J. P. Uhlenbrock, *Herakles: Passage of the Hero through 1000 Years of Classical Art* (New York: A. D. Caratzas, 1986), 1–6; R. Flacelière and P. Devambez, eds., *Héraclès, Images et Récits* (Paris: Boccard, 1966), 17–66; K. Schefold, *Gods and Heroes in Late Archaic Greek Art,* trans. A. Griffiths (Cambridge: Cambridge University Press, 1992), 93–162, and *Die Urkönige, Perseus, Bellerephon, Herakles und Theseus in der klassischen und hellenistischen Kunst* (Munich: Hirmer Verlag, 1988), 128–32, 135–66.
4. Schefold, *Gods and Heroes,* 93 (potential influence of Peisandros's epic on the image of Herakles with a club and lion skin) and 95–100 on Late Archaic iconography of Herakles against the Nemean lion. On the lion skin of Herakles, see also B. Cohen, "The Nemean Lionskin in Athenian Art," in *Le Bestiaire d'Héraclès,* ed. C. Bonnet, C. Jourdain-Annequin, and U. Pirenne-Delforge (Liège: Centre International d'Étude de la Religion Grecque Antique, KERNOS Suppl. 7, 1998), 127–39.
5. S. Deacey, "Herakles and His 'Girl': Athena, Heroism and Beyond," in *Herakles and Hercules. Exploring a Graeco-Roman Divinity,* ed. L. Rawlings and H. Bowden (Swansea: The Classical Press of Wales, 2005), 37–50.

6. On Herakles as guardian and savior, especially his presence on amulets and protection in everyday activities, see A. M. Nicgorski, "The Magic Knot of Herakles, the Propaganda of Alexander the Great and the Tomb II at Vergina," in Rawlings and Bowden, *Herakles and Hercules,* 97–128, 100–101.
7. Herakles was the founder of the Olympic Games (Pindar *Ol.* 2.3–4; 3.10–22; 6.67–9; 10. 43–59), thought to have brought the olive tree to make the crowns (Pindar *Ol.* 3.13–6; Paus. 5.7.7) and measured the length of the Altis stadium (Pindar *Ol.* 10.43–6), while he also participated in the Games (Diod. 4.14.2).
8. E. Stafford, "Vice or Virtue: Herakles and the Art of Allegory," in Rawlings and Bowden, *Herakles and Hercules,* 78–81.
9. Herodotus (2.4.4) describes the ancient roots of Herakles and agrees with the dual worship of Herakles both as a god and as a hero. On the complexity of Herakles's cult, see R. Vollkommer, *Herakles in the Art of Classical Greece* (Oxford: Oxford University Press Monograph no. 25, 1988), 83–86. On the cult of Hercules in Central Italy, see G. Bradley, "Aspects of the Cult of Hercules in Central Italy," in Rawlings and Bowden, *Herakles and Hercules,* 129–151, and for Herakles and the city, see S. Georgoudi, "Héraclès dans les pratiques sacrificielles des cites," in Bonnet et al., *Le Bestiaire d'Héraclès.*
10. On the earliest representations of Herakles in the visual arts from around and after 700 B.C. and the transformation of his image, see Brommer, *Heracles: The Twelve Labors,* 3, 65–66; *LIMC* 5 Herakles 187 (J. Boardman). For an overview of the history of research regarding Herakles, see Padilla, *The Myths of Herakles,* vii–ix, 19–20.
11. *LIMC* 4 Herakles 729–30 (J. Boardman).
12. Ptol. Chenn, in Phot. *Bibl.* 147a–b narrates how Herakles lost a finger to the Nemean Lion.
13. On a statue of a "dwarf Herakles" in Arcadia, see Paus. 8.31.3 (quoting Onomacritus).
14. Pindar *pae.* 20.13 Snell/Maehler. Because of these qualities Herakles assumed in Boeotia the characteristics of a local underworld hero, Charops, i.e., the bright-eyed (Paus. 9.34.5).
15. See, e.g., plastic oinochoe with Herakles's head ca. 480–60 B.C.: *LIMC* 4 Herakles 743, no. 253 (O. Palagia); J.D. Beazley, *Attic Red-Figure Vase-Painters,* 2nd ed. (Oxford: Clarendon Press), 1593.5 (subsequent references to this work are designated *ARV*²). See also E.A. Mackay, "The Hairstyle of Herakles," in *Essays in Honor of Dietrich von Bothmer,* ed. A.J. Clark and J. Gaunt, 203–210 (Amsterdam: Allard Pierson, 2002).
16. N. Loraux, "Herakles: The Super-Male and the Feminine," in *Before Sexuality: The Construction of the Erotic Experience in Ancient Greek World,* ed. D.M. Halperin, J.J. Winkler, and F.I. Zeitlin, 21–52 (Princeton, NJ: Princeton University Press, 1990), and counterargument by L. Llewellyn-Jones, "Herakles Re-Dressed: Gender, Clothing and the Construction of a Greek Hero," in Rawlings and Bowden, *Herakles and Hercules,* 51–69. On his melancholic nature, see Plut. *Lys.* 2.5.
17. M. Jameson, "The Family of Herakles in Attika," in Rawlings and Bowden, *Herakles and Hercules,* 15.
18. E. Stafford, "Vice or Virtue," 82–89.
19. Pindar *Nemean* 1. Uhlenbrock, *Herakles: Passage,* 10, 29–33; Vollkommer, *Herakles in the Art,* 1988, 79–82; B. Effe, "Heroische Grösse: Der Funktionswandel des Herakles-Mythos in des griechisch-römisch Literatur," in *Herakles/Hercules—*

Metamorphosen des Heros in ihrer medialen Vielfalt, vol. 1, ed. R. Kray and S. Oettermann (Basel: Stroemfeld Verlag, 1994), 16–17.

20. Effe, "Heroische Grösse," 16, comments on the marginal references of Herakles in the Homeric epics and the reduced ambiguity and positive view of the hero both in Hesiod and in art of the seventh and later sixth centuries B.C.: *LIMC* 4 Herakles 730–1 (J. Boardman). For Herakles in the literary sources, see also Padilla, *The Myths of Herakles,* 6–10.
21. *LIMC* 5 Geras 180–82 (H. A. Shapiro); Flacelière and Devambez, *Héraclès, Images et Récits,* 113–14, pl. 19; Schefold, *Die Urkönige,* 173–74, figs. 211–12; G. Hafner, "Herakles, Geras, Ogmios," *Jahrbuch des Römisch-germanischen Zentralmuseums, Mainz* 5 (1958): 139–53.
22. Vollkommer, *Herakles in the Art,* 61–78 (theater), 79–82 (philosophy) with more bibliography; Effe, "Heroische Grösse," 17–18; Silk 1985; K. Galinsky, *The Herakles Theme: The Adaptations of the Hero in Literature from Homer to the Twentieth Century* (Oxford: Blackwell, 1972), 40–80 (tragedy), 81–100 (comedy), 101–25 (philosophy). On Herakles and satyr drama, see *Das griechische Satyrspiel,* ed. R. Krumeich, N. Pechstein, and B. Seidensticker (Darmstadt: Wissenschaftliche Buchgesellschaft, 1999), esp. 259–76. Also, on Herakles in Euripides, see Uhlenbrock, *Herakles: Passage,* 23–28.
23. Effe, "Heroische Grösse," 18–20.
24. *LIMC* 4 Herakles 731 (J. Boardman). On the change of function of Herakles's myth in Greek and Roman literature, see Effe, "Heroische Grösse," 15–23.
25. For example, the lost painting by Nearchos depicted Herakles "saddened with remorse": Pliny *NH* 35.141; *LIMC* 4 Herakles 835–36, no. 1685 (J. Boardman).
26. For Herakles's music recitals on Attic vases and a possible connection with the Peisistratids, see J. Boardman, *The History of Greek Vases* (London: Thames & Hudson, 2001), 202–9; H. A. Shapiro, *Art and Cult under the Tyrants in Athens* (Mainz: P. von Zabern, 1989), 158–60. On Herakles and his instructors, see Flacelière and Devambez, *Héraclès, Images et Récits,* 78, pl. 3; Schefold, *Die Urkönige,* 132–34; and Schefold, *Gods and Heroes,* 94; *LIMC* 4 Herakles 810–17 and 833 (J. Boardman). Cf. Padilla, *The Myths of Herakles,* 33.
27. Especially for Herms decorating baths and palaestras, see *LIMC* 4 Herakles (O. Palagia), 781–86. On Herakles as an engineer of waterworks, see C. A. Salowey, "Herakles and the Waterworks. Mycenaean Dams, Classical Fountains, Roman Aqueducts," in *Archaeology in the Peloponnese: New Excavations and Research,* ed. A. K. Sheedy, 77–94 (Oxford: Oxbow, 1994).
28. Xen. *Memorabilia* 2.1.21–34 (Prodikos frag. 2 DK). On Herakles's Choice, see Stafford, "Vice or Virtue," and comments by H. Kloft, "Herakles als Vorbild: Zur politischen Funktion eines griechischen Mythos in Rom," in Kray and Oettermann, *Herakles/Hercules,* 32–34; and Galinsky, *The Herakles Theme,* 101–4.
29. See, e.g., drinking with the centaur Pholos, attacking Auge at Tegea.
30. St. Basil, *On the Value of Greek Literature* 5.55–77. Cf. Herodoros's interpretation of the club of Herakles as "club of the strong soul" that helps him overcome any desires and live a philosopher's life (FHG I nos. 1–4 and 13–37), and the Cynics' praise of Herakles for his virtue: Vollkommer, *Herakles in the Art,* 79–80.
31. Flacelière and Devambez, *Héraclès, Images et Récits,* 111–12, pl. 18 (Louvre G229; *ARV*² 289.3); Schefold, *Die Urkönige,* 189, fig. 231 (Oxford 322, *ARV*² 627.1); *LIMC* 4 Herakles 834–36 (J. Boardman); Vollkommer, *Herakles in the Art,* 1988, 31–39.

32. Boston, Museum of Fine Arts 08.34d, *LIMC* 4 Herakles 824, no. 1557 (J. Boardman). Cf. Roman glass intaglios of Herakles bound by Eros, e.g., Vienna Kunsthist. Mus. IX B 656, first century B.C.; Schefold, *Die Urkönige*, 204, fig. 250.
33. Such an approach explains why the Knot of Herakles was used as a metaphor of fertility and promise of success in sexual unions. The double interlocking loops that secured Herakles's lion skin around his shoulders was thus perceived as a symbol of endurance, and an apotropaic sign of its own, thanks to the hero's bravery, strength, and accomplishments in all fields. For example, a bride's dress is tied with the Herculean knot, so that when untied by the groom they enjoy fertility equal to that of Herakles; similarly, Zeus's and Hera's bodies are bound together in their union, forming the knot of Herakles: *LIMC* 4 Herakles 729 (J. Boardman); Nicgorski, "Magic Knot"; A. Geus, "Patron und Superlativ: Hercules in Biologie, Medizin und Pharmazie," in Kray and Oettermann, *Herakles/Hercules*, 276–78. On Herakles as erastes/eromenos in the literary sources, see Padilla, *The Myths of Herakles*, 28.
34. *LIMC* 4 Herakles 731, 825–26 (J. Boardman); Vollkommer, *Herakles in the Art*, 87–90.
35. Uhlenbrock, *Herakles: Passage*, 8; Brommer, *Heracles: The Twelve Labors*, 55–64. On the cult of Herakles, see 86; for Herakles's cult in Boeotia, see E. Metropoulou, "He latreia tou Herakle ste Boioteia," *Epeterida Boiotikon Meleton* 3 (2000): 624–81.
36. Herodotus's reminder of Leonidas's lineage may have been an intentional attempt to associate Herakles's success in his impossible labors with Leonidas's heroic achievement in the minds of his audience: H. Bowden, "Herakles, Herodotos and the Persian Wars," in Rawlings and Bowden, *Herakles and Hercules*, 9; Vollkommer, *Herakles in the Art*, 88. On Laconian vases depicting Herakles and possibly reflecting Bathykles's throne for Apollo Amyklaios, see J. M. Woodward, "Bathycles and the Laconian Vase-Painters," *Journal of Hellenistic Studies* 52 (1992): 27–31; J. Boardman, "For You Are the Progeny of Unconquered Herakles," in *Philolakon: Lakonian Studies in Honour of Hektor Catling*, ed. H. W. Catling and J. M. Sanders, 25–29 (London: BSA, 1992). On Laconian iconography, see M. Pipili, *Laconian Iconography of the Sixth Century* (Oxford: Oxford University Press, 1987).
37. On Herakles as ancestor of families from Corinth, Thespiai, Sparta, Thebes, Epeiros, Herakleia Pontou and Lydia, see Vollkommer, *Herakles in the Art*, 88 with primary sources. See also A. Arvanitake, *Heroas kai Pole. To Paradeigma tou Herakles ten Archaike Eikonographia tes Korinthou* (Thessaloniki: University Press, 2006); Padilla, *The Myths of Herakles*, 6–7.
38. F. Brommer, *Vasenlisten zu griechischen Heldensage* (Marburg: N. G. Elwert, 1956), 1–123. On Herakles and the Peisistratids, see J. Boardman, "Herakles, Peisistratos and Sons," *Revue Archéologique* (1972): 52–72; Boardman, 1975, 1–12; Boardman, 1978, 227–34; counterargument by Shapiro, *Art and Cult*, 159–63. See also Kloft, "Herakles als Vorbild," 27. On the Archaic Pediments from the Acropolis, see J. M. Hurwit, *The Athenian Acropolis: History, Mythology and Archaeology from the Neolithic Era to the Present*, repr. (Cambridge: University Press, 2001), 105–21.
39. According to Herodotus (1.60), Peisistratos dressed up a young girl named Phye as Athena as part of his theatrical return to Athens. See comments by Shapiro, *Art and Cult*, 14–15; Padilla, *The Myths of Herakles*, 10–13.
40. Following the sources, Herakles was given by either Athena or the Nymphs springs to wash himself after fighting Geryon (Schol. Pindar *Ol.* 12.27; Hesych. s.v. Her-

akleia Loutra), and he is often associated with hot baths. For example, a spring at Troizen was dedicated to his honor while the baths at Thermopylai were also sacred to Herakles (Paus. 2.32.4; Strabon 9.4.13): *LIMC* 4 Herakles 797–98 (J. Boardman).
41. Plat. *Hipparchos* 228b–c; Shapiro, *Art and Cult,* 159–60; see also here above ns. 25, 39.
42. See, e.g., Foce del Sele and Selinus metopes: J. Boardman, *Greek Sculpture—the Late Classical Period* (London: Thames & Hudson, 1995), 146–61, figs. 157, 160, 162.
43. See, e.g., the Siphnian treasury and the Athenian treasury at Delphi; the altar of Herakles in Olympia, bronze dedications, vases. On the metopes of the temple of Zeus at Olympia: Schefold, *Die Urkönige,* 137–40, figs. 166–77; *LIMC* 5 Herakles no. 1705.
44. J. Pedley, *Sanctuaries and the Sacred in the Ancient Greek World* (Cambridge: University Press, 2005), 140–43; Boardman 1991, 158, figs. 211–12. On the iconography of the struggle, see *LIMC* 5 Herakles 141–43 (J. Boardman, S. Woodford).
45. Cf. similar victory monuments of the fourth century B.C. in Delphi: Vollkommer, *Herakles in the Art,* 90. See also Schefold, *Gods and Heroes,* 153, correlating the pedimental sculpture with the First Sacred War, an idea originally expressed by H. W. Parke and J. Boardman, "The Struggle for the Tripod and the First Sacred War," *Journal of Hellenistic Studies* 77 (1957): 276–82.
46. Pedley, *Sanctuaries,* 143–44; Boardman 1991, 159–60, fig. 213; K. Hoffelner, "Die Metopen des Athener-Schatzhauses. Ein neuer Rekonstruktionsversuch," *Mitteilungen des Deutschen Archaologischen Instituts, Athenische Abteilung* 103 (1988): 77–117.
47. Among the various examples one can single out Onatas's bronze Herakles for the Thasians in Olympia, Myron's Group of Herakles, Athena and Zeus at Samos, and Alkamenes's colossal relief at Thebes. On the role of Herakles as a founder of colonies and his image on west Greek coins, see Vollkommer, *Herakles in the Art,* 88–89.
48. In fifth-century Athens, Herakles loses the monopoly of the single most outstanding hero. Images of Theseus, the Athenian answer to the Peloponnesian hero, steadily gain popularity and enjoy wide circulation, largely because they originate from Heraklean prototypes. During the Periklean building program, Herakles's imagery became part of Athens's facelift. His presence was clearly reduced on the Acropolis, and instead Athena and local Attic heroes were put in the spotlight; however, the deeds of Herakles were still attractive in the fifth century and formed part of the decoration of important buildings such as the Hephaisteion in the heart of Athens: *LIMC* 5 Herakles 181–83, no. 1706 (J. Boardman). On the Hephaisteion metopes, see Boardman 1985, 146, fig. 111. Quite interesting is the case of Herakles on Cyprus, where the strong influence of the Phoenician god Melkart resulted in an abundance of Herakles's statues, all rendered in the characteristic Cypriot style. See, for example, Karageorghis, 2000, 124–27, nos. 190–92, and *LIMC* 5 Herakles (Cypri) 192–96 (A. Hermary). On the relation of sport and Herakles-Melkart, see L. Boutros, *Phoenician Sport: Its Influence on the Origin of the Olympic Games* (Amsterdam: J. C. Gieben, 1981), 31–35. Also cf. Padilla, *The Myths of Herakles,* 3. For Herakles/Melkart in the western Mediterranean, see C. Bonnet, "Melqart in Occidente. Percorsi di Appropriazione e di acculturazione," in *Il Mediterraneo di Herakles. Studi e Ricerche,* ed. P. Bernardini and R. Zucca (Rome: Carocci, 2005), 17–28.

49. Padilla, *The Myths of Herakles,* 16–18; Bowden, "Herakles, Herodotos," esp. 4–9; Jameson, "Family of Herakles," 16–17, 22, on the cult of Herakleidai as an Attic phenomenon. On Marathon and the Herakleion, see Petrakos 1995, 50–52; also, Herodotus 6.108; *IG* I³ 1015bis; Paus. 1.32.4. Notice the time- and case-specific representation of Herakles at the sanctuary of Chryse on the island of Lemnos in the fifth century B.C., coinciding with the Athenian attempt to present Herakles as a founder of that sanctuary and thus gain the support of the island: Vollkommer, *Herakles in the Art,* 89.
50. Paris, Louvre G 341, *ARV*² 601.22, 1661; Boardman, *History of Greek Vases,* 272, fig. 300; idem 2005; D. Castriota, "Feminizing the Barbarian and Barbarizing the Feminine: Amazons, Trojans, and Persians in the Stoa Poikile," in *Periklean Athens and Its Legacy,* ed. J Barringer and J. Hurwit, 89–102 (Austin: University of Texas Press, 2005); M. D. Stansbury-O'Donnel, "Polygnotos's *Ilioupersis*: A New Reconstruction," *American Journal of Archaeology* 93 (1989): 203–15.
51. On Herakles in Classical art, see Vollkommer, *Herakles in the Art,* 1–19 (Labors). On the metopes of the temple of Zeus at Olympia, see Schefold, *Die Urkönige,* 138–39, figs. 166–77; *LIMC* 5 Herakles no. 1705 (J. Boardman); Pedley, *Sanctuaries,* 125–26. Herakles's labors decorated the pediments of the temple of Herakles at Thebes by Praxiteles, now lost (Paus. 9.11.6), while quite famous was the statue group by Lysippos for the city of Alyzia (Akarnania), which was eventually moved to Rome (Strabon 10.2.21): Uhlenbrock, *Herakles: Passage,* fig. 34; *LIMC* 5 Herakles nos. 1709, 1710 (J. Boardman).
52. On Herakles and Athena, see *LIMC* 5 Herakles 143–54 (J. Boardman): his patron and protector helps him in labors, throws a rock to snap him out of madness (Eur. *Herc.* 1001–8; Paus. 9.11.2), and gives him a robe as an apotheosis gift (Diod. 4.14.3; Apoll. *Bibl.* 2[71].4.11). On Herakles and Athena, see Deacey, "Herakles and His 'Girl.'"
53. This wedding demonstrates Herakles's divine status, and it is interesting to note that in fifth-century works Herakles appears explicitly rejuvenated; *LIMC* 5 Herakles 160–65 (A.-F. Laurens).
54. *LIMC* 5 Herakles 100–11 (G. Kokkorou-Aleuras); Schefold, *Die Urkönige,* 163–66.
55. *LIMC* 5 121–32 (J. Boardman). On Herakles iconography in Classical art, see Vollkommer, *Herakles in the Art,* 91–93. On Herakles's apotheosis, see also Schefold, *Die Urkönige,* 221–29.
56. See also Xenophon *Symp.* 8.29, where Herakles is praised for his soul rather than his body.
57. See, e.g., tetradrachms showing Alexander the Great with a lion skin produced posthumously, Vollkommer, *Herakles in the Art,* 88, fig. 103; Uhlenbrock, *Herakles: Passage,* fig. 24; L. Rawlings, "Hannibal and Hercules," in Rawlings and Bowden, *Herakles and Hercules,* 165, fig. 2. On Herakles as political model, see Kloft, "Herakles als Vorbild," 27; Vollkommer, *Herakles in the Art,* 88–89.
58. Nicgorski, "Magic Knot," 105–12.
59. Indicative of this trend is the case of the Carthaginian general Hannibal Barca. Associated with Herakles/Melkart through literary and numismatic examples, Hannibal fashioned himself as a new Herakles-Alexander to promote an image familiar to Greeks and Italians alike: Rawlings, "Hannibal and Hercules," esp. 172; A. Campus, "*Herakles,* Alessandro, Annibale," in Bernardini and Zucca, 201–21.
60. For example, the Pergamon rulers purposefully promoted myths that portrayed Herakles as the founder of their city while weakening the role of Telephos and

Auge. Similarly, the adventures of Herakles that occurred in the East, such as the Amazonomachy and his Service to Omphale, were largely popular in the Hellenistic art: Vollkommer, *Herakles in the Art*, 88–89; Schefold, *Die Urkönige*, 207–17; *LIMC* 5 Herakles 191 (J. Boardman).

61. *LIMC* 4 Herakles 781–86 (O. Palagia).
62. For the statues of Herakles, see recently S. Kansteiner, *Herakles. Die Darstellungen in der Grossplastik der Antike* (Köln: Böhlau, 2000).
63. *LIMC* 4 Herakles 747, no. 319 (O. Palagia); O. Palagia, "The Hope Herakles Reconsidered," *Oxford Journal of Archaeology* 3 (1984): 107–26; Shapiro, *Art and Cult*, 162; Padilla, *The Myths of Herakles*, 22–25.
64. *LIMC* 4 Herakles 747–49 (O. Palagia); Boardman, *Greek Sculpture*, fig. 75.
65. New York, Metropolitan Museum of Art 50.11.4; *RVAp* 1.266, 47, attributed to the Group of Boston 00.348. Schefold, *Die Urkönige*, 228, fig. 280; *LIMC* 4 Herakles (O. Palagia) 745, no. 271.
66. *LIMC* 4 Herakles (O. Palagia) 745–46: Alternatively, the Apulian krater may reflect the Albertini Herakles, a type echoing a fourth-century B.C. statue perhaps set in South Italy, that was widely copied in antiquity and may have been one of the cult statues of Herakles in Rome.
67. *LIMC* 4 Herakles 751 (O. Palagia). Cf. the case of the Chiaramonti Herakles, which had a great appeal among the Roman emperors, 752–53, 753–56. Roman copies of the Copenhagen/Dresden Herakles are indicative of the tendency to represent the hero more as a mortal than as a demigod already in the middle of the fourth century B.C., the date of the original statue this copy imitates. Herakles now possesses a rather flabby body, heavy shoulders, stocky legs, and less-toned muscles. He is aged, with short hair and beard. Leaning against his club and with one hand on his waist, Herakles pauses between labors or relaxes after their completion. This is a commemoration of the hero's rest rather than a proud allusion to his achievements; 762–69.
68. On Lysippos, see P. Moreno, *Lisippo: L'arte e la fortuna* (Milan: Fabbri, 1995); Uhlenbrock, *Herakles: Passage*, 11.
69. R. M. Schneider, "Der Hercules Farnese," in *Meisterwerke der antiken Kunst*, ed. L. Giuliani, 136–57 (Munich: Beck, 2005); Vollkommer, *Herakles in the Art*, 79, fig. 99; Schefold, *Die Urkönige*, 140–41; *LIMC* 4 Herakles 762–70, no. 702 [Farnese] (O. Palagia) for variations of this type, e.g., Herakles holding a cup, fully dressed. Cf. Herakles crowning himself and its impact in later periods: C. C. Vermeule, "Herakles Crowning Himself: New Greek Statuary Types and Their Place in Hellenistic and Roman Art," *Journal of Hellenistic Studies* 77 (1957): 283–99. On the type of Weary Herakles, see Vermeule, "The Weary Herakles of Lysippos," *American Journal of Archaeology* 79 (1975): 323–32.
70. The temple of Herakles near Custos Flaminius is thought to have received the Lysippean bronze from Taras (Livy 38.35.4). *LIMC* 4 Herakles 773–74 [no. 931 casket] (O. Palagia); Boardman, *Greek Sculpture*, fig. 40: bronze statuette of a version of Herakles sitting on upturned basket (Ny Carlsberg 3362). For the fate of the statue in Constantinople and Byzantine sources, see below 29–33. On Herakles statue at Taras, see K. Stähler, "Zu Lysipps Herakles in Tarent," *Boreas* 20 (1997): 43–47.
71. *LIMC* 4 Herakles 774–77 (O. Palagia) with more bibliography and variations. Boardman, *Greek Sculpture*, fig. 41: bronze statuette of a version of Herakles Epitrapezios from Pompeii (Naples 2828); Uhlenbrock, *Herakles: Passage*, fig. 35. On

Herakles as a symposiast, see H. Mommsen, *Der gelagerte Herakles* (Berlin: De Gruyter, 1971); *LIMC* 4 Herakles 777–78 (O. Palagia) and 817–21 (J. Boardman).
72. For Herakles as a stock character in Aristophanes's comedies and his gluttony, see K. Dover, *Aristophanes. Frogs* (Oxford: Clarendon Press, 1993), 10 (*Av.* 1565–1693; *Frogs* 55–65, 105–7).
73. *LIMC* 4 Herakles 770 (O. Palagia).
74. *LIMC* 4 Herakles 770–72 (O. Palagia). Cf. the type of Herakles reclining as a symposiast, represented aged and occasionally accompanied by Dionysos. The banqueting Herakles enjoyed great popularity in ancient Greece and Rome, as the numerous variations attest, and it gained a particularly funerary connotation in the Imperial period: 777–79.
75. According to the sources, the cult of Herakles was introduced in the Ara Maxima at the Forum Boarium by the Greek Evander (Verg. *Aen.* 8, 185–272). The cult of Herakles Invictus was also located near the Circus Maximus (*CIL* I^2 244 = *Inscriptiones Italiae* 13.2.191), while Livy (9.44.16) mentions a statue of Herakles on the Capitoline Hill in the fourth century B.C. See also Kloft, "Herakles als Vorbild," 29, on the Archaic temples and sculptural decoration including terra-cotta statues of Athena and Herakles of ca. 530 B.C. For more comments, see Galinsky, *The Herakles Theme*, 126–27.
76. See, e.g., the sarcophagus from ca. 150 A.D., Velletri Antiquarium: Schefold, *Die Urkönige*, 147–48, fig. 186. Herakles's labors on sarcophagi of the late second to fourth centuries A.D. symbolize the reward of immortality through courageous achievements. See also Uhlenbrock, *Herakles: Passage*, 15.
77. *LIMC* 4 Herakles 790–96 (J. Boardman and O. Palagia); Kloft, "Herakles als Vorbild," 30–32.
78. Uhlenbrock, *Herakles: Passage*, 14–15, e.g., the denarius of Pomponius Musa, 68–66 B.C. (fig. 40). On Hercules for the plebs, see Kloft, "Herakles als Vorbild," 30 and 34–37. On Hercules in Vergil's *Aeneid* and its impact on Roman heritage, see L. Nees, "Theodulf's Mythical Silver Hercules Vase, Poetica Vanitas, and the Augustinian Critique of the Roman Heritage," *Dumbarton Oaks Papers* 41 (1987): 443–51. Also, see U. Huttner, "Marcus Antonius und Herakles," in *Rom und der griechische Osten. Festschrift für Hatto H. Schmitt zum 65. Geburtstag*, ed. Ch. Schubert and K. Brodersen, 103–12 (Stuttgart: F. Steiner Verlag, 1995), for the association of Marcus Antonius and Herakles. For the treatment of Herakles in Roman literature, see Galinsky, *The Herakles Theme*, 126–84.
79. See Caligula's attempt to deification, Kloft, "Herakles als Vorbild," 36 (Philon Alex. *Leg. Ad. Gai* 78f).
80. E.R. OKell, "*Hercules Furens* and Nero: The Didactic Purpose of Senecan Tragedy," in Rawlings and Bowden, 185–204; Kloft, "Herakles als Vorbild," 36; Uhlenbrock, *Herakles: Passage*, 35–38; Galinsky, *The Herakles Theme*, 167–84. On *Hercules stoicus*, see Effe, "Heroische Grösse," 21.
81. O. Hekster, "Propagating Power: Hercules as an Example for Second-Century Emperors," in Rawlings and Bowden, 205, fig. 1 (Palazzo Massimo, Rome).
82. Hekster, "Propagating Power," 205–6.
83. Hekster, "Propagating Power," 206–9: e.g., Antoninus Pius's Dodekathlos series, followed by Gordian 3 (e.g., Herakles faces Eurystheus: Schefold, *Die Urkönige*, 136, fig. 163), and Postumus: *LIMC* 4 Herakles 825–26 (J. Boardman).
84. *LIMC* 4 Herakles 790–96 (J. Boardman and O. Palagia); Kloft, "Herakles als Vorbild," 36–37.

85. Hekster, "Propagating Power," 205–14, figs. 3–4, analyzes the different means Roman emperors used while associating themselves with Hercules and concludes to four large categories: literary and rhetorical, iconographical, symbolic, and ceremonial.
86. Kloft, "Herakles als Vorbild," 41.
87. R. Rees, "The Emperors' New Names: Diocletian Jovius and Maximian Herculius," in Rawlings and Bowden, 223–35.
88. *LIMC* 5 Herakles 168–69, nos. 3381–83, 3385 (J. Boardman); E. Harrison, "The Constantinian Portrait," *Dumbarton Oaks Papers* 21 (1967): 96.
89. See, for example, Maguire's study on profane aesthetic in Byzantium (H. Maguire, "The Profane Aesthetic in Byzantine Art and Literature," *Dumbarton Oaks Papers* 53 [1999]: 189–205), and how the case of Herakles on a secular casket invites questions related to nudity and its significance in Byzantine art. Maguire concludes that nudity was often used as a formula for freakish, grotesque, or humorous characters, and that pagan idols were no less profane than the works produced in their own era.
90. S.G. Bassett, "Historiae Custos: Sculpture and Tradition in the Baths of Zeuxippos," *American Journal of Archaeology* 100 (1996): 491–506, especially 501 for reuse of Herakles statues as a "paragon of physical fitness."
91. S.G. Bassett, "The Antiquities in the Hippodrome of Constantinople," *Dumbarton Oaks Papers* 45 (1991): 87–96.
92. A. Cutler, "The De Signis of Nicetas Choniates: A Reappraisal," *American Journal of Archaeology* 72 (1968): 113–18.
93. *Supra* n. 92.
94. Pliny *NH* 35.57; Ovid (*Fasti* 6.209) mentions Herakles Magnus Custos at the Circus Flaminius, and the association between the hero and the circus was maintained in late antiquity; see, e.g., some fifth-century A.D. coins with "Herakles Hippodromos": Bassett, "Antiquities in the Hippodrome," 91, n. 39.
95. L. James, "Pray Not to Fall into Temptation and Be On Your Guard": Pagan Statues in Christian Constantinople," *Gesta* 35 (1996): 13–18, argues that the Hippodrome in Constantinople was decorated with pagan statues, set next to imperial portraits and Christian symbols.
96. See, e.g., Uhlenbrock, *Herakles: Passage*, figs. 56–61.
97. K. Weitzmann, "The Heracles Plaques of the St. Peter's Cathedra," *Art Bulletin* 55 (1973): 1–37; and Weitzmann, "An Addendum to the Heracles Plaques of the St. Peter's Cathedra," *Art Bulletin* 56 (1974): 248–51.
98. Uhlenbrock, *Herakles: Passage*, 15–16. For Hercules's representations within Christian context, see J. Huskinson, "Some Pagan Mythological Figures and Their Significance in Early Christian Art," *Papers of the British School at Rome* 42 (1974): 68–97, esp. 77, 81–82, 86. On Hercules as a rival of Christ, see Galinsky, *The Herakles Theme*, 189–90.

Chapter 10

1. G. Strawson, "The Sense of Self," in *From Soul to Self*, ed. M.J.C. Crabbe (New York: Routledge, 1999), 126.
2. The definition of the self can be conceived of in the following ways: individual distinctiveness; a typical or individual psychological structure; a psychological

structure made up of parts that carry on intra- and/or interpersonal dialogue; our essence as persons or personalities abstracted from the world; personal identity over time; the possession of a first personal point of view; an autonomous, self-legislating unitary "I"; or grounded on moral agency and responsibility, capable of making decisions that conform to the standards and ideals that define a good human life. See C. Gill, *The Structured Self in Roman and Hellenistic Thought* (Oxford: Oxford University Press, 2006), xiv, and 1993: 1.

3. A. MacIntyre, *After Virtue: A Study in Moral Theory*, 2nd ed. (London: Duckworth, 1985); B. Williams, *Shame and Necessity* (Berkeley: University of California Press, 1993); C. Gill, *Character and Personality in Greek Epic and Tragedy* (Oxford: Oxford University Press, 1996), 6–7, 8–15.
4. Gill 1993: 6.
5. Conybeare 2000: 132, says of Augustine's *Confessions* that "what has come to constitute our vocabulary of personhood is . . nascent in this period." She is following scholars who have concluded that Augustine anticipates the *cogito*-argument of Descartes (Taylor 1989: 130–32); "invents" the will (A. Dihle, *The Theory of the Will in Classical Antiquity* [Berkeley: University of California Press, 1986]; C. H. Kahn, "Discovering the Will: From Aristotle to Augustine," in *The Question of Eclecticism: Studies in Later Greek Philosophy*, ed. J. M. Dillon and A. A. Long [Berkeley: University of California Press, 1988]; J. Rist, *Augustine: Ancient Thought Baptized* [Cambridge: Cambridge University Press, 1994], 148–201); and is the first to display the inward turn (Taylor, 1989, 132; Cary, 2000).
6. C. Gill, "The Ancient Self: Issues and Approaches," in *Ancient Philosophy of the Self*, ed. P. Remes and J. Sihvola (Helsinki: Springer, 2008), 40.
7. H. Chadwick, "Philosophical Tradition and the Self," in *Late Antiquity: A Guide to the Post Classical World*, ed. G. W. Bowersock, P. Brown, and O. Grabar (Cambridge, MA: Belknap Press, 1999), 60; cf. R. Sorabji, *Self* (Chicago: University of Chicago Press, 2006), 5; and "Greco-Roman Varieties of Self," in Remes and Sihvola, *Ancient Philosophy of the Self*, 16.
8. Gill 1993: 2.
9. Gill 1993: 13.
10. Gill 1993: 15–16.
11. Gill 1993: 11–12; M. Mastrangelo, "Oedipus and Polyneices: Characterization and the Self in Sophocles' *Oedipus at Colonus*," *Materiali e Discussioni Per L'Analisi Dei Testi Classici*, 44 (2000): 36, 43–45.
12. Sorabji, *Greco-Roman*, 16–26.
13. Sorabji, *Greco-Roman*, 15.
14. Sorabji, *Greco-Roman*, 15.
15. Gill, *The Structured Self*, xiv.
16. Gill, "The Ancient Self," 45.
17. Snell 1960: 8–14, 26; Adkins: 1960; Taylor: 1989, 20, 37, 44, 115–24.
18. Gill, *Character and Personality*, 31.
19. *Iliad* 11.403–10, 17.90–105, 21.552–70, 22.98–130.
20. Williams, *Shame and Necessity*; R. Padel, *In and Out of the Mind* (Princeton, NJ: Princeton University Press, 1992).
21. Williams, *Shame and Necessity*.
22. D. C. Dennett, "Conditions of Personhood," in *The Identities of Persons*, ed. A. O. Rorty (Berkeley: University of California Press, 1976).
23. Gill, *Character and Personality*, 251.

24. T. Irwin, *Plato's Ethics* (Oxford: Oxford University Press, 1995).
25. Gill, *Character and Personality*, 242–42.
26. 441d–444e and 589c–592b; Gill, *Character and Personality*, 243–46.
27. Gill, *Character and Personality*, 255.
28. *Alcibiades* 1.132c–133c; Gill, *Character and Personality*, 256.
29. J. Sihvola, "Aristotle on the Individuality of Self," in Remes and Sihvola, *Ancient Philosophy of the Self*, 125.
30. Gill, *Character and Personality*, 44.
31. *Nicomachean Ethics* 6.2, 1139b4–5.
32. *Nicomachean Ethics* 10.7–8.
33. Gill, *The Structured Self*, xv.
34. Gill, *The Structured Self*, 32–34.
35. S. Berryman, "Review of C. Gill: *The Structured Self in Hellenistic and Roman Thought*," *Journal of the History of Philosophy* 45, no. 2 (2007): 325.
36. Gill, "The Ancient Self," 48; M. Foucault, *The Care of the Self: History of Sexuality*, vol. 3 (London: Penguin, 1988), 4.
37. Foucault, *Care of the Self*, 42.
38. J. Perkins, "The Self as Sufferer," *Harvard Theological Review* 85, no. 3 (1992): 262; J.G. Fitch and S. McElduff, "Construction of the Self in Senecan Drama," *Mnemosyne* 55, no. 1 (2002): 21.
39. Perkins, "The Self as Sufferer," 262–66.
40. Perkins, "The Self as Sufferer," 261–62.
41. Gill, *The Structured Self*, 389–91.
42. P. Hadot, *Philosophy as a Way of Life: Spiritual Exercises from Socrates to Foucault*, trans. M. Chase (Oxford: Blackwell, 1995), 211–13.
43. Gill, "The Ancient Self," 49.
44. Perkins, "The Self as Sufferer," 262.
45. M. Burnyeat, "Idealism and Greek Philosophy: What Descartes Saw and Berkeley Missed," *Philosophical Review* 90 (1982): 3–40; Gill, "The Ancient Self."
46. C. Morris, *The Discovery of the Individual 1050–1200* (New York: Harper and Row, 1972), 11.
47. *Apotheosis* 309.
48. *Apotheosis* 308.
49. M. Mastrangelo, *The Roman Self in Late Antiquity: Prudentius and the Poetics of the Soul* (Baltimore: Johns Hopkins University Press, 2008), 12, 163.
50. *Hamartigenia* 769–74.
51. *Confessions* 10.17.26.
52. Cary 2000: 4, 10, and 78 argues for the self in Augustine conceived as an "inner space" and a "private world" that stands for a level of being. This metaphorical language of inner space is in Augustine, but it does not exclude the argument that the Greco-Roman self can be autonomous yet relational in its definition and function (*On the True Religion* 39.72; *Confessions* 7.16; Cary 2000: 138 and Taylor 1989: 129).
53. Taylor 1989: 137.
54. *Confessions* 10.8.12.
55. M.L. Humphries, "Michel Foucault on Writing and the Self in the *Meditations* of Marcus Aurelius and *Confessions* of St. Augustine," *Arethusa* 30, no. 1 (1997): 125–38, 131–33.
56. *Poem* 17; Trout 1999: 214; *Letter* 11.5; Conybeare 2000: 67, 87.

57. *Letters* 25.1 and 38.1; Taylor 1989: 131; Conybeare 2000: 140, 157.
58. B. Stock, "Ethical Values and the Literary Imagination in the Later Ancient World," *New Literary History* 29, no. 1 (1996): 248, 256.
59. *Commentary on First Alcibiades*, 20, 51; H. Chadwick, "Philosophical Tradition and the Self," in *Late Antiquity: A Guide to the Post Classical World*, ed. G. W. Bowersock, P. Brown, and O. Grabar (Cambridge, MA: Belknap Press, 1999), 73.
60. K. S.-L. Twu, "This Is Comforting? Boethius's Consolation of Philosophy, Rhetoric, Dialectic, and *Unicum Illud Inter Homines Deumque Commercium*," in *New Directions in Boethian Studies,* ed. N. H. Kaylor Jr. and P. E. Phillips (Kalamazoo, MI: Medieval Institute Publications, 2007), 35.
61. *Consolation of Philosoph.* 4m. 1.23–30; 5m 3.22–31; 5 pr. 46–56.
62. E. C. Sweeney, *Logic, Theology, and Poetry in Boethius, Abelard, and Alan of Lille* (New York: Palgrave Macmillan, 2006), 39.
63. M. McCormick, "Sounding Early Medieval Holiness," in *The Long Morning of Medieval Europe: New Directions in Early Medieval Studies,* ed. J. R. Davis and M. McCormick (Hampshire: Ashgate, 2008), 107.
64. C. Wickham, *Framing the Early Middle Ages: Europe and the Mediterranean, 400–800* (Oxford: Oxford University Press, 2005), 158.
65. G. Philippart and M. Trigalet, "Latin Hagiography before the Ninth Century: A Synoptic View," in *The Long Morning of Medieval Europe: New Directions in Early Medieval Studies,* ed. J. R. Davis and M. McCormick (Hampshire: Ashgate, 2008).
66. A. Angenendt, "*Donationes pro anima*: Gift and Countergift in the Early Medieval Liturgy," in Davis and McCormick, *The Long Morning,* 181, quoting H. Seiwert, "Ofer," in *Handbuch religionswissenschaftlicher Grundbegriffe*, vol. 4, ed. H. Cancik et al. (Stuttgart, 1998), 271.
67. B. Rosenwein, *To Be the Neighbor of St. Peter: The Social Meaning of Cluny's Property, 909–1049* (Ithaca, NY: Cornell University Press, 1989), 136–41, cited by T. Head, "The Early Medieval Transformation of Piety," in Davis and McCormick, *The Long Morning,* 159, rightly argues that clerics and laymen were not only making a financial deal for spiritual benefits. These transactions represent "the salvific effects of charity" (Angenendt, "*Donationes pro anima,*" 132).
68. G. H. Brown, *Bede the Venerable* (Boston: Twayne, 1987), 66.
69. Brown, *Bede the Venerable,* 65.
70. For more of Alcuin's sources such as Cassian, Caesarius of Arles, Gregory the Great, and others, see L. Wallach, *Alcuin and Charlemagne: Studies in Carolingian History and Literature* (Ithaca, NY: Cornell University Press, 1959): 236–47, 252–54.
71. Wallach, *Alcuin and Charlemagne,* 256–61.

BIBLIOGRAPHY

Adams, J. N. *The Latin Sexual Vocabulary*. Baltimore: Johns Hopkins University Press, 1982.
Algra, K., J. Barnes, J. Mansfeld, and M. Schofield, eds. *The Cambridge History of Hellenistic Philosophy*. Cambridge: Cambridge University Press, 1999.
Allbutt, T. C. *Greek Medicine in Rome*. New York: Benjamin Blom, 1970. Reprint of 1921 edition.
Allen, J. *Inference from Signs: Ancient Debates about the Nature of Evidence*. Oxford: Clarendon Press, 2001.
Allen, J. "Pyrrhonism and Medical Empiricism: Sextus Empiricus on Evidence and Inference." *Aufstieg und Niedergang der römischen Welt*, 2nd ser., 37, part 1 (1993): 646–90.
Allen, Th. W., and D. B. Monro, eds. *Homer. Homeri Opera in Five Volumes*. Oxford: Oxford University Press, 1920.
Amundsen, D. W. *Medicine, Society, and Faith in the Ancient and Medieval Worlds*. Baltimore: Johns Hopkins University Press, 1996.
Amundsen, D. W., and G. B. Ferngren. "The Perception of Disease and Disease Causality in the New Testament." *Aufstieg und Niedergang der römischen Welt*, 2nd ser., 37, part 3 (1996): 2934–56.
Angenendt, A. "*Donationes pro anima*: Gift and Countergift in the Early Medieval Liturgy." In *The Long Morning of Medieval Europe: New Directions in Early Medieval Studies*, edited by J. R. Davis and M. McCormick. Hampshire: Ashgate, 2008.
Annas, J. E. "Philosophical Therapy, Ancient and Modern." In *Bioethics: Ancient Themes in Contemporary Issues*, edited by M. G. Kuczewski and R. Polansky. Cambridge, MA: MIT Press, 2000.
Annoni, J. M., and V. Barras. "La découpe du corps humain et ses justifications dans l'Antiquité." *Canadian Bulletin for Medical History* 10 (1993): 185–227.
Anton, J. P. "Dialectic and Health in Plato's *Gorgias*: Presuppositions and Implications." *Ancient Philosophy* 1 (1980): 49–60.

Arbesmann, R. "The Concept of 'Christus medicus' in St. Augustine." *Traditio* 10 (1954): 1–28.
Aristophanes. *Clouds, Wasps, Peace*. Translated by Jeffrey Henderson. Cambridge, MA: Harvard University Press, 1998 (Loeb Classical Library).
Aristotle. *Minor Works*. Translated by W. S. Hett. Cambridge, MA: Harvard University Press, and W. Heinemann, 1936 (Loeb Classical Library).
Arvanitake, A. *Heroas kai Pole. To Paradeigma tou Herakles ten Archaike Eikonographia tes Korinthou*. Thessaloniki: University Press, 2006.
Baader, G., and R. Winau, ed. *Die Hippokratischen Epidemien: Theorie-Praxis-Tradition, Verhandlungen des Ve Colloque International Hippocratique, Berlin 10–15.9.1984, Sudhoffs Archiv*, Heft 27. Stuttgart: Franz Steiner, 1989.
Babbitt, F. C. *Plutarch: Moralia*. Vol. 2. Cambridge, MA: Harvard University Press, 1928.
Babbitt, F. C. *Plutarch: Moralia*. Vol. 5. Cambridge, MA: Harvard University Press, 1936.
Balme, D. M., ed. and trans. *Aristotle. History of Animals VII–X*. Cambridge, MA: Harvard University Press, 1991.
Barnes, J., ed. *The Complete Works of Aristotle*. 2 vols. Princeton, NJ: Princeton University Press, 1995.
Barnes, J., V. Barras, and J. Jouanna, eds. *Galien et la philosophie*. Vandoeuvres-Genève: Fondation Hardt, 2003.
Barringer, J., and J. Hurwit, eds. *Periklean Athens and Its Legacy*. Austin: University of Texas Press, 2005.
Barton, Carlin. *The Sorrows of the Ancient Romans: The Gladiator and the Monster*. Princeton, NJ: Princeton University Press, 1993.
Barton, T. *Power and Knowledge: Astrology, Physiognomics, and Medicine under the Roman Empire*. Ann Arbor: University of Michigan Press, 1994.
Bassett, E. O. "Plato's Theory of Social Progress." *International Journal of Ethics* 38 (1928): 467–77.
Bassett, S. G. "Historiae Custos: Sculpture and Tradition in the Baths of Zeuxippos." *American Journal of Archaeology* 100 (1996): 491–506.
Bassett, S. G. "The Antiquities in the Hippodrome of Constantinople." *Dumbarton Oaks Papers* 45 (1991): 87–96.
Beazley, J. D. *Attic Red-Figure Vase-Painters*. 2nd ed. Oxford: Clarendon Press, 1963.
Behr, C. A. *Aelius Aristides and the Sacred Tales*. Amsterdam: A. M. Hakkert, 1968.
Benveniste, E. "La doctrine medicale des Indo-Européens." *Révue de l'histoire des religions* 130 (1945): 5–12.
Bergren, A. "The Homeric Hymn to Aphrodite: Tradition and Rhetoric, Praise and Blame." *Classical Antiquity* 8 (1989): 1–41.
Bernardini, P., and R. Zucca, eds. *Il Mediterraneo di Herakles. Studi e Ricerche*. Rome: Carocci, 2005.
Berryman, S. "Aristotle on *Pneuma* and Animal Self-Motion." *Oxford Studies in Ancient Philosophy* 23 (2002): 85–97.
Berryman, S. "Review of C. Gill: *The Structured Self in Hellenistic and Roman Thought*." *Journal of the History of Philosophy* 45, no. 2 (2007): 324–25.
Bertier, J. "Enfants malades et maladies des enfants dans le *corpus hippocratique*." In *La maladie et les maladies dans la collection hippocratique*, edited by P. Potter, G. Maloney, and J. Desautels. Quebec: Éditions du Sphinx, 1990.

Bertier, J. "La médecine des enfants à l'époque impériale." *Aufstieg und Niedergang der römischen Welt*, 2nd ser., 37, part 3 (1996): 2147–2227.
Bieber, M. *Alexander the Great in Greek and Roman Art*. Chicago: Argonaut, 1964.
Bloch, E. 2002. "Hemlock Poisoning and the Death of Socrates: Did Plato Tell the Truth?" In *The Trial and Execution of Socrates: Sources and Controversies*, edited by T. Brickhouse and N. Smith. New York: Oxford University Press, 2002.
Boak, A.E.R. "The Theoretical Basis of the Deification of Rulers in Antiquity." *Classical Journal* 11, no. 5 (1916): 293–97.
Boardman, J. "Composition and Content on Classical Murals and Vases." In *Periklean Athens and Its Legacy*, edited by J. Barringer and J. Hurwit. Austin: University of Texas Press, 2005.
Boardman, J. "For You Are the Progeny of Unconquered Herakles." In *Philolakon: Lakonian Studies in Honour of Hektor Catling*, edited by H.W. Catling and J.M. Sanders. London: BSA, 1992.
Boardman, J. 2002. *Greek Sculpture—The Archaic Period*. London: Thames & Hudson, 2002.
Boardman, J. *Greek Sculpture—The Late Classical Period*. London: Thames & Hudson, 1995.
Boardman, J. "Herakles, Peisistratos and Sons." *Revue Archéologique* (1972): 57–72.
Boardman, J. *The History of Greek Vases*. London: Thames & Hudson, 2001.
Boardman, J. *The Oxford History of Classical Art*. New York: Oxford University Press, 1993.
Boardman, J., and H.W. Parke. "The Struggle for the Tripod and the First Sacred War." *Journal of the Hellenistic Society* 77 (1957): 276–82.
Bodel, J. "Dealing with the Dead: Undertakers, Executioners and Potter's Fields in Ancient Rome." In *Death and Disease in the Ancient City*, edited by V. Hope and E. Marshall. London: Routledge, 2000.
Bodel, J. "Death on Display: Looking at Roman Funerals." In *The Art of Ancient Spectacle*, edited by B. Bergmann and C. Kondoleon. New Haven, CT: Yale University Press, 1999.
Bodel, J. "Graveyards and Groves: A Study of the *lex Lucerina*." *American Journal of Ancient History* 11 (1994 [1986]).
Bodel, J. "The Organisation of the Funerary Trade at Puteoli and Cumae." In *Libitina e Dintorni: Atti dell" XI Rencontre franco-italienne sur l"épigraphie (Libitina, 3)*, edited by S. Panciera. Rome: Quasar, 2004.
Bonnet, C. "Melqart in Occidente. Percorsi di Appropriazione e di acculturazione." In *Il Mediterraneo di Herakles. Studi e Ricerche*, edited by P. Bernardini and R. Zucca. Rome: Carocci, 2005.
Bonnet, C., C. Jourdain-Annequin, and U. Pirenne-Delforge, eds. *Le Bestiaire d'Héraclès*. Liège: Centre International d'Étude de la Religion Grecque Antique (KERNOS Suppl. 7), 1998.
Borgeaud, P. *The Cult of Pan in Ancient Greece*. Translated by Kathleen Atlass and James Redfield. Chicago, 1988.
Bos, A.P. "*Pneuma* and Ether in Aristotle's Philosophy of Living Nature." *Modern Schoolman* 79 (2002): 255–76.
Bos, A.P. *The Soul and Its Instrumental Body: A Reinterpretation of Aristotle's Philosophy of Living Nature*. Leiden: Brill, 2003.

Bos, A. P., and R. Ferwerda. *Aristotle, On the Life-Bearing Spirit* (De Spiritu). Leiden: Brill, 2008.

Bosworth, A. B. *Conquest and Empire: The Reign of Alexander the Great*. Cambridge: Cambridge University Press, 1993.

Boutros, L. *Phoenician Sport: Its Influence on the Origin of the Olympic Games*. Amsterdam: J. C. Gieben, 1981.

Bowden, H. "Herakles, Herodotos and the Persian Wars." In *Herakles and Hercules: Exploring a Graeco-Roman Divinity*, edited by L. Rawlings and H. Bowden. Swansea: The Classical Press of Wales, 2005.

Bowersock, G. *Greek Sophists in the Roman Empire*. Oxford: Clarendon Press, 1969.

Boyarin, Daniel. "Paul and the Genealogy of Gender." *Representations* 41 (1993): 1–33.

Boylan, M. "The Digestive and 'Circulatory' Systems in Aristotle's Biology." *Journal of the History of Biology* 15 (1982): 89–118.

Bradley, G. "Aspects of the Cult of Hercules in Central Italy." In *Herakles and Hercules: Exploring a Graeco-Roman Divinity*, edited by L. Rawlings and H. Bowden. Swansea: The Classical Press of Wales, 2005.

Bradley, K. *Discovering the Roman Family*. Oxford: Oxford University Press, 1991.

Braidotti, Rosi. "Mothers, Monsters, and Machines." In *Writing on the Body: Female Embodiment and Feminist Theory*, edited by Katie Conboy, Nadia Medina, and Sarah Stanbury. 7th ed. New York: Columbia University Press, 1997.

Branham, Bracht Robert, ed. *Bakhtin and the Classics*. Chicago: Northwestern University Press, 2002.

Braund, S., and G. Most, ed. *Ancient Anger: Perspectives from Homer to Galen* (Yale Classical Studies, vol. 32). Cambridge: Cambridge University Press, 2003.

Brickhouse, T., and N. Smith, ed. *The Trial and Execution of Socrates: Sources and Controversies*. New York: Oxford University Press, 2002.

Brock, A. J. *Galen: On the Natural Faculties*. Cambridge, MA: Harvard University Press, 1916.

Brock, A. J. *Galen, On the Natural Faculties*. London: W. Heinemann, 1916.

Brock, R. "Sickness in the Body Politic: Medical Imagery in the Greek Polis." In *Death and Disease in the Ancient City*, edited by V. M. Hope and E. Marshall. London: Routledge, 2000.

Brock, S., and S. A. Harvey. *Holy Women of the Syrian Orient*. Berkeley: University of California Press, 1987.

Brommer, F. *Heracles: The Twelve Labors of the Hero in Ancient Art and Literature*. Translated by S. J. Schwarz. New York: A. D. Caratzas, 1986.

Brommer, F. *Vasenlisten zu griechischen Heldensage*. Marburg: N. G. Elwert, 1956.

Brothwell, D., and A. T. Sandison. *Diseases in Antiquity*. Springfield, IL: C. C. Thomas, 1967.

Brown, G. H. *Bede the Venerable*. Boston: Twayne, 1987.

Brown, P. *The Body and Society: Men, Women, and Sexual Renunciation in Early Christianity*. New York: Columbia University Press, 1988. Reprinted with a new introduction, 2008.

Bugh, G. R., ed. *The Cambridge Companion to the Hellenistic World*. Cambridge: Cambridge University Press, 2006.

Burke, P. F. "Malaria in the Greco-Roman World: A Historical and Epidemiological Survey." *Aufstieg und Niedergang der römischen Wel*, 2nd ser., 37, part 3 (1996): 2252–81.

Burkert, Walter. *Greek Religion: Archaic and Classical*, 2nd revised edition. Translated by John Raffan. London: Blackwell, 1987.
Burnet, J. *Platonis Opera: Phaedo*. Oxford: Clarendon Press, 1911.
Burnet, John, ed. *Platonis Opera*. Oxford: Oxford University Press, 1903.
Burnyeat, M. "Idealism and Greek Philosophy: What Descartes Saw and Berkeley Missed." *Philosophical Review* 90 (1982): 3–40.
Bury, J. B., ed. and trans. *Plato Laws Books I–VI*. Cambridge, MA: Harvard University Press, 1926.
Calame, C. *Les choeurs de jeunes flues en Grèce archaique*. 2 vols. Rome, 1992.
Campbell, G., ed. *The Grove Encyclopedia of Classical Art and Architecture*. 2 vols. Oxford: Oxford University Press, 2007.
Campus, A. "*Herakles*, Alessandro, Annibale." In *Il Mediterraneo di Herakles. Studi e Ricerche,* edited by P. Bernardini and R. Zucca. Rome: Carocci, 2005.
Caplan, Harry, trans. [*Cicero*], *Ad C. Herennium (Rhetorica ad Herennium)*. Cambridge, MA: Harvard University Press, 1954.
Carone, G. R. "Mind and Body in Late Plato." *Archiv für Geschichte der Philosophie* 87 (2005): 227–69.
Carratello, U. "Omnis Caesareo cedit labor amphitheatre! (not a Mart. *Spect. Lib.*)." *Giornale italiano di filogia* 18 (1965): 295–99.
Carson, A. *Eros the Bittersweet*. Princeton, NJ: Princeton University Press, 1985.
Carson, A. "Putting Her in Her Place: Woman, Dirt, and Desire." In *Before Sexuality: The Construction of Erotic Experience in the Ancient Greek World*, edited by D.M. Halperin. J.J. Winkler, and F.I. Zeitlin. Princeton, NJ: Princeton University Press, 1990.
Cartledge, Paul. *The Greeks: A Portrait of Self and Others*. Oxford: Oxford University Press, 2002.
Castelli, Elizabeth. "'I Will Make Mary Male': Pieties of the Body and Gender Transformation of Christian Women in Late Antiquity." In *Body Guards: The Cultural Politics of Gender Ambiguity*, edited by J. Epstein and K. Straub. New York: Routledge, 1991.
Castriota, D. "Feminizing the Barbarian and Barbarizing the Feminine: Amazons, Trojans, and Persians in the Stoa Poikile." In *Periklean Athens and Its Legacy,* edited by J. Barringer and J. Hurwit. Austin: University of Texas Press, 2005.
Chadwick, H. "Philosophical Tradition and the Self." In *Late Antiquity: A Guide to the Post Classical World*, edited by G.W. Bowersock, P. Brown, and O. Grabar. Cambridge, MA: Belknap Press, 1999.
Cherniss, H., and W.C. Helmbold. *Plutarch: Moralia*. Vol. 12. Cambridge, MA: Harvard University Press, 1957.
Cicero. *Selected Letters*. Translated by D.R. Shackleton-Bailey. Harmondsworth: Penguin, 1982.
Cixous, Helene. "The Laugh of the Medusa." Translated by Keith and Paula Cohen. *Signs* 1 (1976): 875–93.
Clark, E.A. "Holy Women, Holy Words: Early Christian Women, Social History, and the 'Linguistic Turn.'" *Journal of Early Christian Studies* 6 (1998): 413–30.
Clark, G. *Porphyry: On Abstinence from Killing Animals*. Ithaca, NY: Cornell University Press, 2000.
Clark, Gillian. "'In the Foreskin of Your Flesh': The Pure Male Body in Late Antiquity." In *Roman Bodies: Antiquity to the Eighteenth Century*, edited by A. Hopkins and M. Wyke, 43–53. London: The British School at Rome, 2005.
Clark, Gillian. *Women in Late Antiquity. Pagan and Christian Lifestyles*. Oxford: Oxford University Press, 1993.

Clark, Patricia. "Women, Slaves, and the Hierarchies of Domestic Violence: The Family of St. Augustine." In *Women and Slaves in Greco-Roman Culture: Differential Equation,* edited by S. R. Joshel and Sheila Murnaghan. London: Routledge, 1998.

Clarke, M. *Flesh and Spirit in the Songs of Homer.* Oxford: Clarendon Press, 1999.

Clay, J. *The Politics of Olympus: Form and Meaning in the Major Homeric Hymns.* Princeton, NJ: Princeton University Press, 1989.

Cohen, B. "Divesting the Female Breast of Clothes in Classical Sculpture." In *Naked Truths: Women, Sexuality, and Gender in Classical Art and Archaeology,* edited by A. O. Koloski-Ostrow and C. L. Lyons, 66–92. London: Routledge, 1997.

Cohen, B. "The Nemean Lionskin in Athenian Art." In *Le Bestiaire d'Héraclès,* edited by C. Bonnet, C. Jourdain-Annequin, and U. Pirenne-Delforge. Liège: Centre International d'Étude de la Religion Grecque Antique (KERNOS Suppl. 7), 1998.

Cohen, B., ed. *Not the Classical Ideal: Athens and the Construction of the Other in Greek Art.* Leiden: Brill, 2001.

Cohen, B., ed. "Representations of the "Other" in Athenian Art, c. 510–400 B.C." *Source* 15 (1995).

Cohen, D. *Law, Sexuality and Society.* Cambridge: Cambridge University Press, 1991.

Cohn-Haft, L. *The Public Physicians of Ancient Greece.* Northampton, MA: Department of History, Smith College, 1956.

Coleman, K. M. "Fatal Charades: Roman Executions Staged as Mythological Enactments." *Journal of Roman Studies* 80 (1990): 44–73.

Coleman, K. M. *M. Valerii Martialis Liber Spectaculorum.* Oxford: Oxford University Press, 2006.

Connolly, Joy. "Virile Tongues: Rhetoric and Masculinity." In *Companion to Roman Rhetoric,* edited by William J. Dominik and Jon Hall. Malden, MA: Blackwell, 2007.

Conrad, L., M. Neve, V. Nutton, R. Porter, and A. Wear. *The Western Medical Tradition 800 B.C. to A.D. 1800.* Cambridge: Cambridge University Press, 1995.

Coolidge, F. P. "The Relation of Philosophy to Σωφροσύνη: Zalmoxian Medicine in Plato's *Charmides.*" *Ancient Philosophy* 13 (1993): 23–36.

Cooper, J. M., ed. *Plato: Complete Works.* Indianapolis: Hackett, 1997.

Corbeill, A. *Nature Embodied: Gesture in Ancient Rome.* Princeton, NJ: Princeton University Press, 2004.

Corbier, M. "Child Exposure and Abandonment." In *Childhood, Class and Kin in the Roman World,* edited by S. Dixon. London: Routledge, 2001.

Cornford, F. M. *Plato's Cosmology.* London: Routledge & Kegan Paul, 1937.

Corpus Vasorum Antiquorum. Paris: H. Champion, 1923–.

Cosans, C. E. "Aristotle's Anatomical Philosophy of Nature." *Biology and Philosophy* 13 (1998): 311–39.

Cox Miller, Patricia. "Is There a Harlot in This Text? Hagiography and the Grotesque." *Journal of Medieval and Early Modern Studies* 33 (2003): 419–35.

Craik, E. M. "Diet, Diaita and Dietetics." In *The Greek World,* edited by A. Powell. London: Routledge, 1995.

Craik, E. M. *Hippocrates: Places in Man.* Oxford: Clarendon Press, 1998.

Craik, E. M. "Hippokratic Diaita." In *Food in Antiquity,* edited by J. Wilkins, D. Harvey, and M. Dobson. Exeter: University of Exeter Press, 1995.

Craik, E. M. "Plato and Medical Texts: *Symposium* 185c–193d." *Classical Quarterly* 51 (2001): 109–114.

Craik, E. M. "Thucydides on the Plague: Physiology of Flux and Fixation." *Classical Quarterly* 51 (2001): 102–8.

Craik, E. M. *Two Hippocratic Treatises:* On Sight *and* On Anatomy. Leiden: Brill, 2006.

Cunningham, A. "The Theory/Practice Division of Medicine: Two Late Alexandrian Legacies." In *History of Traditional Medicine*, edited by T. Ogawa. Osaka: Division of Medical History, the Taniguchi Foundation, Tokyo, 1986.

Cunningham, I. C., ed. *Herodas, Mimiambi*. Oxford: Clarendon Press, 1971.

Curd, P., and D. W. Graham, eds. *The Oxford Handbook of Presocratic Philosophy*. New York: Oxford University Press, 2008.

Cutler, A. "The De Signis of Nicetas Choniates: A Reappraisal." *American Journal of Archaeology* 72 (1968): 113–18.

D'Arms, J. "The Roman *Convivium* and the Idea of Equality." In *Sympotica: A Symposium on the* Symposium, edited by O. Murray, 308–20. Oxford: Clarendon Press, 1990.

Dasen, V. *Dwarfs in Ancient Egypt and Greece*. Oxford: Clarendon Press, 1993.

Dasen, V., ed. *Naissance et petite enfance dans l'antiquité. Actes du colloque de Fribourg, 28 novembre—1er décembre 2001*. Fribourg and Göttingen: Academic Press and Vandenhoeck & Ruprecht, 2004.

Davidson, J. N. *Courtesans and Fishcakes: The Consuming Passions of Classical Athens*. New York: HarperPerennial, 1997.

Davidson, J. N. "Dover, Foucault, and Greek Homosexuality: Penetration and the Truth of Sex." *Past and Present* 170 (2001): 3–51.

Davies, G. "What Made the Roman Toga *virilis*?" In *The Clothed Body in the Ancient World*, edited by Liza Cleland, Mary Harlow, and Lloyd Llewellyn-Jones. Oxford: Oxbow Books, 2005.

Davies, M. *Epicorum Graecorum fragmenta*. Göttingen: Vandenhoeck & Ruprecht, 1988.

Davis, Stephen J. "Crossed Texts, Crossed Sex: Intertextuality and Gender in Early Christian Legends of Holy Women Disguised as Men." *Journal of Early Christian Studies* 10 (2002): 1–36.

De Lacy, P. *Galen, On the Doctrines of Hippocrates and Plato*. Corpus Medicorum Graecorum 5.4.1.2. Berlin: De Gruyter, 1978–1985.

De Lacy, P. H., ed. and trans. *Galen. De placitis Hippocratis et Platonis*. Corpus Medicorum Graecorum 5.4.2.1–4. Berlin: Akademie Verlag, 1978.

Deacey, S. "Herakles and his 'Girl': Athena, Heroism and Beyond." In *Herakles and Hercules: Exploring a Graeco-Roman Divinity*, edited by L. Rawlings and H. Bowden. Swansea: The Classical Press of Wales, 2005.

Dean-Jones, L. "Literacy and the Charlatan in Ancient Greek Medicine." In *Written Texts and the Rise of Literate Culture in Ancient Greece*, edited by H. Yunis. Cambridge: Cambridge University Press, 2003.

Dean-Jones, L. "The Politics of Pleasure: Female Sexual Appetite in the Hippocratic Corpus." *Helios* 19 (1992): 72–91.

Dean-Jones, L. *Women's Bodies in Classical Greek Science*. Oxford: Clarendon Press, 1994.

Debru, A. "Les demonstrations médicales à Rome au temps de Galien." In *Ancient Medicine in Its Socio-Cultural Context*, edited by P. van der Eijk, H.F.J. Horstmanshoff, and P.H. Schrijvers. 2 vols. Amsterdam: Rodopi, 1995.

Demand, N. *Birth, Death and Motherhood in Classical Greece*. Baltimore: Johns Hopkins University Press, 1994.

Demont, P. "About Philosophy and Humoural Medicine." In *Ancient Medicine in Its Socio-Cultural Context: Papers Read at the Congress Held at Leiden University 13–15 April 1992*, edited by P. van der Eijk, H.F.J. Horstmanshoff, and P.H. Schrijvers. 2 vols. Amsterdam: Rodopi, 1995.

Dennett, D.C. "Conditions of Personhood." In *The Identities of Persons*, edited by A.O. Rorty. Berkeley: University of California Press, 1976.

Detienne, M. *Dionysos Slain*. Translated by Mireille Muellner and Leonard Muellner. Baltimore: Johns Hopkins University Press, 1979.

Detienne, M. *The Gardens of Adonis: Spices in Greek Mythology*. 2nd ed. Translated by Janet Lloyd. Princeton, NJ: Princeton University Press, 1994.

Dexheimer, D. "Portrait Figures on Funerary Altars of Roman *liberti* in Northern Italy: Romanization or the Assimilation of Attributes Characterizing Higher Social Status?" In *Burial, Society and Context in the Roman World*, edited by J. Pearce, M. Millett, and M. Struck. Oxford: Oxbow, 2001.

Diebold, W.J. *Word and Image: An Introduction to Early Medieval Art*. Boulder, CO: Westview Press, 2000.

Dihle, A. *The Theory of the Will in Classical Antiquity*. Berkeley: University of California Press, 1986.

Dover, K. *Aristophanes. Frogs*. Oxford: Clarendon Press, 1993.

Dover, K.J. *Greek Homosexuality*. London: Duckworth. Cambridge, MA: Harvard University Press, 1978 (reprint 1989).

Dover, K.J. *Greek Popular Morality in the Time of Plato and Aristotle*. Oxford: Blackwell, 1974.

Dover, K.J. "Two Women of Samos." In *The Sleep of Reason: Erotic Experience and Sexual Ethics in Ancient Greece and Rome*, edited by M.C. Nussbaum and J. Sihvola. Chicago: University of Chicago Press, 2002.

Downing, E. "Anti-Pygmalion: The *Praeceptor* in *Ars Amatroria*, Book 3." In *Constructions of the Classical Body*, edited by. J.I. Porter. Ann Arbor: University of Michigan Press, 1999.

Drabkin, I. *Caelius Aurelianus: On Acute Diseases and On Chronic Diseases*. Chicago: University of Chicago Press, 1950.

Drescher, J. *Three Coptic Legends: Hilaria, Archellites, the Seven Sleepers*. Supplément aux Annales du Service des Antiquités de l'Égypte, no. 4. Cairo: Institut Français d'Archéologie Orientale, 1947.

duBois, P. *Sowing the Body: Psychoanalysis and Ancient Representations of Women*. Chicago: University of Chicago Press, 1988.

Duff, J.D., ed. and trans. *Lucan: The Civil War*. Cambridge, MA: Harvard University Press, 1928.

Durkheim, Emile. *The Rules of Sociological Method*. Translated by W.D. Halls. New York: The Free Press, 1982. First published 1938.

Dutsch, D. "*Nenia*: Gender, Genre and Lament in Ancient Rome." In *Lament: Studies in the Ancient Mediterranean and Beyond*, edited by A. Suter. Oxford: Oxford University Press, 2008.

Eck, W., Caballos, A., and Fernández, F. *Das Senatus Consultum de Cn. Pisone Patre.* Munich: C.H. Beck, 1996.

Edelstein, E.J., and L. Edelstein. *Asclepius: Collection and Interpretation of the Testimonies.* 2 vols. Baltimore: Johns Hopkins University Press, 1945.

Edelstein, L. *Ancient Medicine: Selected Papers of Ludwig Edelstein,* edited by O. Temkin and C.L. Temkin. Baltimore: Johns Hopkins University Press, 1967.

Edelstein, L. "Hippocratic Prognosis." In *Ancient Medicine: Selected Papers of Ludwig Edelstein,* edited by O. Temkin and C.L. Temkin. Baltimore: Johns Hopkins University Press, 1967.

Edelstein, L. "The History of Anatomy in Antiquity." In *Ancient Medicine: Selected Papers of Ludwig Edelstein,* edited by O. Temkin and C.L. Temkin. Baltimore: Johns Hopkins University Press, 1967.

Edelstein, L. "The Methodists." In *Ancient Medicine: Selected Papers of Ludwig Edelstein,* edited by O. Temkin and C.L. Temkin. Baltimore: Johns Hopkins University Press, 1967.

Edwards, C. "Archetypally Roman? Representing Seneca's Ageing Body." In *Roman Bodies: Antiquity to the Eighteenth Century,* edited by A. Hopkins and M. Wyke. London: The British School at Rome, 2005.

Edwards, C. *Death in Ancient Rome.* New Haven, CT: Yale University Press, 2007.

Edwards, C. *The Politics of Immorality in Ancient Rome.* Cambridge: Cambridge University Press, 1993.

Edwards, C. "The Suffering Body: Philosophy and Pain in Seneca's *Letters.*" In *Constructions of the Classical Body,* edited by J.I. Porter. Ann Arbor: University of Michigan Press, 1999.

Edwards, C. "Unspeakable Professions: Public Performance and Prostitution in Ancient Rome." In *Roman Sexualities,* edited by J.P. Hallet and M.B. Skinner. Princeton, NJ: Princeton University Press, 1997.

Effe, B. "Heroische Grösse: Der Funktionswandel des Herakles-Mythos in des griechisch-römisch Literatur." In *Herakles/Hercules—Metamorphosen des Heros in ihrer medialen Vielfalt,* edited by R. Kray and S. Oettermann. Vol. 1. Basel: Stroemfeld Verlag, 1994.

Effros, Bonnie. "Dressing Conservatively: Women's Brooches as Markers of Ethnic Identity?" In *Gender and the Early Medieval World, East and West, 300–900,* edited by L. Brubaker and J. Smith. Cambridge: Cambridge University Press, 2004.

Evans, Elizabeth C. *Physiognomics in the Ancient World.* Transactions of the American Philosophical Society, vol. 59, part 5.

Evans, Rhiannon. "Ethnography's Freak Show: The Grotesques at the Edges of the Roman Earth." *Ramus* 28 (1999): 54–73.

Fantham, E. "*Stuprum*: Public Attitudes and Penalties for Sexual Offences in Republican Rome." *Échos du Monde Classique/Classical Views* 35 n.s. 10 (1991): 267–91.

Faraone, Christopher. "Binding and Burying the Forces of Evil: The Defense Use of 'Voodoo' Dolls in Ancient Greece." *Classical Antiquity* 10 (1991): 165–205.

Fehr, B. "Entertainers at the *Symposium*: The *akletoi* in the Archaic Period." In *Sympotica: A Symposium on the* Symposium, edited by O. Murray. Oxford: Clarendon Press, 1990.

Felton, D. *Haunted Greece and Rome. Ghost Stories from Classical Antiquity.* Austin: University of Texas Press, 1999.

Ferngren, G. B. "Roman Lay Attitudes towards Medical Experimentation." *Bulletin of the History of Medicine* 59 (1985): 495–505.
Ferngren, G. B., and D. W. Amundsen. "Medicine and Christianity in the Roman Empire: Compatibilities and Tensions." *Aufstieg und Niedergang der römischen Welt*, 2nd ser., 37, part 3 (1996): 2957–80.
Ferrari, G. *Figures of Speech: Men and Maidens in Ancient Greece*. Chicago: University of Chicago Press, 2002.
Ferris, I. M. *Enemies of Rome: Barbarians through Roman Eyes*. Stroud, UK: Sutton, 2000.
Fields, N. "Headhunters in the Roman army." In *Roman Bodies: Antiquity to the Eighteenth Century*, edited by A. Hopkins and M. Wyke. London: The British School at Rome, 2005.
Fisher, Charles Dennis, ed. *Annales ab excessu divi Augusti. Cornelius Tacitus*. Oxford: Clarendon Press, 1906.
Fitch, J. G., and S. McElduff. "Construction of the Self in Senecan Drama." *Mnemosyne* 55 (2002): 18–40.
Flacelière, R., and P. Devambez, eds. *Héraclès, Images et Récits*. Paris: Boccard, 1966.
Flemming, R. "Empires of Knowledge: Medicine and Health in the Hellenistic World." In *A Companion to the Hellenistic World*, edited by A. Erskine. Malden, MA: Blackwell, 2003.
Flemming, R. *Medicine and the Making of Roman Women*. Oxford: Clarendon Press, 2000.
Flemming, R., and A. Hanson. "'Hippocrates' *Peri parthenôn (Diseases of Young Girls)*: Text and Translation." *Early Science and Medicine* 3 (1998): 241–52.
Flower, H. I. *Ancestor Masks and Aristocratic Power in Roman Culture*. Oxford: Clarendon Press, 1996.
Flower, H. I. *The Art of Forgetting: Disgrace and Oblivion in Roman Political Culture*. Chapel Hill: University of North Carolina Press, 2006.
Foerster, R., ed. *Scriptores physiognomonici Graeci et Latini*. 2 vols. Leipzig: Teubner, 1893.
Fortenbaugh, W. W., R. W. Sharples, and M. G. Sollenberger, eds. *Theophrastus of Eresus: On Sweat, On Dizziness, and On Fatigue*. Leiden: Brill, 2003.
Foucault, M. *The History of Sexuality*. Vol. 2, *The Use of Pleasure*. Translated by R. Hurley. New York: Pantheon, 1985.
Foucault, M. *The History of Sexuality*. Vol. 3, *The Care of the Self*. Translated by R. Hurley. New York: Pantheon, 1986.
Foucault, M. *The History of Sexuality*. Vol. 3, *The Care of the Self*. Translated by R. Hurley. New York: Vintage, 1988.
Foucault, M. "On the Genealogy of Ethics: An Overview of Work in Progress." In *The Foucault Reader*, edited by Paul Rabinow. New York: Pantheon, 1984.
Fox, R. L. *Pagans and Christians*. New York: Alfred A. Knopf, 1987.
Frampton, M. F. "Aristotle's Cardiocentric Model of Animal Locomotion." *Journal of the History of Biology* 24 (1991): 291–330.
Frazer, J. *The Fear of the Dead in Primitive Religion*. 3 vols. London: Macmillan, 1934–1936.
Frede, M. *Essays on Ancient Philosophy*. Minneapolis: University of Minnesota Press, 1987.
Frede, M. "The Method of the So-Called Methodical School of Medicine." In *Science and Speculation: Studies in Hellenistic Theory and Practice*, edited by J. Barnes

et al. Cambridge: Cambridge University Press, 1982. Reprinted in *Essays on Ancient Philosophy*. Minneapolis: University of Minnesota Press, 1987.

Frede, M., and R. Walzer. *Galen: Three Treatises on the Nature of Science*. Indianapolis: Hackett, 1985.

French, V. "Midwives and Maternity Care in the Roman World." In *Rescuing Creusa: New Methodological Approaches to Women in Antiquity*, Helios, New Series 13, no. 2 (1986): 69–84.

Freudenthal, G. *Aristotle's Theory of Material Substance: Heat and Pneuma, Form and Soul*. Oxford: Clarendon Press, 1995.

Galinsky, K. *The Herakles Theme: The Adaptations of the Hero in Literature from Homer to the Twentieth Century*. Oxford: Blackwell, 1972.

Garcia Ballester, L. "Soul and Body: Disease of the Soul and Disease of the Body in Galen's Medical Thought." In *Le opere psicologiche di Galeno: Atti del terzo colloquio Galenico internazionale*, edited by P. Manuli and M. Vegetti. *Pavia, 10–12 settembre 1986*, Naples. Reprinted in *Galen and Galenism: Theory and Medical Practice from Antiquity to the European Renaissance*, edited by Jon Arrizabalaga, M. Cabre, L. Cifuentes, and F. Salmon. Aldershot: Ashgate, 2002.

Garland, R. *The Eye of the Beholder: Deformity and Disability in the Greco-Roman World*. Ithaca, NY: Cornell University Press, 1995.

Garland, R. *The Greek Way of Death*. London: Duckworth, 1985.

Garofalo, I. *Erasistrati Fragmenta*. Pisa: Giardini, 1988.

Garrison, D.H. *Sexual Culture in Ancient Greece*. Norman: University of Oklahoma Press, 2000.

Geddes, A.G. "Rags and Riches: The Costume of Athenian Men in the Fifth Century." *Classical Quarterly* 37 (1987): 307–31.

George, Michele. "Slave Disguise in Ancient Rome." In *Representing the Body of the Slave*, edited by T.E.J. Wiedemann and J.F. Gardner. London: Routledge, 2002.

Georgoudi, S. "Héraclès dans les pratiques sacrificielles des cités." In *Le Bestiaire d'Héraclès*, edited by C. Bonnet, C. Jourdain-Annequin, and U. Pirenne-Delforge. Liège: Centre International d'Étude de la Religion Grecque Antique, KERNOS Suppl. 7 (1998): 301–17.

Gerber, D.E. "Semonides, FR. 7.62." *Phoenix* 28 (1974): 251–53.

Gero, S. "Galen on the Christians: A Reappraisal of the Arabic Evidence." *Orientalia Christiana Periodica* 56 (1990): 371–411.

Geus, A. "Patron und Superlativ: Hercules in Biologie, Medizin und Pharmazie." In *Herakles/Hercules—Metamorphosen des Heros in ihrer medialen Vielfalt*, edited by R. Kray and S. Oettermann. Vol. 1. Basel: Stroemfeld Verlag, 1994.

Ghersetti, A. "The Semiotic Paradigm: Physiognomy and Medicine in Islamic Culture." In *Seeing the Face, Seeing the Soul: Polemon's Physiognomy from Classical Antiquity to Medieval Islam*, edited by S. Swain. Oxford: Clarendon Press, 2007.

Giacomelli, A. "The Justice of Aphrodite in Sappho fr. 1." *Transactions of the American Philological Association* 110 (1980): 135–42.

Gill, C. "The Ancient Self: Issues and Approaches." In *Ancient Philosophy of the Self*, edited by P. Remes and J. Sihvola. Helsinki: Springer, 2008.

Gill, C. *Character and Personality in Greek Epic and Tragedy*. Oxford: Oxford University Press, 1996.

Gill, C. *The Structured Self in Roman and Hellenistic Thought*. Oxford: Oxford University Press, 2006.

Gill, M. L. "Plato's *Phaedrus* and the Method of Hippocrates." *Modern Schoolman* 80 (2003): 295–314.

Gill, M. L., and P. Pellegrin, ed. *A Companion to Ancient Philosophy*. Malden, MA: Blackwell, 2006.

Gleason, M. "The Semiotics of Gender: Physiognomy and Self-Fashioning in the Second Century C.E." In *Before Sexuality: The Construction of Erotic Experience in the Ancient Greek World*, edited by D. M. Halperin. J. J. Winkler, and F. I. Zeitlin. Princeton, NJ: Princeton University Press, 1990.

Gleason, Maud. *Making Men: Sophists and Self-Presentation in Ancient Rome*. Princeton, NJ: Princeton University Press, 1995.

Gleason, Maud. "Truth Contests and Talking Corpses." In *Constructions of the Classical Body*, edited by J. I. Porter. Ann Arbor: University of Michigan Press, 1999.

Golden, M. "Change or Continuity? Children and Childhood in Hellenic Historiography." In *Inventing Ancient Culture: Historicism, Periodization and the Ancient World*, edited by Mark Golden and Peter Toohey. London: Routledge, 1997.

Golden, M. *Children and Childhood in Classical Athens*. Baltimore: Johns Hopkins University Press, 1993.

Golden, M. "Did the Ancients Care When Their Children Died?" *Greece and Rome* 35 (1988): 152–63.

Golden, M. "Slavery and Homosexuality at Athens." *Phoenix* 38 (1984): 308–24.

Gorrini, M. E. "The Hippocratic Impact on Healing Cults: The Archaeological Evidence in Attica." In *Hippocrates in Context: Papers Read at the XIth International Hippocrates Colloquium, University of Newcastle upon Tyne, 27–31 August 2002*, edited by P. van der Eijk. Leiden: Brill, 2005.

Gotthelf, A., ed. *Aristotle on Nature and Living Things*. Pittsburgh: Mathesis, 1985.

Gotthelf, A., and J. G. Lennox, ed. *Philosophical Issues in Aristotle's Biology*. Cambridge: Cambridge University Press, 1987

Gottschall, Jonathan. "Homer's Human Animal: Ritual Combat in the *Iliad*." *Philosophy and Literature* 25 (2001): 278–94.

Gourevitch, D. "La gynécologie et l'obstétrique." *Aufstieg und Niedergang der römischen Welt*, 2nd ser., 37, part 3 (1996): 2083–2146.

Gourevitch, D., ed. *Maladie et maladies: Historie et conceptualization (mélanges en l'honneur de Mirko Grmek)*. Geneva: Librairie Droz S.A., 1992.

Gradel, Ittai. *Emperor Worship and Roman Religion*. Oxford: Oxford University Press, 2004.

Graf, Fritz. "Gestures and Conventions: The Gestures of Roman Actors and Orators." In *A Cultural History of Gesture*, edited by J. N. Bremmer and H. Roodenberg. Ithaca, NY: Cornell University Press,1992.

Graham, E.-J. *Death, Disposal and the Destitute: The Burial of the Urban Poor in Italy in the Late Roman Republic and Early Empire*. British Archaeological Reports International Series 1565. Oxford: Archaeopress, 2007.

Graham, E.-J. "Memory and Materiality: Re-embodying the Roman Funeral." In *Memory and Mourning: Studies on Roman Death*, edited by V. M. Hope and J. A. R. Huskinson. Oxford: Oxbow, forthcoming.

Graver, M. *Cicero on the Emotions: Tusculan Disputations 3 and 4*. Chicago: University of Chicago Press, 2002.

Greek Ministry of Culture. *The Human Figure in Early Greek Art*. Washington, DC: National Gallery of Art, 1988.

Greenough, J. B., ed. *Bucolics, Aeneid, and Georgics of Vergil*. Boston: Ginn & Co., 1900.
Gregoric, P. *Aristotle on the Common Sense*. Oxford: Clarendon Press, 2007.
Griffith, Mark. "Slaves of Dionysos: Satyrs, Audience, and the Ends of the *Oresteia*." *Classical Antiquity* 21 (2002): 195–258.
Grmek, M. D. *Diseases of the Ancient World*. Translated by M. Muellner and L. Muellner. Baltimore: Johns Hopkins University Press, 1989.
Grmek, M. D., ed. *Western Medical Thought from Antiquity to the Middle Ages*. Translated by A. Shugaar. Cambridge, MA: Harvard University Press, 1999.
Gunderson, E. *Staging Masculinity: The Rhetoric of Performance in the Roman World*. Ann Arbor: University of Michigan Press, 2000.
Gundert, B. "Soma and Psyche in Hippocratic Medicine." In *Psyche and Soma: Physicians and Metaphysicians on the Mind-Body Problem from Antiquity to Enlightenment*, edited by J. P. Wright and P. Potter. Oxford: Clarendon Press, 2000.
Guthrie, R. D. *The Nature of Paleolithic Art*. Chicago: University of Chicago Press, 2005.
Hadley, D. M. "Negotiating Gender, Family and Status in Anglo-Saxon Burial Practices, c. 600–950." In *Gender and the Early Medieval World, East and West, 300–900*, edited by L. Brubaker and J. Smith. Cambridge: Cambridge University Press, 2004.
Hadot, P. *Philosophy as a Way of Life: Spiritual Exercises from Socrates to Foucault*. Translated by M. Chase. Oxford: Blackwell, 1995.
Hadot, P. *The Veil of Isis: An Essay on the History of the Idea of Nature*. Cambridge, MA: Harvard University Press, 2006.
Hafner, G. "Herakles, Geras, Ogmios." *Jahrbuch des Römisch-germanischen Zentralmuseums, Mainz* 5 (1958): 139–53.
Hall, Edith. "Asia Unmanned: Images of Victory in Classical Athens." In *War and Society in the Greek World*, edited by J. Rich and G. Shipley. London: Routledge, 1993.
Hall, Edith. *Inventing the Barbarian*. Oxford: Oxford University Press, 1989.
Hall, Edith, and Amanda Wrigley, eds. *Aristophanes in Performance, 421 BC–AD 2007: Peace, Birds and Frogs*. London: MHRA, 2007.
Hallett, J. P. "Women as Same and Other in Classical Roman Elite." *Helios* 16 (1989): 59–78.
Hallett, J. P., and M. B. Skinner, eds. *Roman Sexualities*. Princeton, NJ: Princeton University Press, 1997.
Halperin, D. *One Hundred Years of Homosexuality*. London: Routledge, 1990.
Halperin, D. M., J. Winkler, and F. I. Zeitlin, eds. *Before Sexuality: The Construction of Erotic Experience in the Ancient Greek World*. Princeton, NJ: Princeton University Press, 1990.
Hamilakis, Y., M. Pluciennik, and S. Tarlow. *Thinking through the Body: Archaeologies of Corporeality*. New York: Kluwer/Plenum, 2002.
Hampe, R., and Simon, E. *The Birth of Greek Art: From the Mycenaean to the Archaic Period*. New York: Oxford University Press, 1981.
Hankinson, R. J. *Cause and Explanation in Ancient Greek Thought*. Oxford: Clarendon Press, 1998.
Hankinson, R. J. "Galen and the Best of All Possible Worlds." *Classical Quarterly* 39 (1987): 206–27.
Hankinson, R. J. *Galen: On the Therapeutic Method*. Oxford: Clarendon Press, 1991.

Hankinson, R. J. "Galen's Anatomy of the Soul." *Phronesis* 36 (1991): 197–233.
Hankinson, R. J. "Galen's Philosophical Eclecticism." *Aufstieg und Niedergang der römischen Welt,* 2nd ser., 36, part 5 (1992): 3505–22.
Hankinson, R. J. "The Growth of Medical Empiricism." In *Knowledge and the Scholarly Medical Traditions*, edited by D. Bates. Cambridge: Cambridge University Press, 1995.
Hankinson, R. J., ed. *The Cambridge Companion to Galen.* Cambridge: Cambridge University Press, 2008.
Hänninen, M. L. "From Womb to Family: Rituals and Social Conventions Connected to Roman Birth." In *Hoping for Continuity: Childhood, Education and Death in Antiquity and the Middle Ages*, edited by K. Mustakallio, J. Hanska, H.-L. Sainio, and V. Vuolanto. *Acta Instituti Romani Finlandiae* 33 (2005): 49–59.
Hansen, William F. *Handbook of Classical Mythology.* Santa Barbara: ABC-CLIO, 2004.
Hanson, A. "Conception, Gestation, and the Origin of Female Nature." *Helios* 19 (1992): 31–71.
Hanson, A. E. "Diseases of Women I." *Signs* 1 (1975): 567–84.
Hanson, A. E. "Diseases of Women in the *Epidemics.*" In G. Baader and R. Winau 1989: 38–51.
Hanson, A. E. *Studies in the Textual Tradition and the Transmission of the Gynecological Treatises of the Hippocratic Corpus.* Ph.D. dissertation, University of Pennsylvania, 1971.
Hanson, A. E. "'Your Mother Nursed You with Bile': Anger in Babies and Small Children." In *Ancient Anger: Perspectives from Homer to Galen.* Yale Classical Studies vol. 32, edited by S. Braund and G. Most. Cambridge: Cambridge University Press, 2003.
Hanson, Ann Ellis. "Continuity and Change: Three Case Studies in Hippocratic Gynecological Therapy and Theory." In *Women's History and Ancient History*, edited by S. B. Pomeroy. Chapel Hill: University of North Carolina Press, 1991.
Harkins, P. W. *Galen on the Passions and the Errors of the Soul.* Columbus: Ohio State University Press, 1963.
Harlow, M. "Clothes Maketh Man: Power Dressing in the Later Roman Empire." In *Gender and the Early Medieval World, East and West, 300–900*, edited by L. Brubaker and J. Smith. Cambridge: Cambridge University Press, 2004.
Harlow, M., and Laurence, R. *Growing Up and Growing Old in Ancient Rome: A Life Course Approach.* London: Routledge, 2002.
Harris, C.R.S. *The Heart and the Vascular System in Ancient Greek Medicine.* Oxford: Clarendon Press, 1973.
Harris, W. V. "Child-Exposure in the Roman Empire." *Journal of Roman Studies* 54 (1994): 1–22.
Harrison, E. "The Constantinian Portrait." *Dumbarton Oaks Papers* 21 (1967): 76–96.
Head, T. "The Early Medieval Transformation of Piety." In *The Long Morning of Medieval Europe: New Directions in Early Medieval Studies*, edited by J. R. Davis and M. McCormick. Hampshire: Ashgate, 2008.
Heath, John. *The Talking Greeks: Speech, Animals, and the Other in Homer, Aeschylus, and Plato.* Cambridge: Cambridge University Press, 2005.
Heeßel, N. P. "Diagnosis, Divination and Disease: Towards an Understanding of the *Rationale* behind the Babylonian *Diagnostic Handbook.*" In *Magic and Rational-*

ity in Ancient Near Eastern and Graeco-Roman Medicine, edited by H.F.J. Horstmanshoff and M. Stol. Leiden: Brill, 2004.

Heiberg, I. L., ed. *Paulus Aegineta.* 2 vols. Corpus Medicorum Graecorum 9.1–2. Berlin: Teubner, 1921–1924.

Hekster, O. "Propagating Power: Hercules as an Example for Second-Century Emperors." In *Herakles and Hercules: Exploring a Graeco-Roman Divinity,* edited by L. Rawlings and H. Bowden. Swansea: The Classical Press of Wales, 2005.

Henderson, J. "Greek Attitudes toward Sex." In *Civilization of the Ancient Mediterranean,* edited by M. Grant and R. Kitzinger. New York: Scribner's, 1988.

Henderson, Jeffrey. *The Maculate Muse: Obscene Language in Attic Comedy.* New Haven, CT: Yale University Press, 1975.

Henderson, Jeffrey. *The Maculate Muse: Obscene Language in Attic Comedy.* 2nd rev. ed. New York: Oxford University Press, 1991.

Henry, D. "How Sexist Is Aristotle's Developmental Biology?" *Phronesis* 52 (2007): 251–69.

Herington, J. *Poetry into Drama: Early Tragedy and the Greek Poetic Tradition.* Berkeley: University of California Press, 1985.

Herodotus. *The Histories.* Translated by Robin Waterfield. Oxford: Oxford University Press, 1998.

Hicks, R. D. *Diogenes Laertius: Lives of Eminent Philosophers.* Vol. 2. Cambridge, MA: Harvard University Press, 1931.

Hobart, W. K. *The Medical Language of St. Luke.* Piscataway, NJ: Gorgias Press, 2004. Reprint of 1954 edition.

Hoffelner, K. "Die Metopen des Athener-Schatzhauses. Ein neuer Rekonstruktionsversuch." *Mitteilungen des Deutschen Archaologischen Instituts, Athenische Abteilung* 103 (1988): 77–117.

Hoffmann-Axthelm, W. *History of Dentistry.* Translated by H. M. Koehler. Chicago: Quintessence, 1981.

Holmes, B. "Aelius Aristides' Illegible Body." In *Aelius Aristides between Greece, Rome, and the Gods,* edited by W. V. Harris and B. Holmes. Leiden: Brill, 2008.

Holmes, B. "Body, Soul, and Medical Analogy in Plato." In *When Worlds Elide: Classics, Politics, Culture,* edited by J. P. Euben and K. Bassi. Lanham, MD: Rowman and Littlefield, forthcoming.

Holmes, B. "The *Iliad*'s Economy of Pain." *Transactions of the American Philological Association* 137 (2007): 45–84.

Holmes, B. "Plato's Uncanny Soul: Medical Analogy and Ethical Subjectivity." In *When Worlds Elide: Classics, Politics, Culture,* edited by J. P. Euben and K. Bassi. Lanham, MD: Rowman & Littlefield, forthcoming.

Holmes, B. *The Symptom and the Subject: The Emergence of the Physical Body in Ancient Greece.* Princeton, NJ: Princeton University Press, 2010.

Hope, V. M. "Contempt and Respect: The Treatment of the Corpse in Ancient Rome." In *Death and Disease in the Ancient City,* edited by V. M Hope and E. Marshall. London: Routledge, 2000.

Hope, V. M. *Death in Ancient Rome: A Sourcebook.* London: Routledge, 2007.

Hope, V. M. *Roman Death: The Dying and the Dead in Ancient Rome.* London: Hambledon Continuum, 2009.

Hope, V. M., and E. Marshall, eds. *Death and Disease in the Ancient City.* London: Routledge, 2000.

Hopkins, K. *Death and Renewal*. Cambridge: Cambridge University Press, 1983.
Horstmanshoff, H.F.J. "The Ancient Physician: Craftsman or Scientist?" *Journal of the History of Medicine and Allied Sciences* 45 (1990): 176–97.
Horstmanshoff, H.F.J. "Did the God Learn Medicine?: Asclepius and Temple Medicine in Aelius Aristides '*Sacred Tales*.'" In *Magic and Rationality in Ancient Near Eastern and Graeco-Roman Medicine*, edited by H.F.J. Horstmanshoff and M. Stol. Leiden: Brill, 2004.
Hubbard, T.K. "Pederasty and Democracy: The Marginalization of a Social Practice." In *Greek Love Reconsidered*, edited by T.K. Hubbard. New York: Wallace Hamilton Press, 2000.
Hubbard, T.K. "Popular Perceptions of Elite Homosexuality in Classical Athens." *Arion* ser. 3, 6, no. 1 (1998): 48–78.
Hude, Carolus, ed. *Herodati Historiae, Volume I: Books I–IV*. 3rd ed. Oxford: Oxford University Press, 1927.
Hude, Carolus, ed. *Herodati Historiae, Volume II: Books V–IX*. 3rd ed. Oxford: Oxford University Press, 1927.
Humphrey, N. "Great Expectations: The Evolutionary Psychology of Faith Healing and the Placebo Effect." In *The Mind Made Flesh: Essays from the Frontiers of Evolution and Psychology*. Oxford: Oxford University Press, 2002.
Humphries, M.L. "Michel Foucault on Writing and the Self in the *Meditations* of Marcus Aurelius and *Confessions* of St. Augustine." *Arethusa* 30 (1997): 125–38.
Hunter, V. "Constructing the Body of the Citizen: Corporal Punishment in Classical Athens." *Echos du monde classique* 36 (1992): 271–91.
Hurwit, J.M. *The Athenian Acropolis: History, Mythology and Archaeology from the Neolithic Era to the Present*. Reprint. Cambridge: University Press, 2001.
Huskinson, J. "Representing Women on Roman Sarcophagi." In *The Material Culture of Sex, Procreation, and Marriage in Premodern Europe*, edited by Anne L. McClanan and Karen Rosoff Encarnación. New York: Palgrave, 2001.
Huskinson, J. "Some Pagan Mythological Figures and Their Significance in Early Christian Art." *Papers of the British School at Rome* 42 (1974): 68–97.
Huttner, U. "Marcus Antonius und Herakles." In *Rom und der griechische Osten. Festschrift für Hatto H. Schmitt zum 65. Geburtstag*, edited by C. Schubert and K. Brodersen. Stuttgart: F. Steiner Verlag, 1995.
Ilberg, I., ed. *Soranus*, Corpus Medicorum Graecorum 4. Leipzig: Teubner, 1927.
Inwood, B., and L.P. Gerson, ed. *Hellenistic Philosophy: Introductory Readings*. Indianapolis: Hackett, 1997.
Irmscher, J., and R. Müller, ed. *Aristotelis als Wissenschaftstheoretiker (Schriften zur Geschichte und Kultur der Antike 22)*. Berlin: Akademie-Verlag, 1983.
Irwin, T. *Plato's Ethics*. Oxford: Oxford University Press, 1995.
Isaac, Benjamin. *The Invention of Racism in Antiquity*. Princeton, NJ: Princeton University Press, 2004.
Jackson, R. *Doctors and Diseases in the Roman Empire*. London: British Museum Publications, 1987.
Jackson, Ralph. "Circumcision, De-circumcision and Self-Image: Celsus' 'Operations on the Penis.'" In *Roman Bodies: Antiquity to the Eighteenth Century*, edited by A. Hopkins and M. Wyke. London: The British School at Rome, 2005.
Jacoby, F., ed. *Die Fragmente der griechischen Historiker*. Vol. 1. 2nd ed. Leiden: Brill, 1957.

Jaeger, W. "Aristotle's Use of Medicine as Model of Method in His Ethics." *Journal of Hellenic Studies* 77 (1957): 54–61.
James, L. "Pray Not to Fall into Temptation and Be On Your Guard: Pagan Statues in Christian Constantinople." *Gesta* 35 (1996): 12–20.
James, Liz, and Shaun Tougher. "Get Your Kit On! Some Issues in the Depiction of Clothing in Byzantium." In *The Clothed Body in the Ancient World*, edited by Liza Cleland, Mary Harlow, and Lloyd Llewellyn-Jones. Oxford: Oxbow, 2005.
Jameson, M. "The Family of Herakles in Attika." In *Herakles and Hercules: Exploring a Graeco-Roman Divinity*, edited by L. Rawlings and H. Bowden. Swansea: The Classical Press of Wales, 2005.
Johnson, L. T. *The Gospel of Luke*. Collegeville, MN: Liturgical Press, 1991.
Johnston, S. I. *Restless Dead: Encounters between the Living and the Dead in Ancient Greece*. Berkeley: University of California Press, 1999.
Jones, B. W. *Suetonius: Domitian: Edited with Introduction, Commentary and Bibliography*. London: Bristol Classical Press, 1996.
Jones, C. P. "Stigma: Tattooing and Branding in Graeco-Roman Antiquity." *Journal of Roman Studies* 77 (1987): 139–55.
Jones, Horace Leonard, ed. and trans. *Strabo: Geography, Books X–XII*. Cambridge, MA: Harvard University Press, 1928.
Jones, W.H.S. *Hippocrates*. Vol. 1. Cambridge, MA: Harvard University Press, 1923.
Jones, W.H.S. *Hippocrates*. Vol. 2. Cambridge, MA: Harvard University Press, 1923.
Jones, W.H.S. *Hippocrates*. Vol. 4. Cambridge, MA: Harvard University Press, 1931.
Jones, W.H.S. *Malaria: A Neglected Factor in the History of Greece and Rome*. Cambridge: Macmillan and Bowes, 1907.
Jones, W.H.S. *The Medical Writings of Anonymous Londinensis*. Cambridge: Cambridge University Press, 1947.
Jones, W.H.S. *Pliny: Natural History*. Vol. 7. Cambridge, MA: Harvard University Press, 1956.
Jouan, F., and H. van Looy, eds. *Euripide. Fragments 1re partie*. Paris: Les Belles Lettres, 1998.
Jouanna, J. *Hippocrates*. Translated by M. B. DeBevoise. Baltimore: Johns Hopkins University Press, 1999.
Joyce, R. A. "Archaeology of the Body." *Annual Review of Anthropology* 34 (2005): 139–58.
Just, R. *Women in Athenian Law and Life*. London: Routledge, 1989.
Kahn, C. H. "Discovering the Will: From Aristotle to Augustine." In *The Question of Eclecticism: Studies in Later Greek Philosophy*, edited by J. M. Dillon and A. A. Long. Berkeley: University of California Press, 1988.
Kalkavage, P. *Plato's Timaeus*. Newburyport, MA: Focus, 2001.
Kansteiner, S. *Herakles. Die Darstellungen in der Grossplastik der Antike*. Köln: Böhlau, 2000.
Katz, M. A. "Sexuality and the Body in Ancient Greece." *Mltis* 4 (1989): 155–79.
Kaufmann, W. *The Portable Nietzsche*. New York: Penguin, 1976.
Kee, H. C. "Self-Definition in the Asclepius Cult." In *Jewish and Christian Self-Definition*. Vol. 3, *Self-Definition in the Graeco-Roman World*, edited by B. F. Meyer and E. P. Sanders. London: Fortress, 1982.
Keuls, Eva. *The Reign of the Phallus: Sexual Politics in Ancient Athens*. Berkeley: University of California Press, 1985.

King, H. *The Disease of Virgins: Green Sickness, Chlorosis and the Problems of Puberty*. London: Routledge, 2004.
King, H. *Greek and Roman Medicine*. London: Bristol Classical Press, 2003.
King, H., ed. *Health in Antiquity*. London: Routledge, 2005.
King, H. *Hippocrates' Woman*. London: Routledge, 1998.
King, H. "Women's Health and Recovery in the Hippocratic Corpus." In *Health in Antiquity*, edited by H. King. London: Routledge, 2005.
King, J. E. *Cicero: Tusculan Disputations*. Cambridge, MA: Harvard University Press, 1927.
King, R.A.H. *Aristotle on Life and Death*. London: Duckworth, 2001.
King, R.A.H., ed. *Common to Body and Soul: Philosophical Approaches to Explaining Living Behaviour in Greco-Roman Antiquity*. Berlin: Walter de Gruyter, 2006.
Kiple, K., ed. *The Cambridge World History of Human Disease*. Cambridge: Cambridge University Press, 1993.
Kirk, G.S., J.E. Raven, and M. Schofield, eds. *The Presocratic Philosophers: A Critical History with a Selection of Texts*. Cambridge: Cambridge University Press, 1983.
Kloft, H. "Herakles als Vorbild: Zur politischen Funktion eines griechischen Mythos in Rom." In *Herakles/Hercules—Metamorphosen des Heros in ihrer medialen Vielfalt*, edited by R. Kray and S. Oettermann. Vol. 1. Basel: Stroemfeld Verlag, 1994.
Koch, B. *Philosophie als Medizin für die Seelen: Untersuchungen zur den Tusculanae Disputationes*. Stuttgart: Steiner, 2006.
Koenen, L. "The Ptolemaic King as a Religious Figure." In *Images and Ideologies: Self-Definition in the Hellenistic World*, edited by A. Bulloch, E.S. Gruen, A.A. Long, and A. Stewart. Berkeley: University of California Press, 1993.
Konstan, D. *Sexual Symmetry: Love in the Ancient Novel and Related Genres*. Princeton, NJ: Princeton University Press, 1994.
Kosak, J.C. "*Polis Nosousa*: Greek Ideas about the City and Disease in the Fifth Century BC." In *Death and Disease in the Ancient City*, edited by V.M. Hope and E. Marshall. London: Routledge, 2000.
Kray, R., and S. Oettermann, eds. *Herakles/Hercules—Metamorphosen des Heros in ihrer medialen Vielfalt*. Vol. 1. Basel: Stroemfeld Verlag, 1994.
Kron, U. "Priesthoods, Dedications and Euergetism: What Part Did Religion Play in the Political and Social Status of Greek Women?" In *Religion and Power in the Ancient Greek World: Proceedings of the Uppsala Symposium 1993*, edited by P. Hellström and B. Alroth. Uppsala: Acta Universitatis Upsalensis, 1996.
Krumeich, R., N. Pechstein, and B. Seidensticker, eds. *Das griechische Satyrspiel*. Darmstadt: Wissenschaftliche Buchgesellschaft, 1999.
Kuczewski, M.G., and R. Polansky, ed. *Bioethics: Ancient Themes in Contemporary Issues*. Cambridge, MA: MIT Press, 2000.
Kudlien, F. "The Old Greek Concept of 'Relative Health.'" *Journal of the History of Behavioral Sciences* 9 (1973): 53–59.
Kühn, C.G. *Claudii Galeni opera omnia*. 20 vols. Leipzig: C. Cnobloch, 1821–1833.
Kuriyama, S. *The Expressiveness of the Body and the Divergence of Greek and Chinese Medicine*. New York: Zone, 1999.
Kurtz, D., and Boardman J. *Greek Burial Customs*. London: Thames & Hudson, 1971.
Kyle, D. *Spectacles of Death in Ancient Rome*. London: Routledge, 1998.
Lamb, W.R.M., ed. *Lysias*. Cambridge: Cambridge University Press, 1930.
Lamb, W.R.M., trans. *Plato*, vol. 2. Cambridge, MA: Harvard University Press, 1924.

Lang, P. "Medical and Ethnic Identities in Hellenistic Egypt." In *Re-inventions: Essays on Hellenistic and Early Roman Science*, edited by P. Lang. Special Issue of *Apeiron* 37 (2004): 107–31.

Langholf, V. "L'air (*pneuma*) et les maladies." In *La maladie et les maladies dans la collection hippocratique,* edited by P. Potter, G. Maloney, and J. Desautels. Quebec: Éditions du Sphinx, 1990.

Langholf, V. *Medical Theories in Hippocrates: Early Texts and the "Epidemics."* Berlin: De Gruyter, 1990.

Laqueur, T. *Making Sex: Body and Gender from the Greeks to Freud.* Cambridge, MA: Harvard University Press, 1990.

Laskaris, J. *The Art Is Long: On the Sacred Disease and the Scientific Tradition.* Leiden: Brill, 2002.

Lattimore, Richmond, trans. *The Iliad of Homer.* Chicago: University of Chicago Press, 1951.

Lattimore, Richmond, trans. *The Odyssey of Homer.* Chicago: University of Chicago Press, 1967.

Laurence, Ray. "Health and the Life Course at Herculaneum and Pompeii." In *Health in Antiquity*, edited by H. King. London: Routledge, 2005.

Le Blay, F. "Microcosm and Macrocosm: The Dual Direction of Analogy in Hippocratic Thought and the Meteorological Tradition." In *Hippocrates in Context: Papers Read at the XIth International Hippocrates Colloquium, University of Newcastle upon Tyne, 27–31 August 2002*, edited by P. van der Eijk. Leiden: Brill, 2005.

Lefort, L. T. "Saint Athanase: Sur la Virginité." *Le Muséon* 42 (1929): 197–274.

Lennox, J. G. *Aristotle: On the Parts of Animals I–IV.* Oxford: Clarendon Press, 2001.

Lennox, J. G. *Aristotle's Philosophy of Biology: Studies in the Origins of Life Science.* Cambridge: Cambridge University Press, 2001.

Lesher, J., D. Nails, and F. Sheffield, ed. *Plato's Symposium: Issues in Interpretation and Reception.* Cambridge, MA: Harvard University Press, 2006.

Lexicon Iconographicum Mythologiae Classicae. Zurich: Artemis, 1974.

LiDonnici, L. R. *The Epidaurian Miracle Inscriptions: Text, Translation, and Commentary.* Atlanta: Scholars Press, 1995.

Liebeschuetz, J.H.W.G. *Continuity and Change in Roman Religion.* Oxford: Clarendon Press, 1979.

Lilja, Saara. *Dogs in Ancient Greek Poetry.* Helsinki: Commentationes Humanarum Litterarum, 56 (1976). Societas Scientiarum Fennica.

Lindsay, H. "Death-Pollution and Funerals in the City of Rome." In *Death and Disease in the Ancient City,* edited by V. M. Hope and E. Marshall. London: Routledge, 2000.

Lissarrague, François. "Aesop, between Man and Beast: Ancient Portraits and Illustrations." In *Not the Classical Ideal: Athens and the Construction of the Other in Greek Art*, edited by Beth Cohen. Leiden: Brill, 2001.

Lissarrague, François. "The Sexual Life of Satyrs." In *Before Sexuality: The Construction of Erotic Experience in the Ancient Greek World*, edited by David M. Halperin, John J. Winkler, and Froma I. Zeitlin. Princeton, NJ: Princeton University Press, 1990.

Littré, Émile. *Œuvres complètes d'Hippocrate.* 10 vols. Paris: Baillière, 1839–1861.

Llewellyn-Jones, L. "Herakles Re-dressed: Gender, Clothing and the Construction of a Greek Hero." In *Herakles and Hercules: Exploring a Graeco-Roman Divinity,* edited by L. Rawlings and H. Bowden. Swansea: The Classical Press of Wales, 2005.

Lloyd, G.E.R. *Aristotelian Explorations*. Cambridge: Cambridge University Press, 1996.

Lloyd, G.E.R. "The Hippocratic Question." *Classical Quarterly* 25 (1975): 171–92. Reprinted in *Methods and Problems in Greek Science*. Cambridge: Cambridge University Press, 1990.

Lloyd, G.E.R., ed. *Hippocratic Writings*. New York: Penguin, 1978.

Lloyd, G.E.R. *In the Grip of Disease*. Oxford: Oxford University Press, 2003.

Lloyd, G.E.R. *Magic, Reason, and Experience: Studies in the Origin and Development of Greek Science*. Cambridge: Cambridge University Press, 1979.

Lloyd, G.E.R. *Science, Folklore, and Ideology*. Cambridge: Cambridge University Press, 1983.

Lloyd, Genevieve. 1993. *The Man of Reason: "Male" and "Female" in Western Philosophy*. 2nd ed. London: Routledge, 1993.

Lloyd-Jones, Hugh, and N.G. Wilson, eds. *Sophoclis Fabulae*. Oxford: Oxford University Press, 1990.

Lobel, E., and D.L. Page. *Poetarum Lesbiorum fragmenta*. Oxford: Clarendon Press, 1955.

Lones, T.E. *Aristotle's Researches in Natural Science*. London: West, Newman & Co., 1912.

Longrigg, J. *Greek Medicine from the Heroic to the Hellenistic Age: A Source Book*. New York: Routledge, 1998.

Longrigg, J. *Greek Rational Medicine: Philosophy and Medicine from Alcmaeon to the Alexandrians*. London: Routledge, 1993.

Longrigg, J. "Medicine and the Lyceum." In *Ancient Medicine in Its Socio-Cultural Context: Papers Read at the Congress Held at Leiden University 13–15 April 1992*, edited by P. van der Eijk, H.F.J. Horstmanshoff, and P.H. Schrijvers. Vol. 2. Amsterdam: Rodopi, 1995.

Lonie, I.M. *The Hippocratic Treatises "On Generation," "On the Nature of the Child," "Diseases IV."* Berlin: Walter de Gruyter, 1981.

Loraux, N. *The Children of Athena: Athenian Ideas about Citizenship and the Division between the Sexes*. 2nd ed. Translated by Caroline Levine. Princeton, NJ: Princeton University Press, 1993.

Loraux, N. "Herakles: The Super-male and the Feminine." In *Before Sexuality: The Construction of Erotic Experience in the Ancient Greek World*, edited by D.M. Halperin, J.J. Winkler, and F.I. Zeitlin. Princeton, NJ: Princeton University Press, 1990.

Loraux, Nicole. *The Experiences of Tiresias: The Feminine and the Greek Man*. Translated by P. Wissig. Princeton, NJ: Princeton University Press, 1995.

Lozano, Fernando. "*Divi Augusti* and *Theoi Sebastoi*: Roman Initiatives and Greek Answers." *Classical Quarterly* 57 (2007): 139–52.

Macfarlane, P. *A Philosophical Commentary on Aristotle's De spiritu*. Ph.D. dissertation, Duquesne University, 2007.

Macfarlane, P., and R. Polansky. "Disability in Earlier Greek Philosophers." *Skepsis* 15 (2004): 25–41.

MacIntyre, A. *After Virtue: A Study in Moral Theory*. 2nd ed. London: Duckworth, 1985.

Mackay, E.A. "The Hairstyle of Herakles." In *Essays in Honor of Dietrich von Bothmer*, edited by A.J. Clark and J. Gaunt. Amsterdam: Allard Pierson, 2002.

Mackie, C. "The Earliest Jason: What's in a Name?" *Greece and Rome* 48 (2001): 1–17.
Maguire, H. "The Profane Aesthetic in Byzantine Art and Literature." *Dumbarton Oaks Papers* 53 (1999): 189–205.
Majno, G. *The Healing Hand: Man and Wound in the Ancient World*. Cambridge, MA: Harvard University Press, 1975.
Malinowski, B. "Magic, Science and Religion." In *Science, Religion and Reality*, edited by J. Needham. London: MacMillan, 1925.
Manetti, D. "Aristotle and the Role of Doxography in the Anonymous Londiniensis (PBRLIBR INV. 137)." In *Ancient Histories of Medicine: Essays in Medical Doxography and Historiography in Classical Antiquity*, edited by P. van der Eijk. Leiden: Brill, 1999.
Manuli, P. "Donne mascoline, femminine sterile, vergine perpetue. La ginecologia greca tra Ippocrate e Sorano." In *Madre materia: Sociologica e biologia della donna greca*, edited by S. Campese, P. Manuli, and G. Sissa. Turin: Boringhieri, 1983.
Marchant, E. C., and O. J. Todd. *Xenophon*. Vol. 4. Cambridge, MA: Harvard University Press, 1923.
Mastrangelo, M. "Oedipus and Polyneices: Characterization and the Self in Sophocles' *Oedipus at Colonus*." *Materiali e Discussioni Per L'Analisi Dei Testi Classici* 44 (2000): 35–81.
Mastrangelo, M. *The Roman Self in Late Antiquity: Prudentius and the Poetics of the Soul*. Baltimore: Johns Hopkins University Press, 2008.
Mastronarde, Donald J., ed. *Euripides: Medea*. Cambridge: Cambridge University Press, 2002.
Matheson, S. "The Elder Claudia: Older Women in Roman Art." In *I Claudia II*, edited by Diana E. E. Kleiner and Susan B. Matheson. Austin: University of Texas Press, 2000.
May, M. T. *Galen on the Usefulness of the Parts of the Body*. 2 vols. Ithaca, NY: Cornell University Press, 1968.
Mayhew, R. *The Female in Aristotle's Biology*. Chicago: University of Chicago Press, 2004.
McCarthy, Kathleen. *Slaves, Masters, and the Art of Authority in Plautine Comedy*. Princeton, NJ: Princeton University Press, 2000.
McCormick, M. "Sounding Early Medieval Holiness." In *The Long Morning of Medieval Europe: New Directions in Early Medieval Studies*, edited by J. R. Davis and M. McCormick. Hampshire: Ashgate, 2008.
McDonald, Mary Francis, trans. *Lactantius, Minor Works*. Washington, DC: Catholic University of America Press, 1965.
McEvilley, Thomas. *The Shape of Ancient Thought: Comparative Studies in Greek and Indian Philosophies*. New York: Allworth Communications, 2002.
McNiven, Timothy J. "Behaving Like an Other: Telltale Gestures in Athenian Vase Painting." In *Not the Classical Ideal: Athens and the Construction of the Other in Greek Art*, edited by Beth Cohen. Leiden: Brill, 2001.
McPherran, M. "Medicine, Magic, and Religion in Plato's *Symposium*." In *Plato's Symposium: Issues in Interpretation and Reception*, edited by J. Lesher, D. Nails, and F. Sheffield. Cambridge, MA: Harvard University Press, 2006.
McPherran, M. "Socrates, Crito, and Their Debt to Asclepius." *Ancient Philosophy* 23 (2003): 71–92.

McVaugh, M. "Therapeutic Strategies: Surgery." In *Western Medical Thought from Antiquity to the Middle Ages.* Edited by M.D. Grmek. Translated by A. Shugaar. Cambridge, MA: Harvard University Press, 1999.

Mendelsohn, E. *Heat and Life: The Development of the Theory of Animal Heat.* Cambridge, MA: Harvard University Press, 1964.

Meskell, L. "Archaeologies of Identity." In *Archaeological Theory Today,* edited by I. Hodder, 187–213. Cambridge: Cambridge University Press, 2001.

Meskell, L. "Writing the Body in Archaeology." In *Reading the Body: Representations and Remains in the Archaeological Record,* edited by A.E. Rautman. Philadelphia: University of Pennsylvania Press, 2000.

Metropoulou, E. "He latreia tou Herakle ste Boioteia." *Epeterida Boiotikon Meleton* 3 (2000): 624–81.

Migne, J.-P. ed. *Patrologiae Cursus Completus, Series Graeca.* Paris: J.-P. Migne, 1857–1866.

Migne, J.-P., ed. *Patrologiae Cursus Completus, Series Latina.* Paris: J.-P. Migne, 1844–1864.

Miller, H.W. "The Aetiology of Disease in Plato's Timaeus." *Transactions of the American Philological Association* 93 (1962): 175–87.

Miller, S.G. *Ancient Greek Athletics.* New Haven, CT: Yale University Press, 2004.

Mommsen, H. *Der gelagerte Herakles.* Berlin: De Gruyter, 1971.

Montserrat, D. "Experiencing the Male Body in Roman Egypt." In *When Men Were Men: Masculinity, Power and Identity in Classical Antiquity,* edited by L. Foxhall and J. Salmon. London: Routledge, 1998.

Moreno, P. *Lisippo: L'arte e la fortuna.* Milan: Fabbri, 1995.

Moretti, G. "L'arena, Cesare e il mito: appunti sul *De Spectaculis* di Marziale." *Maia* 44 (1992): 55–63.

Morris, C. 1973. *The Discovery of the Individual 1050–1200.* New York: Harper and Row, 1973.

Morris, I. *Death-Ritual and Social Structure in Classical Antiquity.* Cambridge: Cambridge University Press, 1992.

Morris, Ian. "Remaining Invisible: The Archaeology of the Excluded in Classical Athens." In *Women and Slaves in Greco-Roman Culture: Differential Equation,* edited by S.R. Joshel and Sheila Murnaghan. London: Routledge, 1998.

Murgatroyd, Paul. *Mythical Monsters in Classical Literature.* London: Duckworth, 2007.

Mustakallio, K. "Roman Funerals: Identity, Gender and Participation." In *Hoping for Continuity: Childhood, Education and Death in Antiquity and the Middle Ages,* edited by K. Mustakallio, J. Hansks, H.L. Sanio, and V. Vuolanto. Rome: *Acti Instituti Romani Findlandiae* 33 (2005): 179–90.

Musurillo, H. "Euripides' *Medea*: A Reconsideration." *American Journal of Philology* 87 (1966): 52–74.

Naussbaum, Martha Craven. *Aristotle's De motu animalium.* (Text with translation, commentary, and interpretive essays.) Princeton, NJ: Princeton University Press, 1978.

Naussbaum, Martha Craven. *The Fragility of Goodness: Luck and Ethics in Greek Tragedy and Philosophy.* Cambridge: Cambridge University Press, 1986.

Nees, L. "Theodulf's Mythical Silver Hercules Vase, Poetica Vanitas, and the Augustinian Critique of the Roman Heritage." *Dumbarton Oaks Papers* 41 (1987): 443–51.

Newmyer, S. T. *Animals, Rights, and Reason*. London: Routledge, 2006.
Nicgorski, A. M. "The Magic Knot of Herakles, the Propaganda of Alexander the Great and the Tomb II at Vergina." In *Herakles and Hercules: Exploring a Graeco-Roman Divinity*, edited by L. Rawlings and H. Bowden. Swansea: The Classical Press of Wales, 2005.
Nietzsche, Friedrich. "The Greek Woman." Translated by M. A. Mügge. In *The Complete Works of Friedrich Nietzsche*. Vol. 2. London: T. N. Foulis, 1871 (reprinted 1911).
Nock, A. D. 1932. "Cremation and Burial in the Roman Empire." *Harvard Theological Review* 25 (1932): 321–59; reprinted in *Essays on Religion and the Ancient World*, edited by Z. Stewart. 2 vols. Oxford: Clarendon Press, 1972.
Noy, D. "Building a Roman Funeral Pyre." *Antichthon* 34 (2000): 30–45.
Noy, D. "'Goodbye Livia': Dying in the Roman Home." In *Memory and Mourning: Studies on Roman Death*, edited by V. M. Hope and J. Huskinson. Oxford: Oxbow, forthcoming.
Nussbaum, M. *The Therapy of Desire: Theory and Practice in Hellenistic Ethics*. Princeton, NJ: Princeton University Press, 1994.
Nutton, V. *Ancient Medicine*. London: Routledge, 2004.
Nutton, V. "Ancient Medicine: Asclepius Transformed." In *Science and Mathematics in Ancient Greek Culture*, edited by C. J. Tuplin and T. E. Rihll. Oxford: Oxford University Press, 2002.
Nutton, V. "Erasistratus." In *Brill's New Pauly Encyclopedia of the Ancient World*. Vol. 5. 2004.
Nutton, V. *From Democedes to Harvey: Studies in the History of Medicine*. London: Ashgate, 1988.
Nutton, V. "From Galen to Alexander: Aspects of Medicine and Medical Practice in Late Antiquity." *Dumbarton Oaks Papers* 38 (1984): 1–14.
Nutton, V. "Galen at the Bedside: The Methods of a Medical Detective." In *Medicine and the Five Senses*, edited by W. F. Bynum and R. Porter. Cambridge: Cambridge University Press, 1993.
Nutton, V. "God, Galen, and the Depaganisation of Ancient Medicine." In *Religion and Medicine in the Middle Ages*, edited by P. Biller and J. Ziegler. Woodbridge, Suffolk: York Medieval Press, 2001.
Nutton, V. "Healers in the Medical Marketplace: Towards a Social History of Graeco-Roman Medicine." In *Medicine in Society: Historical Essays*, edited by A. Wear. Cambridge: Cambridge University Press, 1992.
Nutton, V. "The Medical Meeting Place." In *Ancient Medicine in Its Socio-Cultural Context*, edited by P. van der Eijk, H.F.J. Horstmanshoff, and P. H. Schrijvers. 2 vols. Rodopi: Amsterdam, 1995.
Nutton, V. "Medicine in Late Antiquity and the Early Middle Ages." In *The Western Medical Tradition: 800 B.C.–1800 A.D.*, edited by L. I. Conrad. Cambridge: Cambridge University Press, 1995.
Nutton, V. "Murders and Miracles: Lay Attitudes to Medicine in Classical Antiquity." In *Patients and Practitioners: Lay Perceptions of Medicine in Pre-Industrial Society*, edited by R. Porter. Cambridge: Cambridge University Press, 1985.
Nutton, V. "Roman Medicine: Tradition, Confrontation, Assimilation." *Aufstieg und Niedergang der Römischen Welt*. Band 2.37.1, Berlin: De Gruyter, 1993.
Nutton, V. "The Seeds of Disease: An Explanation of Contagion and Infection from the Greeks to the Renaissance." *Medical History* 27 (1983): 1–34.

O'Connor, J. T. *The Hidden Manna: A Theology of the Eucharist*. San Francisco: Ignatius Press, 1988.

Ochs, D. J. *Consolatory Rhetoric. Grief, Symbol and Ritual in the Greco-Roman Era*. Columbia: University of South Carolina Press, 1993.

Ogden, Daniel. *Magic, Witchcraft, and Ghosts in the Greek and Roman Worlds: A Sourcebook*. Oxford: Oxford University Press, 2002.

OKell, E. R. "*Hercules Furens* and Nero: The Didactic Purpose of Senecan Tragedy." In *Herakles and Hercules: Exploring a Graeco-Roman Divinity*, edited by L. Rawlings and H. Bowden. Swansea: The Classical Press of Wales, 2005.

Oldfather, C. H., ed. and trans. *Diodorus Siculus: Library of History, Volume 1, Books 1–2.34*. Cambridge, MA: Harvard University Press, 1933.

Oldfather, C. H., ed. and trans. *Diodorus Siculus: Library of History, Volume 2, Books 2.35–4.58*. Cambridge, MA: Harvard University Press, 1935.

Oser-Grote, C. *Aristoteles und das Corpus Hippocraticum*. Stuttgart: Steiner, 2004.

Padel, R. *In and Out of the Mind*. Princeton, NJ: Princeton University Press, 1992.

Padilla, M. W. *The Myths of Herakles in Ancient Greece*. Lanham, MD: University Press of America, 1998.

Page, D. L., ed. *Poetae melici graeci*. Oxford: Clarendon Press, 1962.

Palagia, O. "The Hope Herakles Reconsidered." *Oxford Journal of Archaeology* 3 (1984): 107–26.

Parker, R. *Miasma: Pollution and Purification in Early Greek Religion*. Oxford: Clarendon Press, 1983.

Parsons, Mikeal C. *Body and Character in Luke and Acts. The Subversion of Physiognomy in Early Christianity*. Grand Rapids, MI: Baker Academic, 2006.

Paster, Gail Kern. *The Body Embarrassed: Drama and the Disciplines of Shame in Early Modern England*. Ithaca, NY: Cornell University Press, 1993.

Patterson, J. R. "Patronage, Collegia and Burial in Imperial Rome." In *Death in Towns*, edited by S. Bassett. Leicester: Leicester University Press, 1992.

Peck, A. L. *Aristotle: Generation of Animals*. Cambridge, MA: Harvard University Press, 1942.

Peck, A. L., ed. and trans. *Aristotle: On the Generation of Animals*. Cambridge, MA: Harvard University Press, 1979.

Pedley, J. *Sanctuaries and the Sacred in the Ancient Greek World*. Cambridge: Cambridge University Press, 2005.

Pellegrin, P. "Ancient Medicine and Its Contribution to the Philosophical Tradition." In *A Companion to Ancient Philosophy*, edited by M. L. Gill and P. Pellegrin. Malden, MA: Blackwell, 2006.

Perkins, J. "The Self as Sufferer." *Harvard Theological Review* 85 (1992): 245–72.

Perkins, J. *The Suffering Self: Pain and Narrative Representation in the Early Christian Era*. London: Routledge, 1995.

Pestilli, Livio. "Disabled Bodies: The (Mis)Representation of the Lame in Antiquity and Their Reappearance in Early Christian and Medieval Art." In *Roman Bodies: Antiquity to the Eighteenth Century*, edited by A. Hopkins and M. Wyke. London: The British School at Rome, 2005.

Petrakos, V. *Marathon*. Athens: Archaiologike Hetaireia, 1995.

Petronius. *Satyrica*. Edited and translated by R. B. Branham and D. Kinney. London: J. M. Dent, 1996.

Pfeiffer, R., ed. *Callimachus*. 2 vols. Oxford: Oxford University Press, 1949–1953.

Philippart, G., and Trigalet, M. "Latin Hagiography before the Ninth Century: A Synoptic View." In *The Long Morning of Medieval Europe: New Directions in Early Medieval Studies*, edited by J. R. Davis and M. McCormick. Hampshire: Ashgate, 2008.

Phillips, E. D. *Greek Medicine*. London: Thames & Hudson, 1973.

Pindar. *Odes*. Translated by J. Sandys. Cambridge, MA: Harvard University Press, 1937.

Pipili, M. *Laconian Iconography of the Sixth Century*. Oxford: Oxford University Press, 1987.

Plato. *Collected Dialogues*. Edited by E. Hamilton and H. Cairns. Princeton, NJ: Princeton University Press, 1961.

Pohl, W. "Gender and Ethnicity in the Early Middle Ages." In *Gender and the Early Medieval World, East and West, 300–900*, edited by L. Brubaker and J. Smith. Cambridge: Cambridge University Press, 2004.

Pohl, W. "Introduction: Strategies of Distinction." In *Strategies of Distinction: The Construction of Ethnic Communities, 300–800*, edited by W. Pohl and H. Reimitz. Leiden, 1998.

Pohl, W. "Telling the Difference—Signs of Ethnic Identity." In *Strategies of Distinction: The Construction of Ethnic Communities, 300–800*, edited by W. Pohl and H. Reimitz. Leiden, 1998.

Polansky, R. *Aristotle's* De anima. Cambridge: Cambridge University Press, 2007.

Polansky, R. "Is Medicine Art, Science, or Practical Wisdom? Ancient and Contemporary Reflections." In *Bioethics: Ancient Themes in Contemporary Issues*, edited by M. G. Kuczewski and R. Polansky. Cambridge, MA: MIT Press, 2000.

Polansky, R. "The Unity of Plato's *Crito*." *Scholia* 6 (1997): 49–67.

Pollini, J. "Ritualizing Death in Republican Rome: Memory, Religion, Class Struggle and the Wax Ancestral Mask Tradition's Origins and Influence on Veristic Portraiture." In *Performing Death. Social Analyses of Funerary Traditions in the Ancient Near East and Mediterranean*, edited by N. Laneri. Chicago: University of Chicago Press, 2007.

Pomeroy, S. B. *Families in Classical and Hellenistic Greece: Representations and Realities*. Oxford: Clarendon Press, 1997.

Pomeroy, S. B. "TECHNIKAI KAI MOUSIKAI: The Education of Women in the Fourth Century and in the Hellenistic Period." *American Journal of Ancient History* 2 (1977): 51–68.

Pomeroy, S. B. "Women's Identity and the Family in the Classical *polis*." In *Women in Antiquity: New Assessments*, edited by R. Hawley and B. Levick. London: Routledge, 1995.

Porter, James I., ed. *Constructions of the Classical Body*. Ann Arbor: University of Michigan Press, 2002.

Potter, David S. "Odor and Power in the Roman Empire." In *Constructions of the Classical Body*, edited by J. I. Porter. Ann Arbor: University of Michigan Press, 1999.

Potter, P. *Hippocrates*. Vol. 5. Cambridge, MA: Harvard University Press, 1988.

Potter, P. *Hippocrates*. Vol. 6. Cambridge, MA: Harvard University Press, 1988.

Potter, P. *Hippocrates*. Vol. 8, Cambridge, MA: Harvard University Press, 1995.

Potter, P. "Some Principles of Hippocratic Nosology." In *La maladie et les maladies dans la collection hippocratique*, edited by P. Potter, G. Maloney, and J. Desautels. Quebec: Éditions du Sphinx, 1990.

Potter, P., G. Maloney, and J. Desautels, ed. *La maladie et les maladies dans la collection hippocratique*. Quebec: Éditions du Sphinx, 1990.

Powell, A., ed. *The Greek World*. London: Routledge, 1995.

Pratt, Louise. "The Parental Ethos of the *Iliad*." In *Constructions of Childhood in Ancient Greece and Italy*. Hesperia Supplement 41, edited by Ada Cohen and Jeremy B. Rutter. Athens: ASCSA, 2007.

Preus, A. "Aristotle and Hippocratic Gynecology." In *Aristotelis als Wissenschaftstheoretiker (Schriften zur Geschichte und Kultur der Antike 22)*, edited by J. Irmscher and R. Müller. Berlin: Akademie-Verlag, 1983.

Price, S. "From Noble Funerals to Divine Cult: The Consecration of Roman Emperors." In *Rituals of Royalty: Power and Ceremonial in Traditional Societies*, edited by D. Cannadine and S. Price. Cambridge: Cambridge University Press, 1987.

Rabel, Robert J. "The Harpies in the '*Aeneid*.'" *Classical Journal* 80 (1985): 317–25.

Rackham, H., ed. and trans. *Aristotle*. Vol. 21. Cambridge, MA: Harvard University Press, 1944.

Raval, Shilpa. "Cross-Dressing and 'Gender Trouble' in the Ovidian Corpus." *Helios* 29 (2002): 149–72.

Rawlings, L. "Hannibal and Hercules." In *Herakles and Hercules: Exploring a Graeco-Roman Divinity*, edited by L. Rawlings and H. Bowden. Swansea: The Classical Press of Wales, 2005.

Rawlings, L., and H. Bowden, eds. *Herakles and Hercules: Exploring a Graeco-Roman Divinity*. Swansea: The Classical Press of Wales, 2005.

Reed, J. D. "Arsinoe's Adonis and the Poetics of Ptolemaic Imperialism." *Transactions of the American Philological Association* 130 (2000): 319–51.

Reeder, Ellen D. *Pandora: Women in Classical Greece*. Princeton, NJ: Princeton University Press, 1995.

Rees, R. "The Emperors' New Names: Diocletian Jovius and Maximian Herculius." In *Herakles and Hercules: Exploring a Graeco-Roman Divinity*, edited by L. Rawlings and H. Bowden. Swansea: The Classical Press of Wales, 2005.

Reeve, C.D.C., trans. *Plato, Republic*. Indianapolis: Hackett, 2004.

Rengakos, A., and A. Tsakmakis, ed. *Brill's Companion to Thucydides*. Leiden: Brill, 2006.

Repath, Ian, trans. "The *Physiognomy* of Adamantius the Sophist." In *Seeing the Face, Seeing the Soul: Polemon's Physiognomy from Classical Antiquity to Medieval Islam*, edited by S. Swain. Oxford: Clarendon Press, 2007.

Reynolds, P. L. *Food and the Body: Some Peculiar Questions in High Medieval Theology*. Leiden: Brill, 1999.

Richlin, A. "Cicero's Head." In *Constructions of the Classical Body*, edited by J. Porter. Ann Arbor: University of Michigan Press, 1999.

Richlin, A. "Emotional Work: Lamenting the Roman Dead." In *Essays in Honor of Gordon Williams: Twenty-Five Years at Yale*, edited by E. Tywalsky and C. Weiss. New Haven, CT: Schwab, 2001.

Richlin, A. *The Garden of Priapus: Sexuality and Aggression in Roman Humor*. New Haven, CT: Yale University Press, 1983.

Richlin, A. "Making Up a Woman: The Face of Roman Gender." In *Off with Her Head! The Denial of Women's Identity in Myth, Religion, and Culture*, edited by H. Eilberg Schwartz and Wendy Doniger. Berkeley: University of California Press, 1995.

Richlin, A. "Towards a History of Body History." In *Inventing Ancient Culture: Historicism, Periodization, and the Ancient World*, edited by M. Golden and P. Toohey. London: Routledge, 1997.

Richlin, A. "Zeus and Metis: Foucault, Feminism, Classics." *Helios* 18 (1991): 160–80.

Riddle, J. *Dioscorides on Pharmacy and Medicine*. Austin: University of Texas Press, 1985.

Rist, J. *Augustine: Ancient Thought Baptized*. Cambridge: Cambridge University Press, 1994.

Roller, M. B. *Dining Posture in Ancient Rome: Bodies, Values, and Status*. Princeton, NJ: Princeton University Press, 2006.

Roselli, A. *[Aristotele] de spiritu*. Pisa: ETS Editrice, 1992.

Rosenthal, Franz. *The Classical Heritage in Islam*. Berkeley: University of California Press, 1975.

Rosenwein, B. *To Be the Neighbor of St. Peter: The Social Meaning of Cluny's Property, 909–1049*. Ithaca, NY: Cornell University Press, 1989.

Rösler, W. *Dichter und Gruppe: Eine Untersuchung zu den Bedingungen und zur historischen Funktion fruher griechischer Lyrik am Beispiel Alkaios*. Munich: W. Fink, 1980.

Rothwell, Kenneth S. *Nature, Culture, and the Origins of Greek Comedy: A Study of Animal Choruses*. Cambridge: Cambridge University Press, 2007.

Russell, D. A., trans. *Quintilian, the Orator's Education, Books 1–2*. Cambridge, MA: Harvard University Press, 2001.

Rusten, J., and I. C. Cunningham. *Theophrastus, Characters; Herodas, Mimes, Sophron and Other Mime Fragments*. Cambridge, MA: Harvard University Press, 2002 (Loeb Classical Library).

Salazar, C. F. *The Treatment of War Wounds in Graeco-Roman Antiquity*. Leiden: Brill, 2000.

Sallares, R. *The Ecology of the Ancient Greek World*. Ithaca, NY: Cornell University Press, 1991.

Sallares, R. *Malaria in Ancient Rome*. Oxford: Oxford University Press, 2003.

Salowey, C. A. "Herakles and the Waterworks: Mycenaean Dams, Classical Fountains, Roman Aqueducts." In *Archaeology in the Peloponnese: New Excavations and Research*, edited by A. K. Sheedy. Oxford: Oxbow, 1994.

Sassi, M. M. *The Science of Man in Ancient Greece*. Translated by P. Tucker. Chicago: University of Chicago Press, 2001.

Saunders, K. B. "The Wounds in *Iliad* 13–16." *Classical Quarterly* 49 (1999): 345–63.

Saxonhouse, Arlene W. *Free Speech and Democracy in Ancient Athens*. Cambridge: Cambridge University Press, 2006.

Scanlon, T. *Eros and Greek Athletics*. New York: Oxford University Press, 2002.

Scarborough, J. "Diphilus of Siphnos and Hellenistic Medical Dietetics." *Journal of the History of Medicine and Allied Sciences* 25 (1970): 194–201.

Scarborough, J. *Roman Medicine*. Ithaca, NY: Cornell University Press, 1969.

Scarborough, J. "Theoretical Questions in Hippocratic Pharmacology." In *Formes de pensée dans la collection hippocratique: Actes du IVᵉ Colloque international hippocratique, Lausanne, 21–26 septembre 1981*, edited by F. Lasserre and P. Mudry. Geneva: Droz, 1983.

Schefold, K. *Die Urkönige, Perseus, Bellerephon, Herakles und Theseus in der klassischen und hellenistischen Kunst*. Munich: Hirmer Verlag, 1988.

Schefold, K. *Gods and Heroes in Late Archaic Greek Art.* Translated by A. Griffiths. Cambridge: University Press, 1992.

Schiefsky, M. *Hippocrates, On Ancient Medicine.* Leiden: Brill, 2005.

Schneider, R.M. "Der Hercules Farnese." In *Meisterwerke der antiken Kunst,* edited by L. Giuliani. Munich: Beck, 2005.

Schulz-Flügel, Eva, trans. *Tertullien, De virginibus velandis.* Sources chrétiennes 424. Paris: Les Éditions du Cerf, 1997.

Scott, E. "Unpicking a Myth: The Infanticide of Female and Disabled Infants in Antiquity." In *Proceedings of the Theoretical Roman Archaeology Conference 2000,* edited by G. Davies, A. Garner, and K. Lockyear. Oxford: Oxbow, 2001.

Sebesta, J.L. "Women's Costume and Feminine Civic Morality in Augustan Rome." In *Gender and the Body in the Ancient Mediterranean,* edited by M. Wyke. Malden, MA: Blackwell, 1998.

Seidler, Victor. "Reason, Desire, and Male Sexuality." In *The Cultural Construction of Sexuality,* edited by Patricia Caplan. London: Routledge, 1987.

Seiwert, H. "Ofer." In *Handbuch religionswissenschaftlicher Grundbegriffe,* edited by H. Cancik et al. Vol. 4. Stuttgart, 1998.

Sellars, J. *Stoicism.* Berkeley: University of California Press, 2006.

Settis, S. *I Greci: Storia, cultura, arte, societa.* Torino: G. Einaudi, 1996.

Shackleton Bailey, D.R., ed. and trans. *Martial Epigrams.* Vol. 1. Cambridge, MA: Harvard University Press, 1993.

Shapiro, H.A. *Art and Cult under the Tyrants in Athens.* Mainz: P. von Zabern, 1989.

Shapiro, H.A. "Courtship Scenes in Attic Vase Painting." *American Journal of Archaeology* 85 (1981): 133–43.

Sharples, R.W. "Common to Body and Soul: Peripatetic Approaches After Aristotle." In *Common to Body and Soul: Philosophical Approaches to Explaining Living Behaviour in Greco-Roman Antiquity,* edited by R.A.H. King. Berlin: Walter de Gruyter, 2006.

Sharples, R.W. "Philosophy for Life." In *The Cambridge Companion to the Hellenistic World,* edited by G.R. Bugh. Cambridge: Cambridge University Press, 2006.

Sharples, R.W. *Stoics, Epicureans, and Sceptics: An Introduction to Hellenistic Philosophy.* London: Routledge, 1996.

Sharples, R.W., and P. van der Eijk, trans. *Nemesius, On the Nature of Man.* Liverpool: Liverpool University Press, 2008.

Shaw, B. "Raising and Killing Children: Two Roman Myths." *Mnemosyne* 54 (2001): 31–77.

Shaw, Brent D. "Body, Power, Identity: The Passions of the Martyrs." *Journal of Early Christian Studies* 4 (1996): 269–312.

Siegel, R.E. *Galen's System of Physiology and Medicine.* Basel: Karger, 1968.

Sifakis, G.M. *Parabasis and Animal Choruses: A Contribution to the History of Attic Comedy.* London: Athlone Press, 1971.

Sihvola, J. "Aristotle on the Individuality of Self." In *Ancient Philosophy of the Self,* edited by P. Remes and J. Sihvola. Helsinki: Springer, 2008.

Silver, Morris. *Taking Ancient Mythology Economically.* Leiden: Brill, 1992.

Simon, B. *Mind and Madness in Ancient Greece.* Ithaca, NY: Cornell University Press, 1978.

Singer, P.N. *Galen: Selected Works.* New York: Oxford University Press, 1997.

Sissa, G. *Greek Virginity.* Translated by A. Goldhammer. Cambridge, MA: Harvard University Press, 1990.

Sissa, G. "Sexual Bodybuilding: Aeschines against Timarchus." In *Constructions of the Classical Body*, edited by J. I. Porter. Ann Arbor: University of Michigan Press, 1999.

Skinner, M. B. "Alexander and Ancient Greek Sexuality: Some Theoretical Considerations." In *Responses to Alexander*, edited by F. Greenland. Cambridge: Cambridge University Press, forthcoming.

Skinner, M. B. *Sexuality in Greek and Roman Culture*. Oxford: Blackwell, 2005.

Smith, W. D. "Erasistratus's Dietetic Medicine." *Bulletin of the History of Medicine* 56 (1982): 398–409.

Smith, W. D. *Hippocrates*. Vol. 7, Cambridge, MA: Harvard University Press, 1994.

Smith, W. D. *The Hippocratic Tradition*, Ithaca, NY: Cornell University Press, 1979.

Smith, W. D. "So-Called Possession in Pre-Christian Greece." *Transactions of the American Philological Association* 96 (1965): 403–26.

Snell, B. *The Discovery of the Mind: The Greek Origins of European Thought*. Translated by T. G. Rosenmeyer. Cambridge, MA: Harvard University Press, 1953.

Solmsen, F. "Greek Philosophy and the Discovery of the Nerves." *Museum Helveticum* 18 (1961): 169–97. Reprinted in *Kleine Schriften*. 3 vols. Hildesheim: G. Olms, 1968–1982.

Solmsen, F. "Tissues and the Soul: Philosophical Contributions to Physiology." *Philosophical Review* 59 (1950): 435–68.

Solmsen, F. "The Vital Heat, the Inborn Pneuma and the Aether." *Journal of Hellenic Studies* 77 (1957): 119–23. Reprinted in *Kleine Schriften*. 3 vols. Hildesheim: G. Olms, 1968–1982.

Sorabji, R. *Animal Minds and Human Morals: The Origins of the Western Debate*. Ithaca, NY: Cornell University Press, 1995.

Sorabji, R. *Emotion and Peace of Mind: From Stoic Agitation to Christian Temptation*. Oxford: Oxford University Press, 2002.

Sorabji, R. "Greco-Roman Varieties of Self." In *Ancient Philosophy of the Self*, edited by P. Remes and J. Sihvola. Helsinki: Springer, 2008.

Sorabji, R. "The Mind-Body Relation in the Wake of Plato's *Timaeus*." In *Plato's Timaeus as Cultural Icon*, edited by G. J. Reydams-Schils. Notre Dame, IN: University of Notre Dame Press, 2003.

Sorabji, R. *Self*. Chicago: University of Chicago Press, 2006.

Spencer, W. G. *Celsus: De Medicina*. Vol. 1. Cambridge, MA: Harvard University Press, 1935.

Stafford, E. "Vice or Virtue: Herakles and the Art of Allegory." In *Herakles and Hercules: Exploring a Graeco-Roman Divinity*, edited by L. Rawlings and H. Bowden. Swansea: The Classical Press of Wales, 2005.

Stähler, K. "Zu Lysipps Herakles in Tarent." *Boreas* 20 (1997): 43–47.

Stansbury-O'Donnel, M. D. "Polygnotos's *Ilioupersis*: A New Reconstruction." *American Journal of Archaeology* 93 (1989): 203–15.

Stears, K. "Death Becomes Her: Gender and Athenian Death Ritual." In *The Sacred and the Feminine in Ancient Greece*, edited by S. Blundell and M. Williamson. London: Routledge, 1998.

Steckerl, F. *The Fragments of Praxagoras of Cos and His School*. Leiden: Brill, 1958.

Steiner, Gary. *Anthropocentrism and Its Discontents: The Moral Status of Animals in the History of Western Philosophy*. Pittsburgh: University of Pittsburgh Press, 2005.

Stewart, A. *Art, Desire, and the Body in Ancient Greece*. Cambridge: Cambridge University Press, 1997.

Stewart, Andrew, and Celina Gray. "Confronting the Other: Childbirth, Aging, and Death on an Attic Tombstone at Harvard." In *Not the Classical Ideal: Athens and the Construction of the Other in Greek Art*, edited by Beth Cohen. Leiden: Brill, 2001.

Stewart, P. *Statues in Roman Society: Representation and Response.* Oxford: Oxford University Press, 2003.

Stock B. "Ethical Values and the Literary Imagination in the Later Ancient World." *New Literary History* 29 (1998): 1–13.

Stone, S. "The Toga: From National to Ceremonial Costume." In *The World of Roman Costume*, edited by J. Sebesta and L. Bonfante. Madison: University of Wisconsin Press, 1994.

Storace, P. "Marble Girls of Athens." *New York Review of Books*, 1996 (Oct. 3), 7–11.

Strawson, G. "The Sense of Self." In *From Soul to Self*, edited by M.J.C. Crabbe. New York: Routledge, 1999.

Strohmaier, G. "Reception and Tradition: Medicine in the Byzantine and Arab World." In *Western Medical Thought from Antiquity to the Middle Ages*. Edited by M.D. Grmek. Translated by A. Shugaar. Cambridge, MA: Harvard University Press, 1999.

Studtmann, P. "Living Capacities and Vital Heat in Aristotle." *Ancient Philosophy* 24 (2004): 365–79.

Suter, A. *Lament: Studies in the Ancient Mediterranean and Beyond.* Oxford: Oxford University Press, 2008.

Sutton, R. F., Jr. "Pornography and Persuasion on Attic Pottery." In *Pornography and Representation in Greece and Rome*, edited by A. Richlin. New York: Oxford University Press, 1992.

Sweeney, E. C. *Logic, Theology, and Poetry in Boethius, Abelard, and Alan of Lille.* New York: Palgrave Macmillan, 2006.

Tanner, J. "Nature, Culture and the Body in Classical Greek Religious Art." *World Archaeology* 33 (2001): 257–276.

Tarlow, S. "The Aesthetic Corpse in Nineteenth-Century Britain." In *Thinking Through the Body: Archaeologies of Corporeality*, edited by Y. Hamilakis, M. Pluciennik, and S. Tarlow. New York: Kluwer/Plenum, 2002.

Tecusan, M. *The Fragments of the Methodists.* 2 vols. Leiden: Brill, 2004.

Temkin, O. "Byzantine Medicine: Tradition and Empiricism." *Dumbarton Oaks Papers* 16 (1962): 97–115.

Temkin, O. *The Double Face of Janus and Other Essays in the History of Medicine.* Baltimore: Johns Hopkins University Press, 1977.

Temkin, O. *Galenism: Rise and Decline of a Medical Philosophy.* Ithaca, NY: Cornell University Press, 1973.

Temkin, O. "Greek Medicine as Science and Craft." *Isis* 44 (1953): 213–25.

Temkin, O. "Studies on Late Alexandrian Medicine." *Bulletin of the History of Medicine* 3 (1935): 405–30.

Thivel, A., and A. Zucker, ed. *Le normal et la pathologique dans la collection hippocratique.* Nice: Université de Nice, 2002.

Thomas, R. *Herodotus in Context: Ethnography, Science, and the Art of Persuasion.* Cambridge: Cambridge University Press, 2000.

Thomas, R. "Thucydides' Intellectual Milieu and the Plague." In *Brill's Companion to Thucydides*, edited by A. Rengakos and A. Tsakmakis. Leiden: Brill, 2006.

Thumiger, Chiara. "ἀνάγκης ζεύγματ' ἐμπεπτώκαμεν: Greek Tragedy between Human and Animal." *Leeds International Classical Studies* 7 (2008): 1–21.

Totelin, L.M.V. *Hippocratic Recipes: Oral and Written Transmission of Pharmacological Knowledge in Fifth- and Fourth-Century Greece.* Leiden: Brill, 2009.

Toynbee, J.M.C. *Death and Burial in the Roman World.* London: Thames & Hudson, 1971.

Tracy, T. *Physiological Theory and the Doctrine of the Mean in Plato and Aristotle.* Chicago: Loyola University Press, 1969.

Trendall, A.D., and A. Cambitoglou. *The Red-Figured Vases of Apulia.* Oxford: Clarendon Press, 1978–1982.

Tress, D.M. "Aristotle against the Hippocratics on Sexual Generation: A Reply to Coles." *Phronesis* 44 (1999): 228–41.

Tsekourakis, D. "Pythagoreanism or Platonism and Ancient Medicine? The Reasons for Vegetarianism in Plutarch's 'Moralia.'" *Aufstieg und Niedergang der römischen Welt,* 2nd ser., 36, part 1 (1987): 366–93.

Tuck, R., ed. *Thomas Hobbes: Leviathan.* Cambridge: Cambridge University Press, 1996.

Turcan, Marie, trans. *Tertullien, De cultu feminarum,* Sources chrétiennes 173. Paris: Les Éditions du Cerf, 1971.

Twu, K.S.-L. "This Is Comforting? Boethius's Consolation of Philosophy, Rhetoric, Dialectic, and *Unicum Illud Inter Homines Deumque Commercium.*" In *New Directions in Boethian Studies,* edited by N.H. Kaylor Jr. and P.E. Phillips. Kalamazoo, MI: Medieval Institute Publications, 2007.

Tyrrell, W.M. Blake. *Amazons: A Study in Athenian Mythmaking.* Baltimore: Johns Hopkins University Press, 1984.

Uhlenbrock, J.P. *Herakles. Passage of the Hero through 1000 Years of Classical Art.* New York: A.D. Caratzas, 1986.

Vallance, J. *The Lost Theory of Asclepiades of Bithynia.* Oxford: Oxford University Press, 1990.

Vallance, J. "The Medical System of Asclepiades of Bithynia." *Aufstieg und Niedergang der Römischen Welt,* Band 2.37.1, Berlin: De Gruyter, 1993.

van der Eijk, P. "Aristotle on 'Distinguished Physicians' and on the Medical Significance of Dreams." In *Ancient Medicine in Its Socio-Cultural Context: Papers Read at the Congress Held at Leiden University 13–15 April 1992,* edited by P. van der Eijk, H.F.J. Horstmanshoff, and P.H. Schrijvers. Vol. 2. Amsterdam: Rodopi, 1995.

van der Eijk, P.J. "Aristotle's Psycho-Physiological Account of the Soul-Body Relationship." In *Psyche and Soma. Physicians and Metaphysicians on the Mind-Body Problem from Antiquity to Enlightenment,* edited by J.P. Wright and P. Potter. Oxford: Clarendon Press, 2000.

van der Eijk, P. "Galen's Use of the Concept of 'Qualified Experience' in His Dietetic and Pharmacological Works." In *Galen on Pharmacology: Philosophy, History and Medicine,* edited by A. Debru. Leiden: Brill, 1997.

van der Eijk, P. "Historical Awareness, Historiography and Doxography in Greek and Roman Medicine." In *Ancient Histories of Medicine: Essays in Medical Doxography and Historiography in Classical Antiquity,* edited by P. van der Eijk. Leiden: Brill, 1999.

van der Eijk, P. *Medicine and Philosophy in Classical Antiquity: Doctors and Philosophers on Nature, Soul, Health and Disease.* Cambridge: Cambridge University Press, 2005.

van der Eijk, P. "The Role of Medicine in the Formation of Early Greek Thought." In *The Oxford Handbook of Presocratic Philosophy*, edited by P. Curd and D.W. Graham. New York: Oxford University Press, 2008.

van der Eijk, P. "Towards a Rhetoric of Ancient Scientific Discourse: Some Formal Characteristics of Greek Medical and Philosophical Texts (Hippocratic Corpus, Aristotle)." In *Grammar as Interpretation: Greek Literature in Its Linguistic Contexts*, edited by E.J. Bakker. Leiden: Brill, 1997.

van der Eijk, P., ed. *Ancient Histories of Medicine: Essays in Medical Doxography and Historiography in Classical Antiquity.* Leiden: Brill, 1999.

van der Eijk, P., ed. *Hippocrates in Context: Papers Read at the xith International Hippocrates Colloquium, University of Newcastle upon Tyne, 27–31 August 2002.* Leiden: Brill, 2005.

van der Eijk, P., ed. and trans. *Diocles of Carystus: A Collection of Fragments with Testimonia.* 2 vols. Leiden: Brill, 2000–2001.

van der Eijk, P., H.F.J. Horstmanshoff, and P.H. Schrijvers, ed. *Ancient Medicine in Its Socio-Cultural Context: Papers Read at the Congress Held at Leiden University 13–15 April 1992.* 2 vols. Amsterdam: Rodopi, 1995.

Van Hooff, A.J.L. "Ancient Euthanasia: 'Good Death' and the Doctor in the Graeco-Roman World." *Social Science and Medicine* 58 (2004): 975–85.

Van Hooff, A.J.L. *From Autothanasia to Suicide: Self-Killing in Classical Antiquity.* London: Routledge, 1990.

van Wees, Hans. "Trailing Tunics and Sheepskin Coats: Dress and Status in Early Greece." In *The Clothed Body in the Ancient World*, edited by Liza Cleland, Mary Harlow, and Lloyd Llewellyn-Jones. Oxford: Oxbow, 2005.

Varner, E. "Execution in Effigy: Severed Heads and Decapitated Statues in Imperial Rome." In *Roman Bodies: Antiquity to the Eighteenth Century*, edited by A. Hopkins and M. Wyke. London: The British School at Rome, 2005.

Varner, E.R. *Mutilation and Transformation: Damnatio memoriae and Roman Imperial Portraiture.* Leiden: Brill, 2004.

Vegetti, M. "Between Knowledge and Practice: Hellenistic Medicine." In *Western Medical Thought from Antiquity to the Middle Ages.* Edited by M.D. Grmek. Translated by A. Shugaar. Cambridge, MA: Harvard University Press, 1999.

Vegetti, M. "L'Épistémologie d'Érasistrate et la technologie hellénistique." In *Ancient Medicine in Its Socio-Cultural Context*, edited by P. van der Eijk, H.F.J. Horstmanshoff, and P.H. Schrijvers. 2 vols. Amsterdam: Rodopi, 1995.

Vermeule, C.C. "Herakles Crowning Himself: New Greek Statuary Types and Their Place in Hellenistic and Roman Art." *Journal of the Hellenistic Society* 77 (1957): 283–99.

Vermeule, C.C. "The Weary Herakles of Lysippos." *American Journal of Archaeology* 79 (1975): 323–32.

Vermeule, E. *Greece in the Bronze Age.* Chicago: University of Chicago Press, 1964.

Vernant, J.P. *Mortals and Immortals: Selected Essays*, edited by F.I. Zeitlin. Princeton, NJ: Princeton University Press, 1991.

Vernant, Jean-Pierre. "Dim Body, Dazzling Body." Translated by Anne M. Wilson. In *Fragments for a History of the Human Body*, edited by Michel Feher, with Ramona Naddaff and Nadia Tazi. Vol. 3, part 1. New York: Zone, 1986, 1989.

Vince, J.H., trans. *Demosthenes*, vol. 5. Cambridge, MA: Harvard University Press, 1935.

Vizgin, V.P. "Hippocratic Medicine as a Historical Source for Aristotle's Theory of the Dynameis." *Studies in the History of Medicine* 4 (1980): 1–12.

Voisin, J. L. "Les Romains, chasseurs de têtes." In *Du Châtiment dans la Cité: Supplices Corporels et Peine de Mort dans le Monde Antique*. Collection de l'École Française de Rome 79, 241–93. Rome: l'École Française de Rome, 1984.

Vollkommer, R. *Herakles in the Art of Classical Greece*. Oxford: Oxford University Press Monograph no. 25, 1988.

Von Hesberg, H. *Römische Grabbauten*. Darmstadt: Wissenschaftliche Buchgesellschaft, 1992.

von Staden, H. "Anatomy as Rhetoric: Galen on Dissection and Persuasion." *Journal of the History of Medicine and Allied Sciences* 50 (1995): 47–66.

von Staden, H. "Body and Machine: Interactions between Medicine, Mechanics, and Philosophy in Early Alexandria." In *Alexandria and Alexandrianism*, edited by M. True and K. Hamma. Malibu, CA: The J. Paul Getty Museum, 1996.

von Staden, H. "Body, Soul, and Nerves: Epicurus, Herophilus, Erasistratus, the Stoics, and Galen." In *Psyche and Soma: Physicians and Metaphysicians on the Mind-Body Problem from Antiquity to the Enlightenment*, edited by J. P. Wright and P. Potter. Oxford: Clarendon Press, 2000.

von Staden, H. "The Discovery of the Body: Human Dissection and Its Cultural Contexts in Ancient Greece." *Yale Journal of Biology and Medicine* 65 (1992): 223–41.

von Staden, H. "Hairesis and Heresy: The Case of the *haireseis iatrikai*." In *Jewish and Christian Self-Definition*. Vol. 3, *Self-Definition in the Graeco-Roman World*, edited by B. F. Meyer and E. P. Sanders. London: S.C.M., 1982.

von Staden, H. *Herophilus: The Art of Medicine in Early Alexandria*. Cambridge: Cambridge University Press, 1989.

von Staden, H. "Lexicography in the Third Century B.C.: Bacchius of Tanagra, Erotian, and Hipócrates." In *Tratados hipocráticos: Estudios acerca de su contenido, forma e influencia; actas del VIIe Colloque international hippocratique, Madrid, 24–29 de septiembre de 1990*, edited by J. A. López Férez. Madrid: Universidad Nacional de Educación a Distancia, 1992.

von Staden, H. "Teleology and Mechanism: Aristotelian Biology and Early Hellenistic Medicine." In *Aristotelische Biologie*, edited by W. Kullmann and S. Föllinger. Stuttgart: Steiner Verlag, 1997.

von Staden, H. "Women and Dirt." *Helios* 19 (1992): 7–30.

Wallach, L. *Alcuin and Charlemagne: Studies in Carolingian History and Literature*. Ithaca, NY: Cornell University Press, 1959.

Walters, J. "Invading the Roman Body: Manliness and Impenetrability in Roman Thought." In *Roman Sexualities*, edited by J. P. Hallett and M. B. Skinner. Princeton, NJ: Princeton University Press, 1997.

Walzer, R. *Galen on Jews and Christians*. Oxford: Clarendon Press, 1949.

Walzer, R., and M. Frede, ed. *Galen: Three Treatises on the Nature of Science*. Indianapolis: Hackett, 1985.

Weiler, Ingomar. "Inverted *Kalokagathia*." In *Representing the Body of the Slave*, edited by T.E.J. Wiedemann and J. F. Gardner. London: Routledge, 2002.

Weinrich, Otto. *Studien zu Martial*. Stuttgart: W. Kohlhammer, 1928.

Weissenrieder, A. *Images of Illness in the Gospel of Luke*. Tübingen: Mohr Siebeck, 2003.

Weitzmann, K. "An Addendum to the Heracles Plaques of the St. Peter's Cathedra." *Art Bulletin* 56 (1974): 248–51.

Weitzmann, K. "The Heracles Plaques of the St. Peter's Cathedra." *Art Bulletin* 55 (1973): 1–37.

Wessner, P., ed. *Aeli Donati quod fertur Commentum Terenti*. 3 vols. Stuttgart: Teubner, 1902–1908.
West, M. L., ed. *Iambi et elegi Graeci ante Alexandrum cantati: Callinus. Mimnermus. Semonides. Solon. Tyrtaeus.* Rev. ed. Oxford: Clarendon Press, 1992.
West, M. L., ed. and trans. *Greek Epic Fragments*. Cambridge, MA: Harvard University Press, 2003.
Westerink, L. G., ed. and trans. *Stephanus, Commentary on Hippocrates' Aphorisms.* 3 vols. Corpus Medicorum Graecorum 11.1.3. Berlin: De Gruyter, 1985–1995.
Wickham, C. *Framing the Early Middle Ages: Europe and the Mediterranean, 400–800*. Oxford: Oxford University Press, 2005.
Wickkiser, B. L. *Asklepios, Medicine, and the Politics of Healing in Fifth-Century Greece*. Baltimore: Johns Hopkins University Press, 2008.
Wiles, D. *The Masks of Menander: Sign and Meaning in Greek and Roman Performance*. Cambridge: Cambridge University Press, 1991.
Wilfong, T. "Reading the Disjointed Body in Coptic: From Physical Modification to Textual Fragmentation." In *Changing Bodies, Changing Meanings: Studies on the Human Body in Antiquity*, edited by D. Montserrat. London: Routledge, 1998.
Wilkins, J., D. Harvey, and M. Dobson, ed. *Food in Antiquity*. Exeter: University of Exeter Press, 1995.
Williams, B. *Shame and Necessity*. Berkeley: University of California Press, 1993.
Williams, C. A. *Roman Homosexuality: Ideologies of Masculinity in Classical Antiquity*. New York: Oxford University Press, 1999.
Wills, Garry. *What Paul Meant*. New York: Viking Adult, 2006.
Winkler, J. J. *The Constraints of Desire: The Anthropology of Sex and Gender in Ancient Greece*. New York: Routledge, 1990.
Withington, E. T. *Hippocrates*. Vol. 3. Cambridge, MA: Harvard University Press, 1928.
Wohl, Victoria. *Love among the Ruins: The Erotics of Democracy in Classical Athens*. Princeton, NJ: Princeton University Press, 2002.
Woodford, S. "Cults of Heracles in Attica." In *Studies Presented to George M.A. Hanfmann*, edited by D. G. Mitten and J. Pedley. Mainz: P. von Zabern, 1971.
Woodward, J. M. "Bathycles and the Laconian Vase-Painters." *Journal of the Hellenistic Society* 52 (1932): 25–41.
Worman, Nancy. *The Cast of Character: Style in Greek Literature*. Austin: University of Texas Press, 2002.
Worthington, Ian, ed. *Alexander the Great: A Reader*. London: Routledge, 2003.
Wright, J. P., and P. Potter, ed. *Psyche and Soma: Physicians and Metaphysicians on the Mind-Body Problem from Antiquity to Enlightenment*. Oxford: Clarendon Press, 2000.
Wyke, Maria. "Woman in the Mirror: The Rhetoric of Adornment in the Roman World." In *Women in Ancient Societies: An Illusion of the Night*, edited by Léonie J. Archer, Susan Fischler, and Maria Wyke. Basingstoke: Macmillan, 1994.
Zeitlin, F. I. *Playing the Other: Gender and Society in Classical Greek Literature*. Chicago: University of Chicago Press, 1996.

CONTRIBUTORS

Amalia Avramidou is a researcher at the Centre de Recherches Archéologiques—Université Libre de Bruxelles and a visiting lecturer at the University of Crete. She specializes in the art and archeology of Greece and Etruria, and her monograph *The Codrus Painter: Iconography and Reception of Attic Vases in the Age of Pericles* will appear this fall published by the University of Wisconsin Press.

Page duBois is Distinguished Professor of Classics and Comparative Literature at the University of California, San Diego. Some of her earlier books include *Sappho is Burning* (University of Chicago Press) and *Torture and Truth* (Routledge). More recent publications include *Slavery: Antiquity and Its Legacy* (I.B. Tauris), and *Out of Athens: New Ancient Greeks* (Harvard University Press), both published in 2010. She is currently working on ancient polytheism.

Brooke Holmes is an Assistant Professor of Classics at Princeton University. Her primary areas of interest are in the history of Greco-Roman medicine and the life sciences, Greek tragedy, literary and critical theory, and the history of philosophy. Her monograph *The Symptom and the Subject: The Emergence of the Physical Body in Ancient Greece* was published by Princeton University Press in 2010. A short book, *Gender: Antiquity and Its Legacy*, is forthcoming in the series "Ancients and Moderns" (I. B. Tauris and OUP-USA).

Valerie M. Hope is a University Lecturer in Classical Studies at the Open University. Her main area of research is Roman social history, focusing in particular on funeral rites and funerary monuments. She is the author of *Constructing*

Identity: the Funerary Monuments of Aquileia, Mainz and Nîmes (2001); *Death in Ancient Rome: A Sourcebook* (2007); and *Roman Death* (2009). She also co-edited *Death and Disease in the Ancient City* (2000) and *Memory and Mourning: Studies on Roman Death* (forthcoming).

Marguerite Johnson is Senior Lecturer in Classics at The University of Newcastle, Australia. Her primary research interests are in ancient literature with a focus on cultural representations of gender, sexualities, and the body. Related to these areas is her work on later engagements with ancient texts and traditions. She is the co-author (with Terry Ryan) of *Sexuality in Greek and Roman Society and Literature: A Sourcebook* (Routledge, 2005) and author of *Sappho* for Duckworth's *Ancients in Action* series (2006).

Patrick Macfarlane is an Assistant Professor of Philosophy at Providence College, Rhode Island. His research interests include ancient philosophy and ancient medicine. His most recent publication is an article he co-authored with Ronald Polansky (Duquesne University), entitled "God, the Divine, and Νους in Relation to the De anima" in *Ancient Perspectives on Aristotle's De anima*, G. van Riel and P. Destrée (eds.), Leuven University Press (2009).

Marc Mastrangelo teaches Classics at Dickinson College in Carlisle, Pennsylvania. His chief areas of interest are Late Antique Latin poetry and Greek intellectual history. His most recent publications include articles on the decline of fourth-century Latin poetry, the Judith story in fourth-century Latin literature, and a book, entitled, *The Roman Self in Late Antiquity: Prudentius and the Poetics of the Soul* (2007).

Marilyn B. Skinner is Professor of Classics at the University of Arizona in Tucson. Her primary research specialization is Roman literature of the Republican and Augustan eras. She has authored two monographs in that field, *Catullus' Passer: The Arrangement of the Book of Polymetric Poems* (1981) and *Catullus in Verona* (2003), and has edited or co-edited two collections of scholarly essays, *Vergil, Philodemus, and the Augustans* (2004) and the Blackwell *Companion to Catullus* (2007). Dr. Skinner is also well known for her work on sexuality and gender in antiquity. She has published numerous articles on the Greek female poetic tradition and has recently co-edited *The New Sappho on Old Age: Textual and Philosophical Issues* (2009).

Froma I. Zeitlin is Ewing Professor of Greek Language and Literature and Professor of Comparative Literature at Princeton University. She has published extensively in the field of Greek literature (epic, tragedy and comedy, and prose fiction). Author of *Under the Sign of the Shield: Semiotics and Aeschylus' Seven*

Against Thebes (1982; 2nd ed. 2009) and *Playing the Other: Gender and Society in Classical Greek Literature* (1996), she also edited *Mortals and Immortals: Selected Essays of Jean-Pierre Vernant* (1991) and co-edited *Nothing to Do with Dionysos?* (1990) and *Before Sexuality* (1990).

INDEX

Abassid patronage, 20
abortion, 12
abscesses, 57
abstinence, sexual, 168
Achaeans, 8, 86
   in Homer, 89
Acharnia, 195
Achilles, 8, 11, 86, 87, 125, 145, 162, 186, 210, 242
Acropolis, 15
acropolis of the body, 58
Actaeon, 188
actors, 37, 38, 81, 171
acupuncture, 2
Adam & Eve, 237
adultery, 71, 80, 81
Aeëtes, 210
Aelius Aristides *Sacred Tales*, 99
Aeschines, 167
   Against Timarchus, 169
Aeschylus
   Agamemnon, 179
   Eumenides, 68
Aesculapius, 18
Aesop, 177
Aëtius of Amida, 19, 103
Agamemnon, 8, 165, 242
Agrippina, 127
air, 46, 48, 53, 61, 100, 125
Ajax, 87
aklêtoi, 180
akrateia, 73
Alcmaeon of Croton, 47, 48, 53, 88
Alcuin
   On the Virtues and Vices, 252
Alexander I, 228

Alexander III, 200
Alexander the Great, 15, 60, 75, 112, 228, 231, 232, 234
Alexander/Herakles, 229
Alexandria, Egypt, 11, 15, 16, 19, 22, 94, 97, 98, 235
   Hellenistic, 94, 95, 100, 169
   in late antiquity, 102, 103
   medical school in, 100, 102
Alexandrian period, 84
Alkmene, 218
Amazons, 204, 215
Ambrose, 250
ameleia, 92
Ammon, 200
Amphiaraos, 219
amphidromia, 29
Anacreon, 175
anal sex, 73
Anaxagoras, 88, 89
Anaximander, 88
Anaximenes, 88
Anchises, 148
Andromeda, 171
Antioch, 22
Antonine smallpox plague, 19
Antoninus Pius, 234
anus, 72, 73
aorta, 56
Apate, 144
apes, 162
Aphrodite, 76, 141, 143
Apollo, 11, 14, 18, 86, 89, 237
Apollo Sauroktonos, 15
Apollonius of Rhodes, 97
   Argonautica, 75, 204

apoplexy, 48
apples of the Hesperides, 227
Apuleius
    Golden Ass, 114
Apulia, 229
Arabic world, 100, 102, 103
Arabic-Islamic medicine, 20
arch, anatomical, 14
Archaic Greece, 20, 32, 192
Archaic period, late, 145
Archangel Michael, 182
architecture, 14
Ares, 208
aretê, 7
Ariadne, 5
Aristides *Sacrae Orationes*, 245
Aristophanes, 13, 74, 118, 121, 122
    Acharnians, 195
    Birds, 197
    Ecclesiazusae, 122
    Frogs, 175
    in Plato, 77
    Thesmophoriazusae, 122, 140
    Wasps, 119, 131
Aristotelian school, 113
Aristotelianism, 100
Aristotelians, 103
Aristotle, 12, 15, 16, 17, 20, 45, 47, 56, 57, 58, 59, 60, 61, 63, 68, 69, 71, 88, 93, 103, 138, 143, 183, 186, 188, 190, 205, 244
    Generation of Animals, 191, 203
    History of Animals, 57
    Metaphysics, 203
    Motion of Animals, 58
    Nicomachaean Ethics, 73
    Parts of Animals, 58, 62
    Physics, 56
    Politics, 189
arms, 10
Arsinoë II, 76
Artemidorus of Ephesus, 25
Artemis/Diana, 148, 188
arteries, 51, 94, 95, 100, 104
Artimision Zeus, 14
ascetic traditions, 104
Asclepiades (epigrammatist), 77
Asclepiades of Bithynia, 17, 61, 96
Asclepieia, 99
Asclepius, 27, 99, 127
    cult, 84
askêsis, 108, 168
asymmetry, sexual, 139
Athena, 144, 148, 227
Athena/Minerva, 188
Athenaeus, 108
Athens, 15, 16, 28, 31, 68, 71, 77, 142, 223, 227

Atrahasis, 10
Attic comedy, 169
Attis, 20
Auge, 223, 235
Augean stables, 231, 236
Augustus Caesar, 17, 18, 80, 81, 112, 174, 201, 212
Aulus Gellius, 104
Aurelian, 177

Bacchanalian rites, 209
Baghdad, 19, 103
barbarians, 162, 208
Barbary apes, 100
baths, 229
Baths of Zeuzippos, 235
Bathyllus, 81
beard, 163, 219, 225
beauty, 71
    female, 75
Bede, 252
    Life of Felix, 251
Berenice II of Cyrene, 76
Bergren, Ann, 150
Bes, 228
Bible, 247
bile, 50, 51, 53
    black, 2, 48, 53
    green, 2, 53
    yellow, 47, 48
    yellowish orange, 53
birth, 26
Bitinna, 159, 173, 181, 183
bitter, 88
Bitto, 77
black box, 11, 15, 89
Black Sea, 208
bladder, 51
blood, 2, 47, 48, 53, 58, 59, 61, 69, 70, 95, 113, 162
Boardman, John, 22, 224
Bodel, John, 36
body, Alexandrian, 100
body care, 104
Boeotia, 223
Boethius, 246
bonding, social, 72
bone, 51, 53, 61
Boreas, 144
boundaries, gender, 79
Bourdieu, Pierre, 161, 169
bowels, 20, 51
brain, 2, 51, 56, 62, 95, 102
breast, 40, 75, 121
    male, 169
brides, 74
Brown, Peter, 22
bulla, 29, 174

INDEX

burial clubs, 36
Butler, Judith, 169, 212
Byzantine world, 100, 102, 104
Byzantium, 235

Caesar, Julius, 17, 26
Calchas, 86
Caligula (Gaius), 35, 78, 176
Callimachus
   Hymn 5, 188
   Lock of Berenice, 76
Calypso, 146
Campbell, Gordon, 22
Cannae, battle of, 40
Canon (Polycleitus), 10, 14
care
   of the self, 98
   of the soul, 92
Carolingian dynasty, 250
Carratello, U., 201
Carson, A., 214
Carthage, 173
caskets, Byzantine ivory, 232
Cassandra, 178
castration, 182
Cato, 162
Catullus, 20, 76, 78
celibacy, 170
Celsus, 86, 98, 108
Celts, 38
Ceramicus cemetery, Athens, 129
cervix, 69
Chaos, 12
charis, 144, 145
Charites, 145
Charlemagne, 240, 250
chastity, 82
   of boys, 80
   marital, 76
cheeks, 40
Cherchel Herakles, 230
chest, 20
childbearing, 100
childbirth, 12, 27, 28, 70
child care, 27
children, 67, 70, 76, 165, 190
Choice of Herakles, 221
choler, 2
Christ, 45, 181, 247, 250
Christianity, 16, 18, 19, 20–1, 45, 104, 112, 160, 172, 210, 235
   early, 178
Christianization, 250
Christian martyrs, 180
Christians, 64, 66, 99, 102, 202, 246, 250
Christian world, west, 100
Chrysostom, John, 171

Church, Christian, 251
Cicero, 26, 38, 45, 131, 168
   De Officiis, 221
   Pro Caelio, 81
cinaedus, 167
Circe, 146, 210
circumcision, 169, 182
Circus Maximus, 236
Cixous, Hélène, 207
class
   distinctions, 108, 112
   divisions, 80
Classical period, 10, 14, 20
cleanliness, 108
Cleanthes, 167
Clement of Alexandria, 171
club of Herakles, 230
Clytemnestra, 146
Cnidus, 15
Codex Theodosianus, 173
coition, 72
Colchis, 210
cold, 61, 88, 100, 170
Coleman, K. M., 199, 201
Colosseum, Roman, 196, 198, 215
column, anatomical, 14
comedy
   New, 75, 174, 177
   Old, 73, 186
Comedy, Attic, 180
Commodus-Hercules, 234
community of conditions, 97
Compitalia festival, 164
complexion, 113
concoction (*pepsis*, digestion), 54, 57
Conrad, Lawrence, 22
Constantine, 175
Constantinople, 235
Constantinus Herculius, 235
cooking, 50
Corinth, 172, 223
Cornelia, 33
Cornelius Nepos, 68
corpse abuse, 37
corpses, 34, 35, 38, 85
Cos, 15
courtship, 72, 74
cremation, 32
Creon, 211
Crete, Minoan, 4, 5
Cronus, 153
crucifixion, 37
culture, 70
cupping, 2
curses, 127
Cybele, 20
Cycladic, 4, 6
Cynics, 220

Daedalic, 9
Damascus, 19
damnatio memoriae, 38
dance, 13
Dawn (Eos), 143
death, 30
deathbed, 31
decapitation, 38
decorum, medical, 12
Deianera, 221
delicati, 81
Delphi, 224
demas, 85
Demeter, 140
Democritus of Abdera, 63, 129
demons, 104
Demosthenes, 137, 160, 165
Dennett, Daniel, 244
density of body, 89
deontology, medical, 12
depilation, pubic, 120
Descartes, R., 105, 244
Devil, 211
diagnosis, medical, 11
Dicaeopolis, 195
Diebold, William J., 22
diet, 11, 21, 48, 49, 50, 98, 108
digestion, 50
Dikaiarchos, 219
Diocles of Carystus, 12, 15, 17, 45, 92, 93
Diocletian, 235
Diogenes of Apollonia, 58
Dionysiac/Bacchic cults, 233
Dionysus, 144, 145, 175, 209, 230, 237
Dioscorides, 105
Dioskouroi, 236
Dipylon Amphora, 10
disability, 57
dislocations, 11, 90
dissection, 98
Divus Iulius, 200
Dogmatist sect, 17, 97, 101
Dolos, 144
dominance, male, 78
Domitian, 234
Domitius Tullus, 162
Doryphoros, 14
Dover, Kenneth J., 13, 22
drapery, 21
dreams, 25
Dreyfus, Herbert, 110
dropsy, 48
drugs, 85
Drusus, 31
dry, 50, 61, 70, 100, 170
dryness, 69
Dying Gaul, 176
dynameis ("powers"), 105

Early Modern period, 253
earth, 46, 53, 61, 67, 100
Edelstein, Ludwig, 12
effeminacy, 171
Egypt, 76
   Roman, 176
Eileithyia, 27
ejaculation, 70
ekphora, 32
Elagabalus, 179
elastico-fluid systems, 95
Empedocles of Acragas, 46, 53, 88
Empiricists (medical sect), 17, 60, 96, 97, 101, 105
environment, 62
Eos, 151
Ephesus, 37
Epictetus, 246
Epicureanism, 20, 60, 245
Epicureans, 246
Epicurean school, 100
Epicurus, 20
Epidemics (Hippocratic), 12, 30
epigram, Hellenistic, 75
epilepsy, 48, 57
Erasistratus of Ceos, 16, 95, 96, 101
erastês, 72, 74, 81
Erinyes, 143
Eristrateans, 101
erômenos, 72, 81
Eros, 12, 97, 144
erôs, 74
Erotes, 145
eructation, 50
ethics, medical, 12
Ethiopia, 177
Ethiopians, 124
etiology, 11
Eucharist, 45
eunuchs, 182
Euripides, 140
   Alcestis, 232
   Bacchae, 209
   Hippolytus, 73
   Medea, 210
Europa, 5
Eurylochus, 206
Eurystheus, 218, 223
Eutropius, Byzantine eunuch, 182
evil eye, 127
exercise, 11, 49, 55, 98
eyes, 31, 113

facies Hippocratica, 90
familia funesta, 31, 39
farming, 67
Farnese Herakles, 231
fat, 51

Favorinus, 171
feces, 50
feet, 116
fertility goddesses, 68
fetus, 69, 70
fever, 48, 57, 60, 95
fingers, 59
fire, 46, 53, 61, 100
flatulence, 50
Flavian Amphitheater (Colosseum), 199
Flemming, Rebecca, 94
flesh, 61
Flexner Report, 17
floor, anatomical, 14
fluxes, 56
folk medicine, 104
food, bitter, 88
Forestier's disease, 183
Forum (Rome), 32
Forum Boarium, Rome, 229
Foucault, Michel, 130, 161, 169
Fox, Robin Lane, 22
fractures, 90
Freud, Sigmund, 137, 244
Fronto, 99
Furies, 218
Fury, 210

Gaia, 12, 143, 153
Galatea, 207
Galba, 38
Galenism, 19, 91
Galen of Pergamon, 16, 19, 20, 45, 47, 59, 69, 83, 85, 91, 96, 162, 171, 189
   The Best Doctor is Also a Philosopher, 60
Games, Olympic, 7, 9
gangrene, 57
Ganymede, 143
garden of the Hesperides, 228, 237
Garrison, Daniel H., 23
Gastron, 159, 169, 173, 181, 183
Gauls, 176
Gela Painter, 198
gender in language, 160
gender, 69
gendered roles, 43
Genesia (Gk.), 33
Genesis, biblical, 247
genitalia, female, 120
Genius Augusti, 234
George, Michele, 176
Gêras, 220
Germans, 176
gestation, 69
gesture, 168
ghost stories, 35
Gigantomachies, 227

Gilgamesh, 228
Gilgamesh Epic, 10
Gill, Christopher, 240, 243
gladiators, 37, 171
Gleason, Maud, 166, 178
Glindoni, Henry Gillard, 198
Gomorrah, 247
Gorgons, 204, 208
Gospel of John, 65
Gospel of Luke, 18, 19
Great Britain, 19
Gregory the Great (pope), 21
gums, 59
Guthrie, R. Dale, 23
gymnasia, 229
gynaikerastria, 77
gynecology, 11, 28, 57

Hades, 8, 85, 126
Hadot, Pierre, 54
Hadrian, 234
Hageladas, 229
hair, 219
Haly Abbas, 22
Hampe, Roland, 23
Hannibal, 40, 232
Hanson, Ann Ellis, 164
Harpies, 208, 215
Harvey, William, 16
headache, 48
heart, 2, 58, 62, 102
heat, 56, 69, 81
   vital, 58, 69, 93
Hebe, 221, 223, 227
Hector, 11, 192, 242
Hecuba, 192
hêgemonikon, 93
Helen, 144, 146
Heliodorus *Ethiopian Story*, 171
Helios, 210
Helladic, Middle, 6
Hellas, 210
Hellenistic
   period, 20, 74, 82, 84, 91, 93, 96, 98, 154, 244
   world, 112
helots, 176
Henderson, Jeffrey, 120, 134
Hephaestus, 145, 154
Hera, 12, 144, 147, 154, 194, 218
Heraclitus, 54
Herakleidai, 227
Herakles (Hercules), 12, 31
   twelve labors, 218, 227
Herakles Alexikakos, 229
Herakles Epitrapezios, 232
Herakles vs. Hercules, 233
Herakles of Lysippus, 229

Herakles Mingens, 233
Herakles/Alexander, 234
Herakles-Melkart, 228
Heraklid clan, 223
Heraklids of Argos, 228
herbalism, 17
Herculaneum, 183
Hercules, 200
Hercules Romanus, 234
Herodas, 173, 181
Herodotus, 45, 182, 208, 219
  Histories, 124
Herophilus of Chalcedon, 16, 17, 60, 95, 96
Hesiod, 9, 12, 143, 145, 203, 219
  Theogony, 145, 151, 153, 204
  Works & Days, 9, 70, 145
Hestia, 148
hetaira, hetairai, 13, 71
heterosexuality, 145
hidden causes, 96
Hilaria, 173
Hippias of Elis, 7
Hippocrates
  Airs Waters Places, 48, 62, 124, 165
  Decorum, 63
  Diseases 1, 50
  Diseases of Women, 69
  Epidemics, 65
  Nutriment, 51
  On Ancient Medicine, 49, 62, 88
  On Breaths, 48
  On the Nature of Man, 47, 48, 51, 83, 89, 100
  On Places in Man, 87
  On Prognostic, 90
  On the Sacred Disease, 87
  Prorrhetic I, II, 90
  Regimen in Acute Diseases, 89, 90, 91
  Regimen in Health, 49, 59
Hippocrates of Cos, 2, 6, 11, 16, 20, 58, 59, 61, 83, 85, 100
Hippocratic
  corpus, 138
  works, 12, 45, 46, 51, 84, 93
Hippocratics, 16, 17
Hippolytus (Euripides), 20
Hippolytus, 147
Historia Augustae, 177
Hobbes, Thomas, 49
homeopathy, 17
Homer, 4, 9, 10, 12, 15, 20, 45, 46, 89, 145, 197, 219, 240, 242
  Iliad, 7, 8, 11, 18, 85, 86, 125, 165, 180, 191, 199, 204, 242
  Odyssey, 6, 7, 85, 146, 154, 210, 242
homosexuality, 13
homosexuals, 137

Hope Herakles, 229
Horace (Quintus Horatius Flaccus), 122, 210
  Satire I.8, 212
hospitals, 22
hospitium, 19
hot, 50, 61, 88, 100, 170
humoral
  body, 94, 95
  medicine, 83
humors, 47, 51, 57, 59, 60, 61, 62, 89, 100
hydraulics, Hellenistic, 95
hydrostatics, Hellenistc, 95
hysterical suffocation, 70

iatricê (medicine), 85
iatros, 84, 85, 86, 87
iatrosophists, 103
iconography, 221, 229, 237
  sexual, 145
ideology, sexual, 134
Idomeneus, 165
imagines, 38
incontinence, sexual, 168
indigestion, 50
Indo-European, 6
infants, 162
  mortality and, 26, 67
infidelity, 81
inhumation, 32
intercourse, sexual, 70
Iole, 221
Ionia, 10
Ionian philosophers, 220
Iphigenia, 143
irridology, 17
Isaiah (prophet), 65
Isis, 76
Islam, 19, 21
Ixion, 144

Jason, 211
Jerome, 250
Jerusalem, 22, 64
Jewish
  culture, 12, 19, 21, 102
  people, 172
  Talmud, 18
Julia, 179
Julian, Emperor, 103
Julius Caesar, 78, 200
Juno Lucina, 27
Jupiter/Zeus, 199, 202
Justinian *Digest*, 163

Kafka
  "In The Penal Colony," 179

INDEX

kalokagathia, 165, 180
Kant, E., 241, 244
katapugôn, 73
Kavala Herakles, 232
Keuls, Eva, 120
kharizesthai, 73
kidneys, 51
kinaidos, 73
klinê (Greek couch), 31
Knidian Aphrodite, 15
korê statues, 14
kouros statues, 13, 14
kourotrophos, 4
krasis (blending), 56
krisis, 63
Kritios Boy, 15
Kuriyama, Shigehisa, 94
Kypsileds, 223

labia, genital, 121
labor pains, 27
labyrinth, anatomical, 14
lactation, 70
Lang, Phillipa, 94
latifundia, 18
law, Athenian, 68, 71
lectus funebris, 31
legs, 10, 14, 113
Lenbach Herakles, 229
Leo Bible, 182
Leo I, 182
Leonardo da Vinci, 14
Leonidas, 223
leprosy, 57
Lesbos, 77, 139
Lex Iulia de adulteriis coercendis, 80
lex porcia, 160
liberal education, 76
lion skin, 230
liver, 2, 20, 51, 61, 62, 102
Lloyd, G. E. R., 50, 58, 62
Lochia, 27
logos, medical, 87
looseness, 97
Lot, 247
  wife of, 247
love
  charms of, 139
  conjugal, 141
  romantic, 76
Lucan, 210
  Civil War, 213
Lucius Plotius, 175
Lucretius, 45
Luke (Apostle), 64
luxury, 79
Lyceum, 60
Lysippean Herakles, 236

Lysippos, 232
Lysippos of Sikyon, 231

Macedon, 228
Macedonia, 112
Machaon, 86
Macintyre, Alisdair, 243
Maecenas, 81
Maenads, 204, 209, 207, 210, 215
magico-religious medicine, 89
Magnus of Nisibis, 103
Mamurra, 78
Marcus Aurelius, 16, 21, 99, 245
  Meditations, 246
Mark (Apostle), 64
marriage, 12, 68, 75, 82
marrow, 53
  spinal, 51
masculinity, 73, 77, 168
Mass (Christian eucharist), 45
matrona, 80
Matthew (Apostle), 64
Maximian Herculius, 235
Medea, 75, 97, 141
medicalization, 99
medical marketplace, 84
medicine, 10
medicus, medici, 98
Medieval art, 21
Medusa, 212, 215
Megara, 221
Megarians, 195
Melkart, 228
memory, 62
Menander, 75
Menelaus, 8
menses, 162
menstrual fluid, 69
menstruation, 12, 162
mentula, 122
Methodists (medical sect), 17, 60, 97, 101, 105
microcosm, 10
Middle Ages, 170, 210
midriff, 51
midwives, 27, 28, 30, 42
Miletus, 11
Miller, Stephen G., 23
mimes, 171
mimesis, 172
mind (*nous*), 164
Minoans, 5, 6
Minos, 5
Minotaur, 5
misconduct, sexual, 79
misogyny, 122
mixis, 148
mobility, class, 78

moisture, 69, 70, 100
Moretti, G., 201
motion, voluntary, 94
mourners, 39, 40
mourning, 41
mouth, 31, 51, 72, 122
Muhammad ibn Zakariya' ar-Razi
    (Rhazes), 103
Muses, 145
Mycenaean, 5, 6
    Greece, 192

Nannion, 77
Narcissus, 134, 138
nature (Gk. *physis*), 46, 70, 88
naturopathy, 17
neck, 20
Nemean lion, 236
nenia, 41
Neolithic, 4
neo-Platonists, 102, 103
Neo-Pythagoreanism, 76
Nephele, 144
Nephelokokkygia, 197
Nero, 17, 127, 202, 234
nerves, 95, 100
neura, "sinews," 94
New Testament, 65, 181
Nicetas Choniates, 236
Nietzsche, F., 55
Nike, 15, 75
nosology, 54, 57, 59
nudity, 75
    female, 10, 13
    male, 7, 9
nutritionism, 17
Nutton, Vivian, 23, 84
Nymphs, 209

Oaristus, 144
Oath, Hippocratic, 12
obscenity, comic, 139
obstretics, 11
Oceanid, 210
Odysseus, 8, 125, 146, 165, 180
Oedipus, 134, 219
oesophagus, 51
Ogden, Daniel, 127
oinochoe, 198
old age, 57, 67, 220
Old Oligarch, 176
Old Testament, 248
Olympia, 7
    temple of Zeus, 227
Olympic games, 22
Olympus, 152, 219
Omphale, 219
Onatas, 230

onkoi, 96
organs, female, 69
Oribasius of Pergamum, 19, 103
Orphism, 20
Osiris, 76
osteopathy, 17
Ouranos, 143, 153
ovaries, 69
Ovid, 198, 207, 252
    Baucis & Philemon, 114
    Elegies, 167
    Metamorphoses, 172, 186, 199, 202

paideia, 167
pain, 48, 89
palaestras, 229
Paleolithic age, 3, 4
palla, 174
pallium, 174
pallor, 113
Pandora, 9, 145, 167, 203
Panhellenic, 7
Pantokrator hospital, 22
Parentalis, 33
Parthenon, 15
Pasiphaë, 199
passivity, 73
Paster, Gail Kern, 208
paterfamilias, 98
Patroclus, 125, 145
patronage, 79, 80, 82
Paul (Apostle), 64, 172
Paul of Aegina, 19, 103, 169
Paulinus of Nola, 246, 248, 251
Paulus, 39
Pedanius Dioscorides of Anazarbus, 105
pederasty, 13, 81, 82, 121, 142
Peisandros of Rhodes
    Herakleia, 219
Peisetairos, 197
Peisistratid Athens, 224
Peisistratos, 224
Peitho, 145
Peleus, 193
Peloponnese, 224
Peloponnesian War, 74, 227
Penelope, 146
penetration, sexual, 67, 72, 77, 81
Penia, 143
penis, 68, 69, 73, 122, 195
Pentheus, 73
Pergamon, 235
Perimedes, 206
Peripatetic school, 60, 100
Peripatetics, 100, 114
peritoneum, 51
Perkins, Judith, 245
Perpetua, 173

Perse, 210
Persian Wars, 14
Persians, 165, 208, 227
Pertinax, 36
Petronius, 37
phallic, 4
phallicism, 120
phalluses, padded, 121
Phaon, 77
Pharaohs, 76
pharmaka, 86, 99
Pharsalus, 213
Phasians, 166
Phasis River, 124, 165
Philinus of Cos, 96
Philip II, 200
Philo, 176
Philomela, 187
phlebes, 94
phlebotomy, 2
phlegm, 2, 47, 48, 50, 51, 53, 56, 90
Phoenicians, 7
Phryne, 15
Phylotimus, 59
Physiognomics, 113, 124
physiognomy, 103, 166
physiology, 103
physis, 10, 84
Pindar, 127, 219
Pisistratids, 74
Piso, Cnaeus Calpurnius, 40
placebo effect, 27, 84
plague, 18, 50
  Justinianic, 19
Plato, 45, 58, 61, 63, 77, 81, 100, 103, 138, 169, 171, 186, 188, 190, 197, 243
  Alciabiades I, 249
  Charmides, 52
  Crito, 63
  Gorgias, 52
  Laws, 62, 92, 208
  Phaedo, 92, 125, 245
  Phaedrus, 52, 139
  Philebus, 92
  Protagoras, 168
  Republic, 45, 51, 52, 55, 92, 189, 195, 245
  Socratic dialogues, 92
  Symposium, 13, 20, 52, 139
  Timaeus, 20, 54, 62, 65, 92
  tripartite soul, 240
Platonic school, 100
Platonism, 102, 247
Plautus, 118, 180
  Mostellaria, 126
plethora, 96
pleurisy, 2

Pliny, 175, 177
Pliny the Elder, 27
Pliny the Younger, 26
Plutarch, 28, 110, 179
  Advice about Keeping Well, 99
  Moralia, 129
pneuma, 48, 54, 57, 58, 93, 94, 245
pneumatics, Hellenistic, 95
Podalirius, 86, 87
poetry, Hellenistic, 97
Pohl, Walter, 177
poison, 54
pollution, 30, 31, 34, 35, 40
Polybus, 58
Polycleitus, 10, 14
Polyxena, 143
pompa, 32
Pompey the Great, 180
Pomponius Mela, 177
pornê, pornai, 13
Poros, 143
Posidonius, 189
Postumus, 235
pottery
  black figure, 13, 197
  red figure, 13
Praxagoras of Cos, 45, 59, 60, 94, 95
Praxiteles, 15
  Aphrodite of Cnidos, 75
pregnancy, 27, 70
  in New Comedy, 75
Pre-Socratic philosophers, 10, 45, 46, 58, 63
Priam, 165
Priapus, Priapeia, 123, 212, 221
Proclus, 249
Procne, 187
prognosis, 11
prognostic, 90
Prometheus, 151
proportion, numerical, 10
proskynesis, 201
prostitutes, 37
  male, 81
Protagoras, 10
Prudentius
  Apotheosis, 247
prurient speculation, 77
Pseudo-Aristotle
  Physiognomica, 166
  Problemata, 73, 162
pseudo-Galen, 60
psychê, 69, 85
Ptolemies, 94, 95, 97
Ptolemy I Soter, 15, 76
Ptolemy II Philadelphus, 15, 76
Ptolemy III Euergetes, 76
puberty, 80

pulse, 94, 95
purification, 29, 32
purity, 20, 131
puticuli (pits), 36
Pygmalion, 207
Pythagorean
 philosophers, 12, 47
 philosophy, 203
Pythagoreanism, 20, 187

Quintilian, 15, 168, 171, 179

Rabinow, Paul, 110
rape, 75, 221
 anal, 123
Rationalists (medical sect), 60, 97
Ravenna, 104
Reeder, Ellen D., 209
reflexology, 17
regimen, 11, 70
residues (*perittômata*), 57
Rhazes, 103
rhesus monkeys, 100
rhetoric, 91, 168
Riace bronzes, 14
Richlin, Amy, 122
Roman Catholicism, 45
Roman Christian thought, 250
Roman imperial period, 84
Roman period, 98
Romanitas, 18
romantic attachments, 81
Rome, 22, 71, 235
 imperial, 169
Romulus, 200
roof, anatomical, 14
Roscius Otho, 175
Rothwell, Kenneth, 196
Rufus of Ephesus, 163

Sack of Ilion, 86, 87
Sagana, 212
Samos, 225
sanatoria, 99
Sappho of Lesbos, 77
scent, 175
Scribonia Attice, 29
scrotum, 69
Scylla, 210
Scythia, 219
Scythians, 125, 166, 208
sects, 101
seduction of a boy, 80
seed, female, 170
Sejanus, 38
semen, sperm, 68, 70, 73, 143
semiotics, 183
 medical, 89

Semomides of Amorgos, 190, 195
senatus consultum ultimum, 209
Seneca, 78, 170, 176
 Hercules Furens, 234
sensation, 94
Septimius Severus, 36, 235
Serapis, 18
Serenus of Marseilles, 21
serum, 56
sex as metaphor, 78
sexual
 identity, 137
 positions, 13
sexuality, 71, 72, 74, 108, 137, 153, 173
signs (*sêmeia*), 90, 91
 prognostic, 91
Silens, 209
Silver Latin, 214
sinew, 53, 61
Siphnian treasury, Delphi, 224
Sirens, 204, 206, 208, 215
Sissa, Giulia, 117
skênê, theatrical, 211
Skeptics, 60
skin, 51, 85
Skylla, 236
slave body, 160
slaves, 28, 78, 79, 81, 98, 112, 119,
 132, 159, 171, 175
 black, 179
smallpox, 19
smile, Archaic, 14
snake goddess, 5
Socrates, 20, 45, 55, 63, 72, 73, 92,
 125, 143, 177, 190
Socratic schools, 145
Sodom, 247
Solomon, 64
sôma, 85–9
sôma sêma, 126
Sophists, 220
Sorabji, Richard, 241
Soranus, 17, 27, 28, 163
 Gynecology, 169
soul, 8, 45, 58, 62, 63, 85, 93, 102, 104,
 108, 125, 189
Sparta, 223
Spartacus, 37
Spartans, 28
sphygmology, 94
spleen, 2, 51
St. Augustine, 65, 246, 250
 Confessions, 251
St. Peter's Throne, Vatican, 237
steatopygous, 3
Steiner, 189
stereotypes, gender, 81
Stewart, Andrew, 23

Stoa Poikilê, Athens, 227
Stoic
　philosophy, 215
　school, 100
Stoicism, 20, 21, 245
Stoics, 60, 64, 189, 246
stola, 174
stomach, 20, 51, 61
Storace, Patricia, 144
stratification, social, 78
Strawson, Galen, 239
stuprum, 80
submission, female, 78
Successors, 229, 232
Suetonius, 127
suicides, 37
　assisted, 12
Sulla, 31, 232
surgery, 12
surrender of self, erotic, 82
sweet, 88
Syennesis of Cyprus, 57
symposium, 12
symptoms, 90
Synoptic writers, 64
Syriac translations, 103

Tacitus *Annales*, 202
Taras, 231, 236
Tartaros, 12
taste, 58
tattoo, 35, 159, 178, 181
technê, medical, 87
teeth, 59
teleology, 96, 101, 140
temperaments, 100
tendons, 1
Terence, 118
Tereus, 187, 197
Tertullian, 46, 64, 173
testicles, testes, 68, 73
tetanus, 57
Tetrarchs, 235
Thales, 88
Thanatos, 144
Thecla, 173
Theocritus
　Simaitha, 75
Theodosius I, 22
Theogony (Hesiod), 9, 10
Theonexia, 232
Theophrastus
　Characters, 114
Theriomorphic choruses, 197
Thermopylae, 227
Thersites, 8, 165, 180
Theseus, 5, 219, 226
Thessalus of Tralles, 97

Thessaly, 213
Thetis, 193
Thucydides, 45
thumos, 210
Tiber River, 37
Tiberius, 31, 37, 38
tightness, 97
Timarchus, 167
Tiresias, 187
Tiro, 132
Tithonus, 143
toga praetexta, 174
toga virilis, 80, 174
tongue, 51
torso, 10
touch, 58
tragedy, Attic, 73
Tree of Knowledge, 237
triplokia, 96
Trojan War, 6, 7
Trojans, 8, 194
Troy, 4, 144, 145
Twelve Tables, 40

undertakers, 37, 42
Underworld, 218
unseen things, 96
urination, 232
urine, 50
uterus, womb, 1, 2, 51, 68, 69, 70, 162, 170

vagina, 2, 69, 72, 117, 121, 182
valetudinarium, 18
valves, 96
Varro, 36
vase painting, 72
　Attic, 71, 221, 224
vases
　Athenian black-figure, 224
　Corinthian, 224
　homoerotic, 74
vault, anatomical, 14
vegetarianism, 21
Vegetti, Mario, 95
vein, 2, 51, 61, 95, 100, 104
　basilic, 2
　cephaliic, 2
venesection, 2
Venus figures (paleolithic), 3, 5, 13
Vermeule, Emily, 6, 23
Vernant, Jean-Pierre, 187, 198
Vesalius, Andreas, 1
Vespasian, 36
vessel, 61
Vestal Virgins, 79
Vesuvius, Mt., 183
Virgil, 18, 220, 252

Virgilian Appendix, 123
virgin soil epidemics, 19
virginity, 117
vitalism, 17
   teleological, 96
Vitellius, 37
Vitruvian man, 14

water, 46, 53, 61, 100
wealth, 76
Weary Herakles, 231, 234
Weinrich, Otto, 201
wet, 61, 170

Williams, Bernard, 243
winds, 62
witches, 204
Wohl, Victoria, 165
womb, wandering, 117, 164
wounds, 11, 50

Xenophon, 73, 138

Zeno, Byzantine emperor, 173
Zephyrs, 144
Zeus, 5, 7, 12, 88, 205, 223, 226, 230, 235